ROUTLEDGE RESEARCH COMPANION TO LANDSCAPE ARCHITECTURE

The *Routledge Research Companion to Landscape Architecture* considers landscape architecture's increasingly important cultural, aesthetic, and ecological role. The volume reflects topical concerns in theoretical, historical, philosophical, and practice-related research in landscape architecture—research that reflects our relationship with what has traditionally been called 'nature'. It does so at a time when questions about the use of global resources and understanding the links between human and non-human worlds are more crucial than ever.

The twenty-five chapters of this edited collection bring together significant positions in current landscape architecture research under five broad themes—History, Sites and Heritage, City and Nature, Ethics and Sustainability, Knowledge and Practice—supplemented with a discussion of landscape architecture education. Prominent as well as up-and-coming contributors from landscape architecture and adjacent fields including Tom Avermaete, Peter Carl, Gareth Doherty, Ottmar Ette, Matthew Gandy, Christophe Girot, Anne Whiston Spirn, Ian H. Thompson and Jane Wolff seek to widen, fuel, and frame critical discussion in this growing area.

A significant contribution to landscape architecture research, this book will be beneficial not only to students and academics in landscape architecture, but also to scholars in related fields such as architecture, urban design, planning, history, and social studies.

Ellen Braae has been Professor of Landscape Architecture Theory and Method at the University of Copenhagen, Denmark, since 2009, where she heads the research group 'Landscape Architecture and Urbanism'. She has been Visiting Professor at AHO, Norway (2010), and TU Delft, the Netherlands (2018). Her research bridges design and humanities, with a focus on the transformation of post-industrial and 'welfare' landscapes. This crossover is partly reflected in her recent book *Beauty Redeemed: Recycling Post-Industrial Landscapes* (2015) and partly in her position chairing the Danish Art Council | Architecture (2018–2021) and being a member of the National Independent Research Council for Culture and Communication (2011–2015).

Henriette Steiner is Associate Professor at the Section for Landscape Architecture and Planning at the University of Copenhagen, Denmark, and is currently Visiting Associate Professor at the Department for Urban Studies and Planning at Massachusetts Institute of Technology (MIT).

Her research investigates the cultural role and meaning of architecture, cities, and landscapes. She is author of the book *The Emergence of a Modern City: Golden Age Copenhagen 1800–1850* (Routledge, 2014) and has co-edited six academic volumes, including *Architecture and Control* (Brill, 2018). She obtained her PhD from the University of Cambridge, UK, in 2008, and afterwards held a position as Research Associate in the Department of Architecture at ETH Zurich in Switzerland for five years.

ROUTLEDGE RESEARCH COMPANION TO LANDSCAPE ARCHITECTURE

Edited by Ellen Braae and Henriette Steiner

Routledge
Taylor & Francis Group
LONDON AND NEW YORK

First published 2019 by Routledge

2 Park Square, Milton Park, Abingdon, Oxfordshire OX14 4RN
52 Vanderbilt Avenue, New York, NY 10017

Routledge is an imprint of the Taylor & Francis Group, an informa business

First issued in paperback 2020

British Library Cataloguing-in-Publication Data
A catalogue record for this book is available from the British Library

Library of Congress Cataloging-in-Publication Data
A catalog record has been requested for this book

ISBN: 978-1-4724-8468-0 (hbk)
ISBN: 978-0-367-65973-8 (pbk)

Typeset in Bembo
by Deanta Global Publishing Services, Chennai, India

CONTENTS

FIGURES

CONTRIBUTORS

Tom Avermaete is Professor for the History and Theory of Urban Design at ETH Zurich. He has a special research interest in the public realm and the architecture of the city in Western and non-Western contexts. His research examines precedents, with the explicit ambition to construct a critical base of design knowledge. Avermaete is the author of *Another Modern* (2005), amongst other works. He is co-editor of *Architectural Positions* (2009), *Colonial Modern* (2010), *Architecture of the Welfare State* (2014), *Shopping Towns Europe* (2017), and *OASE Architectural Journal*, as well as curator of exhibitions such as *Casablanca Chandigarh* (Montreal, 2015).

Alan M. Berger is Norman B. and Muriel Leventhal Professor of Advanced Urbanism at Massachusetts Institute of Technology (MIT), USA, where he co-directs LCAU and P-REX lab. His most recent book is *Infinite Suburbia* (2017, with Joel Kotkin and Celina Balderas Guzman), a global perspective on suburban expansion. In addition to his award-winning books *Drosscape: Wasting Land in Urban America* and *Reclaiming the American West*, his other books include *Designing the Reclaimed Landscape* and *The Infrastructural Monument and Scaling Infrastructure* (2006, with Alexander D'Hooghe). Prior to joining MIT, Berger was Associate Professor of Landscape Architecture at Harvard Graduate School of Design. He is a Prince Charitable Trusts Fellow of the American Academy in Rome, Italy.

Inge Bobbink is Associate Professor in Landscape Architecture at Delft University of Technology (TU Delft), the Netherlands, since 2007. She trained as an architect at TU Delft and holds a postgraduate degree from the Berlage Institute. She gained her PhD at TU Delft, allowing her to expand her interest in the form and structure of the Dutch water system. In addition to developing and co-ordinating the master's programme in landscape architecture, since 2010 she has been teaching and supervising master's and PhD theses. Her current research focus is on landscape architectonics and the sustainable characteristics of global water systems. Various publications bear witness to this research agenda: *Land inSight* (2005), *The Polder Atlas of the Netherlands* (2009), *Delta Urbanism* (2010), *Water inSight* (2012), *Blue Bliss* (2016), and *Waterworks* (2017).

Anne Bordeleau is O'Donovan Director of the School of Architecture at the University of Waterloo, Canada. She is an architect and historian with publications on the temporal dimensions of drawings, maps, buildings, and architecture more generally, as well as the author of *Charles Robert Cockerell, Architect in Time: Reflections Around Anachronistic Drawings* (2014). In 2016 she oversaw the creation of plaster casts as co-principal on 'The Evidence Room', an exhibition prepared for the Fifteenth Venice Biennale in 2016 that was also shown at the Royal Ontario Museum in 2017 and selectively as 'Architecture as Evidence' at the Canadian Centre for Architecture.

Ellen Braae has been Professor of Landscape Architecture Theory and Method at the University of Copenhagen, Denmark, since 2009, where she heads the research group 'Landscape Architecture and Urbanism'. She is Visiting Professor at AHO, Norway (2010) and TU Delft, the Netherlands (2018). Her research bridges design and humanities with a focus on the transformation of post-industrial and welfare landscapes. This crossover is partly reflected in her recent book *Beauty Redeemed: Recycling Post-Industrial Landscapes* (2015) and partly in her position chairing the Danish Art Council | Architecture (2018–2021) and being a member of the National Independent Research Council for Culture and Communication (2011–2015).

Christina Capetillo is an architect and photo-based artist. She holds a master's (1995) and PhD (2010) in architecture from the Royal Danish Academy of Fine Arts' School of Architecture in Copenhagen, Denmark. In her photographic practice, Capetillo examines human interventions in nature and records how the contemporary landscape is transformed into a cultural condition where nature appears as traces and vestiges. Her work is represented in private and public collections including the Museum für Photographie Braunschweig, the National Museum of Photography, Denmark, Brandts Museum, Vraa Kunstbygning, Skagens Museum, and the New Carlsberg Foundation.

Peter Carl graduated from Princeton University, USA, and after a stint at the American Academy in Rome, Italy, taught at the University of Kentucky, Lexington, USA, before moving to Cambridge University, UK, where for thirty years he taught design and in the Graduate Programme in the History and Philosophy of Architecture. After developing the PhD Programme in Architecture at the CASS, London Metropolitan University, UK, for seven years, he took up a post as Visiting Professor at the GSD, Harvard, USA. His most recent interest is architectural design and practice as an embodiment of 'practical wisdom', the tension between freedom-from and obligation-to deep contexts.

Christina E. Crawford is Assistant Professor at Emory University, USA, a historian of architecture and urban form, and a licensed architect. Her current book project, *Spatial Revolution*, explores the foundations of early Soviet urban theory and practice through an investigation of specific industrial sites. New research follows the international transfer of ideas about social housing using the first federally funded housing project in the USA—Atlanta's Techwood Homes (1934)—as a principal node. Her scholarly writing can be found in *Future Anterior, Harvard Design Magazine*, and *Journal of Urban History*, among other publications. She received her PhD and MArch from Harvard University, USA.

Lisa Diedrich studied architecture and urbanism in Paris, Marseille, and Stuttgart, and landscape architecture at the University of Copenhagen, Denmark, where she received a doctoral degree for her study of site-specific design approaches in contemporary European harbour transformation. She currently works as Professor of landscape architecture at the Swedish University of Agricultural Sciences (SLU) in Malmö, Sweden, where she directs the research platform SLU Urban Futures. Her research and teaching focuses on critical practices in today's urbanization-induced landscape systems, on transdisciplinary strategies for urban transformation in climate-change affected water landscapes, and on methods for retrieving and communicating site qualities.

Gareth Doherty is Director of the master's programme in Landscape Architecture, Assistant Professor of Landscape Architecture, and Senior Research Associate at Harvard University Graduate School of Design, USA. At the core of his research and teaching is an interest in the relationship between people and the landscapes they inhabit, with the central inquiry into how ethnographic fieldwork methods can inspire and inform design and planning innovations. Doherty's publications include *Roberto Burle Marx Lectures: Landscape as Art and Urbanism* (2018); *Paradoxes of Green: Landscapes of a City-State* (2017); *Is Landscape...? Essays on the Identity of Landscape*, edited with Charles Waldheim (2015); and *Ecological Urbanism*, edited with Mohsen Mostafavi (2010, revised 2016).

Tao DuFour is Assistant Professor at the Department of Architecture at Cornell University, USA. His work explores the overlaps between architecture, anthropology, and philosophy, building on his research on the phenomenology of Edmund Husserl. His current research is concerned with the question of architecture's embeddedness in environmental histories. He holds a PhD and MPhil in the history and philosophy of architecture from the University of Cambridge, UK, and a BArch from the Cooper Union, USA. He is the author of *Husserl and Spatiality: Toward a Phenomenological Ethnography of Space* (forthcoming 2019).

Catharina Dyrssen is Professor Emerita in Architecture and Design Methods at Chalmers University of Technology, Sweden, and has long experience as a practising architect, musicologist, and interdisciplinary scholar, including as head of the Swedish Research Council's Committee for Artistic Research. After her doctoral thesis *Musical Space* she has been extensively engaged in professional training, research, lectures, experimentation, workshops, and PhD supervision that combine explorative spatial-material-bodily approaches with compositional and sociocultural perspectives along three main tracks: (1) applied design thinking in urban landscape contexts, public spaces, and regional development; (2) sonic spaces—intersections between architecture, music, sound, and urban space; (3) artistic research methodology and theory transgressing disciplinary domains.

Shelley Egoz is a landscape architect and Professor at the School of Landscape Architecture, Norwegian University of Life Sciences (NMBU), Norway. Her research and publications focus on landscape justice. She is interested in the political role that landscape architecture plays in society, and in 2014 she founded the Centre for Landscape Democracy at NMBU (www.nmbu.no/clad). One of her current topical interests is the interface between landscape and migration, which investigates how landscape could be employed for the integration of marginalized communities into society. Egoz is the initiator and co-editor of the books *The Right to Landscape* (2011) and *Defining Landscape Democracy* (2018).

Ottmar Ette is Chair of Romance Literatures and Comparative Literature at the University of Potsdam, Denmark. He is an honorary member of the Modern Language Association of America, member of the Berlin-Brandenburg Academy of Sciences, and regular member of Academia Europaea. Ette headed the research project 'Alexander von Humboldt's American Travel Diaries: Genealogy, Chronology, and Epistemology' (2014–2017), and since 2015 has been directing the eighteen-year Academy project 'Travelling Humboldt: Science on the Move', which focuses on editing the manuscripts of Alexander von Humboldt's American and Russian-Siberian travel diaries. His most recent publications available in English are *TransArea: A Literary History of Globalization* (2016) and *Writing-Between-Worlds: TransArea Studies and the Literatures-Without-a-Fixed-Abode* (2016).

Matthew Gandy is Professor of Geography at the University of Cambridge, UK. His publications include *Concrete and Clay: Reworking Nature in New York City* (2002), *The Fabric of Space: Water, Modernity, and the Urban Imagination* (2014), and *Moth* (2016), along with articles in *New Left Review*, *International Journal of Urban and Regional Research*, *Society and Space*, and many other journals. He is currently researching the interface between cultural and scientific aspects of urban biodiversity.

Christophe Girot has been Professor of Landscape Architecture at the Swiss Federal Institute of Technology (ETH) in Zürich, Switzerland, since 2001. He was born in Paris, France, and has a double master's in architecture (1986) and landscape architecture (1988) from the University of California Berkeley, USA. From 1990–2000 he was Chair of Landscape Architecture at the National School of Landscape Architecture in Versailles, and he currently directs the ETH Institute of Landscape Architecture and Landscape Modelling and Visualizing Laboratory. His research and teaching focuses on large-scale landscape design and point cloud modelling. Girot is the author of a major book titled *The Course of Landscape Architecture* (2016).

Clare E.L. Guest is Visiting Research Fellow at Trinity College Dublin, Ireland. Amongst her publications on architectural humanism and the prehistory of aesthetics are *The Understanding of Ornament in the Italian Renaissance* (2016) and the edited volumes *The Muses and Their Afterlife in Post-Classical Europe* (co-edited with K. Christian and C. Wedepohl, 2014), *The Formation of the Genera in Early Modern Culture* (2009), and *Rhetoric, Theatre and the Arts of Design* (2008).

Natalie Marie Gulsrud is Assistant Professor in the Department of Geosciences and Natural Resource Management at the University of Copenhagen, Denmark. Her focus is the governance of urban nature, and through her research, writing, and teaching, she aims to rethink urban nature for human well-being and livelihoods.

Jeanne Haffner on the curatorial staff of the New-York Historical Society. She is the author of The *View from Above: The Science of Social Space* (MIT Press, 2013), *The Environment Built* (Routledge Press, forthcoming), and numerous articles on urbanism, landscape architecture, waterfront development, and techniques of representation. She has been a postdoctoral fellow in urban landscape studies at Dumbarton Oaks (Harvard) and has taught at Brown and Harvard Universities and the ETH in Zürich.

Flavio Janches is a practising architect, researcher, and teacher of architecture and urban design at the University of Buenos Aires, Argentina. He has worked on projects that aim to improve the physical and social conditions of informal urban settlements. In 2012 he published *Public Space in a Fragmented City: Strategy for Socio-Physical Urban Interventions in Marginalized Communities* (Nobuku) and *Urban Interrelation: Work Methodology for the Insertion of Public Space in Informal Settlements* (at ISSUU with Max Rohm). Alongside his research, he opened the Argentine office of the PlaySpace Foundation, a non-profit organization based in the Netherlands, with the mission to use public places to improve the lives of children in slums.

Gini Lee is a landscape architect, interior designer and pastoralist, and Professor at the University of Melbourne (UoM), Australia. She was the Elisabeth Murdoch Chair of Landscape Architecture at UoM from 2011 to 2017. Her academic focus in research and teaching is on cultural and critical landscape architecture and spatial interior design theory and studio practice, to engage with the curation and post-production of complex landscapes. Focusing on arid environments, her multidisciplinary research into the water landscapes of remote territories contributes to the scientific, cultural, and Indigenous understanding of and management strategies for fragile landscapes. Her recent landscape curation and installation practice is an experiment with deep mapping methods to investigate the cultural and scientific landscapes of remote and rural Australia, Scandinavia, global archipelagos, and the arid lands of western USA. Since 2014 she has been an invited researcher at the Swedish University of Agricultural Sciences in Malmö, where she collaborates on fieldwork-based research into transect travel methodologies. She is a registered landscape architect and contributes to the strategic planning, design, and practice of urban and educational landscapes in Melbourne and beyond.

Lilli Lička is a Vienna-based landscape architect who graduated from the University of Natural Resources and Life Sciences (BOKU) in Vienna, Austria, before examining urban green spaces in the Netherlands and collaborating with BplusB in Amsterdam. She was principal of koselička, landscape architecture from 1991 to 2016 and started LL-L Landscape Architecture in 2017. She has been heading the Institute of Landscape Architecture at BOKU since 2003. Her works include *Nextland*, an online collection and book on contemporary Austrian landscape architecture (http://www.nextland.at/, 2005-), and *LArchiv*, an ongoing archival project of Austrian landscape architecture of the twentieth and twenty-first centuries. She conducts projects and research on the public realm, streets, squares, housing, and corporate landscapes. She has been a member of various communal and provincial design boards, juries, and scientific committees.

Sneha Mandhan is Research Associate at the Norman B. Leventhal Center for Advanced Urbanism at the Massachusetts Institute of Technology (MIT), USA. She is an urban planner and architect who has worked on several urban planning, environmental planning, and urban design projects in various international contexts. She has a master's in city planning from MIT, where she received the Flora Crockett Stephenson Writing Prize for her graduate thesis. Her research focuses on the integration of physical form with social and cultural forces within cities, disaster-resilient planning, urban sustainability, public space planning and design, and spatial planning pedagogy.

Samantha Martin-McAuliffe is tenured Assistant Professor in the School of Architecture, Planning and Environmental Policy at University College Dublin, Ireland. Her main research and teaching interests lie in Classical antiquity and the intersections of food and architecture. She is the editor of *Food and Architecture: At the Table* (2016), and her other book projects include *New Directions in Ancient Urban Planning in the Mediterranean* (2018) and a monograph on the urban order of ancient Athens. Martin-McAuliffe completed her PhD in Architecture at the University of Cambridge, UK, in 2007. Before that she received an MPhil in the History and Philosophy of Architecture, also from Cambridge (2003). She is a graduate of Smith College, USA.

Kathryn Moore is immediate past President of the International Federation of Landscape Architects (2014–2018) and Professor of Landscape Architecture at Birmingham City University, UK, where she directs the CATiD Research Centre (Critical Artistic Thinking in Design).

She has developed an approach to teaching, research, and practice that sets the study of landscape at the heart of the design of the built environment. Her book *Overlooking the Visual: Demystifying the Art of Design* (2010) argues that redefining theories of perception provides the basis for critical artistic discourse and has clear implications for architecture, planning, urban design, and other art and design disciplines, in addition to philosophy, aesthetics, and education more generally. A member of the National Independent Design Review Panel of High Speed 2. Her recent study 'Creating a National Park for the West Midlands: A New 21st Century Identity' is being considered by the Combined Authority in the West Midlands in the context of the long-term transformation of the region.

Tom Nielsen is an architect and Full Professor of Urban and Landscape Planning at the Aarhus School of Architecture, Denmark, where he has been teaching landscape architecture, urban design, and urban planning since 2004. His research focuses on the transformation of the Danish welfare city. This has included research into urban landscapes and public space (*Formløs*, 2001), suburban transformation, urbanizing territories (*Det Urbaniserede Territorium*, 2009; *Den Østjyske Millionby* with B. B. Jensen, Dansk Byplanlaboratorium, 2017), and the values and ethics of contemporary models of urban transformation (*Gode Intentioner, Uregerlige Byer*, 2008).

Martin Prominski is Full Professor and Chair in Designing Urban Landscapes at Leibniz University, Hannover, Germany. He studied landscape planning at the Technical University (TU) Berlin, Germany, and received a master's in landscape architecture from Harvard Graduate School of Design, USA. He has a PhD from TU Berlin (2003) and is a registered landscape architect. His current research focuses on design research strategies, qualification of urban landscapes, and integrative concepts of nature and culture. He is a member of Studio Urbane Landschaften, an interdisciplinary platform for research, practice, and teaching on urban landscapes, and a co-founder of the Sino-German Cooperation Group on Urbanization and Locality Research.

Svava Riesto is an art historian and Associate Professor at the Section for Landscape Architecture and Planning, University of Copenhagen, Denmark. She is particularly concerned with how historical perspectives can inform critical action in urban landscapes. A long-standing interest in her research and urban practice is the role of heritage in spatial planning and design. Her publications include *Biography of an Industrial Landscape: Carlsberg's urban spaces retold* (2018); *Forankring i Forandring*, a co-edited volume about alternative preservation in Freetown Christiania (2007); and 'Doing Heritage Together', a critical discussion of collaborative heritage-making (2015).

Maggie Roe is Reader in Landscape Planning Research and Policy Engagement at Newcastle University, UK, a Director of the UK-based international charity Landscape Research Group Ltd, and an editor of the journal *Landscape Research*. She has worked in landscape architecture practice and academia and been a member of the UK Landscape Institute for over thirty years. Her research focuses on participatory landscape planning, new cultural landscapes, and landscape change. She has given numerous lectures and examined in many countries. Funded by national and international government agencies, research councils, and environmental bodies, she has published widely and has extensive reviewing experience including for the UK's Research Excellence Framework 2014.

David Grahame Shane studied at the Architectural Association in London, UK, and received a master's in Architecture and a PhD in Architectural and Urban History, both at Cornell

University, USA, with Professor Colin Rowe. He started teaching at Columbia University, USA, in 1985, becoming Urban Design Programme Co-Director (1991–1997), and still teaches a seminar. He has lectured widely and published in Europe, the USA, and Asia. He co-edited *Sensing the 21st Century City: Close-Up and Remote* (2005). He is the author of *Recombinant Urbanism: Conceptual Modeling in Architecture, Urban Design and City Theory* (2005) and *Urban Design Since 1945: A Global Perspective* (2011).

Kelly Shannon is Professor of Urbanism at the Department of Architecture, Faculty of Engineering Sciences, KU Leuven, Belgium. Her design research is at the intersection of interpretative mapping, projective cartography, urbanism, and landscape. Her research engages numerous contexts (Belgium, Estonia, Vietnam, China, Indonesia, Bangladesh, India, Sri Lanka, Kenya, Morocco, etc.), primarily in the public sector, and focuses on the development of robust landscape structures as a form of adaptation to deal with contemporary design challenges (particularly climate change) at the territorial and urban design scales.

Anne Whiston Spirn is Cecil and Ida Green Professor of Landscape Architecture and Planning at the Massachusetts Institute of Technology, USA. Her books include *The Granite Garden: Urban Nature and Human Design* (1984), *The Language of Landscape* (1998), *Daring to Look* (2008), and *The Eye Is a Door* (2014). Since 1987, Spirn has directed the West Philadelphia Landscape Project, an action research programme integrating research, teaching, and community service. With her first website in 1995, she was an early adopter of the Web as a creative medium. In 2001, Spirn received Japan's International Cosmos Prize for 'contributions to the harmonious coexistence of nature and mankind'. For more, see www.annewhistonspirn.com.

Henriette Steiner is Associate Professor at the Section for Landscape Architecture and Planning at the University of Copenhagen, Denmark, and is currently Visiting Associate Professor at the Department for Urban Studies and Planning at Massachusetts Institute of Technology (MIT). Her research investigates the cultural role and meaning of architecture, cities, and landscapes. She is author of the book *The Emergence of a Modern City: Golden Age Copenhagen 1800–1850* (Routledge, 2014) and has co-edited six academic volumes, including *Architecture and Control* (Brill, 2018). She obtained her PhD from the University of Cambridge, UK, in 2008, and afterwards held a position as Research Associate in the Department of Architecture at ETH Zurich in Switzerland for five years.

Alan Tate is Professor and Head of Landscape Architecture in the Faculty of Architecture at the University of Manitoba, Canada. He has a degree in town and country planning and a graduate diploma in landscape design from the University of Manchester, UK. Tate's doctorate in architecture, from Edinburgh College of Art, UK, examined urban space types. He spent over twenty years running practices in London and Hong Kong before moving to the University of Manitoba in 1998. His ongoing research on city parks includes two editions of the book *Great City Parks* (2001 and 2015). He teaches professional practice and construction, history of landscape and urbanism, and design studio.

Michael Tawa is a registered architect and Professor of Architecture at the University of Sydney, Australia. He completed his BArch in 1980 and PhD in 1992. Tawa has practiced architecture in Alice Springs, Adelaide, and Sydney, and has taught at the University of New South Wales, Australia, the University of South Australia, and the University of Newcastle upon Tyne, UK. He has written extensively on architectural theory and on relationships between architecture, music, cinema, and atmosphere—notably *Agencies of the Frame: Tectonic Strategies in Cinema and Architecture* (2010), *Theorising the Project: A Thematic Approach to Architectural Design* (2011), 'Vaporous Circumambience: Towards an Architectonics of Atmosphere' (2014), and *Recuperative Architectonics: Matter, Memory, Immanence* (2016).

Ian Thompson is Reader in Landscape Architecture at Newcastle University, UK, and a chartered landscape architect. In addition to qualifications in landscape architecture, he holds a BA in philosophy and a master's in photography. His PhD thesis became the book *Ecology, Community and Delight* (1999), which won a Landscape Institute Award in 2001. He wrote *The Sun King's Garden* (2006), the award-winning *The English Lakes: A History* (2010), and *Landscape Architecture: A Very Short Introduction* (2014). Thompson is a former editor of the peer-reviewed journal *Landscape Research* (2004–8).

Anne Tietjen is an architect and Associate Professor at the Section for Landscape Architecture and Planning, University of Copenhagen, Denmark. She specialises in the transformation, preservation and development of existing built environments and landscapes. A long-standing interest in her research and practice concerns the role of heritage in spatial planning and design. Her publications include the book *Towards an Urbanism of Entanglement* (2011); *Forankring i Forandring*, a co-edited volume about alternative preservation in Freetown Christiania (2007); and 'Doing Heritage Together', a critical discussion of collaborative heritage-making (2015).

Jane Wolff is Associate Professor at the University of Toronto, Canada's Daniels Faculty of Architecture, Landscape and Design. Her design research investigates the complicated landscapes that emerge from interactions between natural processes and cultural interventions; its goal is to articulate terms that make these difficult (and often contested) places legible to the wide range of audiences with a stake in the future. Her subjects have ranged from the western Netherlands and the California Delta to post-Katrina New Orleans, as well as the shoreline of San Francisco Bay and the metropolitan landscape of Toronto.

THE ROLE OF LANDSCAPE ARCHITECTURE RESEARCH

Introduction to the volume

Ellen Braae and Henriette Steiner

The *Routledge Research Companion to Landscape Architecture* is a statement! A statement of the increasingly important role that landscape architecture plays in society. This volume reflects a profession that has its own history and is equipped with its own tools and measures for intervention in open spaces, a practice that theoretically and spatially reflects our current relationship with what has traditionally been called nature. It does so at a time when questions about the use of global resources and the way the human and non-human worlds link together are arguably more crucial than ever. From the point of view of our current juncture—which is witnessing the large-scale effects of waning industrial capitalist structures including economic restructuring, urban expansion, climate change and impending technological developments such as increased automatization and the emergence of artificial intelligence—designers from all disciplines are crucial 'first responders', culturally digesting as well as contributing to these shifts and transformations.[1] Being attentive to large-scale design problems as well as to the shifting, conflicting and often non-linear temporalities of the natural world, landscape architects are particularly well positioned to make a significant contribution to this difficult predicament.

In their practice, landscape architects synthesize heterogeneous knowledge derived from a variety of fields—natural, technical, life and social sciences—into spatial entities. The temporal conditions and material contexts in which the landscape architect operates require a constant reinterpretation of the changing nature motifs. If landscape architects therefore inevitably take part in current processes of cultural reorientation, landscape architecture design needs to engage in self-reflexive dialogue concerning the forms of knowledge and the practice ethos it employs. This calls for contributions from research, but concomitantly situates research in landscape architecture as a cultural undertaking as much as an academic one.

This is the fulcrum of this volume. In order to collate a Research Companion to Landscape Architecture, which interrogates these cultural dimensions of the discipline's research and practice, we have included chapters that reflect on landscape architecture topics from various corners of the discipline's theoretical, historical, philosophical and practice-related strands. Our aim with the volume is to create a concrete place for reflection as a dialogue partner for researchers and with a view to current concerns of practictioners. Rather than attempting to understand the full scope of current landscape architecture research by giving a synthetic overview, we focus on themes, issues and materials where we see a need for research to go into more depth. Eschewing metaphors of the frontier or the forefront, this volume instead marks out crucial directions for research which qualitatively add to our knowledge about landscape architecture and the cultural conditions with which it is interrelated. This means that the problems and modes of questioning the volume seeks to evoke cannot necessarily be envisioned as linear progressions. We instead have invited our contributors to move like crabs or lobsters, tripping sideways and backwards, often taking unlikely, slow or difficult routes to approach their material and questions.

Moving backwards into the discipline

As a discipline, landscape architecture has developed beyond its historical origins in horticulture and the design of singular parks and gardens. Today it is a discipline which addresses the spatial, i.e. the material, discursive and practice-related aspects of urban and rural development. It also addresses the fact of an increasing sense of limitation of resources. The concrete, material context therefore needs careful attention as the landscape architect synthesizes various needs, dreams and aims in designs. Landscape architecture today deals with different types of often 'green' but just as often 'blue', 'grey', 'yellow', 'grey' or 'black' open spaces, which are laid out for different cultural activities and for many different participants, human and non-human alike. Taking into account the social and material conditions, climate, terrain, plants, vegetation, water and built contexts of the settings, landscape architecture may play a role in endeavours to find new ways of inhabiting the earth, enabling a turn against the view of natural resources as standing reserves—and instead designing from a perspective of care for the long-lasting future well-being of humans and non-humans.

For one important figure in current landscape architecture and practice, the French landscape architect Gilles Clément, this means that landscape architects may be seen as playing the role of *planetary gardeners*.[2] The motif of the planetary gardener reflects the way landscape architecture has always worked as both a prism and an agent at the intersection of the human and natural worlds, and it marks out a current direction with respect to our changing relationships with and within the latter. The role of *gardener* emphasizes the landscape architect's commitment to the spatial and material context, in the sense of both a concrete engagement with and an attentive pruning of the open spaces—landscapes—in question. This is a process which concomitantly folds backwards and forwards in time, working with concrete material conditions as they have developed over time as well as embodying the orientation towards the future of any design. In the context of landscape architecture, this also requires a commitment to non-linear temporalities, such as the cyclical movement of the seasons or even the promise of eternity, that gardens have historically been seen to make. But the role of a *gardener* oriented towards the particularities of a closed garden space, on which Clément calls, also invokes motifs of non-political inclination and limited commitment to societal concerns. In contrast, Clément's idea of the *planetary gardener* calls attention to the overarching problems pertaining to the planet as our shared living ground, to which landscape architecture is increasingly oriented in light of the current ecological crisis. In the context of the Anthropocene, where the activities of human industrial culture have arguably impacted on the remotest corners of the planet, every intervention—even in an individual garden space—may always be linked to larger effects of human activity. While we might caution against an overly anthropocentric line of argument, the motif of the landscape architect as a gardener who cares for the planet evokes a highly relevant horizon regarding some of the most pressing current concerns of sustainability. Yet, the idea also marks out an extremely distant horizon, and allows us only to establish only highly open and malleable metaphors of the common-to-all. This volume, with its precise studies pertaining to concrete material contexts, therefore emphasizes the need to take into account the concrete context as well as this *planetary* horizon, even as we caution against the gigantism that horizon involves. Needless to say, moreover, in current discourse on landscape architecture and climate, even the planetary image is available to be abstractly moulded and optimized.[3]

The metaphor of the *planetary gardener* therefore entails all the well-known dangers of projection embodied in the suffix '-scape' in landscape, an inbuilt difficulty which landscape architecture needs to carefully consider. That suffix indicates an ontological levelling and a heightened orientation towards the visual field. Insofar as one of the landscape architect's core concerns, in practice as well as research, is with spatial entities and relationships, we therefore need to consider the many different scales and relationships that lie between the horizon of the gardener and that planetary horizon. What we see across the contributions to this volume is a concern to work out these differentiated domains.

For examples of these complex challenges, consider how, in the context of climate change as an effect of changing conditions, emerging motifs of nature are often conceptualized as positive image-related aspects of cities, captured as 'urban nature', 'urban farming', 'urban forestry', 'urban wildering', etc. This phenomenon is bound up with a utopian promise, with the hope that landscape architecture will offer *solutions* to fix it all—addressing environmental problems, as well as making people happier and healthier, and even improving social cohesion at the same time. There is thus an inbuilt risk that landscape architecture design might itself become a plane of projection for visions of betterment. If the role of the landscape architect today does not solely entail a preoccupation with aesthetic concerns, mere stewardship, or fixed and bounded spatial entities, landscape architecture is a practice that requires multiply connected modes of interpretation of material, cultural and historical contexts. This is exactly where landscape architecture research comes into play, offering concrete spatial and scholarly opportunities for reflection. The contributions that make up this volume collectively promote an awareness of the ambiguities of these endeavours, look more deeply into how we came to this impasse, and reflect on continuities as well as current attempts to recalibrate our understanding of fundamental relationships between humans and non-humans, in particular by including a concern for non-binary ways of thinking. A key role of this companion to landscape architecture research, therefore, is to function as a joint hermeneutic exercise across a range of university institutions around the world and involving both established and new voices in the debate, from landscape architecture practice and research in its more traditional definition to related disciplines and debates.

Forms of knowledge

While research on the material and matters of landscape architecture takes place in many different established academic genres, research in landscape architecture itself, as a systematic and organized activity at universities and research institutions around the world, is a relatively new phenomenon. This is partly a result of the profession's recent establishment in the form in which we know it today, and partly a result of the generally increased prioritization of research in the second half of the twentieth century in design educational institutions and beyond. However, the institutional frameworks and relevant research traditions of landscape architecture vary substantially, encompassing universities, schools and colleges rooted in or associated with institutional contexts ranging from the fine arts to the humanities, natural sciences and even agriculture.

Despite such diverse academic settings, as witnessed by this volume, landscape architecture research has developed significant tools, approaches and bodies of knowledge in its own right.

While its methodological apparatus tends to be heterogeneous and remains undertheorized, a significant proportion of it is characterized by a strong interpretive element and is driven by questions of ethical urgency, linking directly to the practical and aesthetic dimensions of the discipline. These methodologies offer ways of providing answers to questions that are unique to landscape architecture, and often do so by mediating between seemingly contradictory spheres, including different academic disciplines, research and practice, academia and everyday life.

Cutting across modern-day disciplinary divisions, we may also understand the varying forms of knowledge with which the work of this volume is in dialogue in terms of ancient Greek philosophy's distinction between *episteme*, *techne* and *phronesis*. Today, we could label *episteme* as scientifically grounded knowledge, while *techne* would entail knowledge about how a thing should be made, formulated within a context that did not separate art from craft or architecture. *Phronesis* is knowledge about the right thing to do, how and why to act in a proper way, and thus designates an ethics or ethos.[4] The contributions to this volume address the intersections of all three knowledge forms, while at the same time keeping an eye out for an embedment of research that takes account of its own relationship to practice and hence to choices regarding concrete problems.

While different constellations of methodologies and guiding themes might lead to conclusions other than those reached in this volume, this volume is grounded in an interpretive humanities tradition. We believe that the humanities, with their concern with the nuanced interpretation of phenomena that may be textual, material and affective as well as spatial, constitute the most useful starting point for engaging with the questions raised in the volume and is the best possible way to enable the volume to form an overarching *companion* to research in landscape architecture.

Configuring the volume

The volume is divided into five parts that suggest different themes for consideration. They each comprise five chapters dealing with thematic aspects that are crucial to research in the discipline, covering the history and historicization of the idea of landscape itself, archives and digital representations, perceived nature/culture dichotomies, the politics of designing for the future, and the philosophy of landscape architecture in relation to both practice and education. The five parts together provide a diverse yet catalysing framework for outlining, deepening and reflecting on central aspects of landscape architecture. Each is accompanied by a photograph by architect and photographer Christina Capetillo, which we have included less as a metaphorical exercise and more as a reminder of the richness of the concrete contexts to which a wordy volume such as this refers, even in the most unexpected places.

The **first part** is titled 'Landscape in the rear-view mirror: Historicizing the field'. The chapters in this part provide a historical perspective on the ways modern ideas of 'landscape' are conceived and landscape architecture is practised. The term 'land-scape' involves a distant way of looking at the world implied in the modern, predominantly Western, embellishment of the notion of 'land' with the abstracting view implied in the suffix '-scape'. We may therefore ask how these ways of looking have developed over time, what ways of knowing they offer, and what their limitations are. In different ways, the chapters in this part consider the epistemological, cultural-political and aesthetic trajectories and implications of how Western understandings of landscape and landscape architecture have developed over time in relation to various cultural and historical contexts. Taking a look at landscapes by looking at historical formations of how landscape is seen—looking at history through the rear-view mirror, we might say—these chapters simultaneously historicize their own ways of considering landscapes of the past. In doing so, they establish that a historicization of the field of landscape architecture must always be accompanied by a critical scrutiny of our ways of looking and of interpreting the past, and of increasingly charged notions central to this undertaking, such as 'history' and 'field'. At the same time, the chapters demonstrate that the interpretation today of past practices allows us to reflect on our current inclinations in fundamental ways.

In the **first chapter**, 'Culture, nature: A punkt in space', Peter Carl opens the volume by identifying the modern understanding of landscape as informed by culturally inherited binaries. But rather than reducing the modern view of landscape to its own vanishing point, he unfolds the richness of the continuum that lies between what we may call nature and culture. According to his argument, the exact sciences' increasing knowledge of nature is accompanied by a diminished capacity to understand human culture as part of nature. By laying bare the characteristically uneven and conflict-ridden relationships between the continuum's fictional end points, he argues, it is possible to call attention to how concrete contexts enact claims on us and provide us with meaning, with a potential sense of direction. In the **second chapter**, 'Renaissance gardens: Topicality and the scene of nature', Clare E.L. Guest considers significant topoi of gardens of the Italian Renaissance as concrete spaces that illuminate the poetic role of coexisting multiple meanings and thereby make gardens places for ethical existential reflection. Yet, positioning the garden as a 'scene of nature' simultaneously allows the identification of an aggressive, distant attitude to nature whose ecological consequences we are facing today. The **third chapter**, 'The birth of landscape from the spirit of theory: Alexander von Humboldt's artistic and scientific *American Travel Journals*' by Ottmar Ette, focuses on the ways of seeing, mapping and knowing landscape presented in the modern ideas of planetary and moving landscapes formulated by the German philosopher, scientist and explorer Alexander von Humboldt at the turn of the nineteenth century. These journals reveal a (constructed) view from above of a landscape that is so dynamic and interconnected that it confronts the 'thinking man' with his own limitations, emphasizing the idea of a discontinuous, broadening present. In the **fourth chapter**, 'Flight from modernity: Historicizing the aerial promise in landscape architecture', Jeanne Haffner also confronts the difficult entrapments of the modern view of landscape 'from above'. She considers how French social scientists in the 1960s insisted on challenging that view with a view 'from below', regarding cities and landscapes as 'social space'. However, as Haffner outlines, not only was this view from below just as constructed as that from above, it was also created in a deep interdependence with the latter, thanks to the role played by the aerial photography of landscapes in post-war French social science. The **fifth chapter**, 'Beyond innocence: The norms and forms of colonial urban landscapes' by Tom Avermaete, further considers the difficult consequences of collapsing the view from above into the view from below, this time in relation to French colonies and the planning practices attached to them. These difficult sites make apparent the contested nature of any view of a landscape imbued with political aspirations, whether from above or from below, since such views rely on discursive mechanisms that construct our historical narrative of them. As this and other chapters in this part demonstrate, the challenge for research is to sound out a more direct understanding of the interrelationships between land and -scape, between the material and the constructed, the embodied and the discursive, without necessarily supplanting one with the other but understanding their distinct—if heavily interrelated—ways of knowing and impacting.

The **second part**, 'The art of archiving landscapes: Tools for capturing moving relationships', considers temporal and spatial notions that confront the standard narratives of linearity, progression and the homogeneity of Cartesian space that are favoured in modern Western culture. In different ways, the chapters challenge such notions. Instead of linearity, we may consider past, present and future as folded into each other; instead of fixed, measurable spaces, we find ourselves in much more unpredictable territory. This broadening present rejects the usual scalar relations; it swallows up both past and future simultaneously, rather than presenting itself as a fleeting moment or point on history's measuring tape. Despite the associated challenges, some possible responses, as some of the chapters show, may lie in digital technology's means for tackling complex scalar and spatio-temporal relations, while other chapters call for heightened self-reflection concerning how we attune ourselves to a landscape's multiple temporalities.

The part opens with the volume's **sixth chapter**, an image essay by Christina Capetillo that is also an expansion of the images by her that we have chosen as accompaniments to the main parts of the book. This contribution is titled 'Scenes from an Anthropocenic Archive' and seeks to represent the Anthropocene as an imprint on concrete landscapes, with human culture becoming visible in new types of objects that challenge our notions of scale in different corners of the planet. Capetillo's work simultaneously also attempts to collect instances where the Anthropocene as a cultural condition becomes visible, to capture its effect in a photographic archive, and thereby to challenge the implicit historicist underpinning of the term as a periodizing exercise. In the **seventh chapter**, 'Transareal excursions into landscapes of fragility and endurance: A contemporary interpretation of Alexander von Humboldt's mobile science', Lisa Diedrich and Gini Lee conduct transareal excursions into waterside landscapes, creating travelling transects in areas as diverse as the Canary Islands. They adopt and reframe Alexander von Humboldt's ideas of mobile science to develop a means of understanding the qualities of landscapes that are becoming increasingly fragile as a result of political, economic and climatic shifts. The authors thus propose tools to capture large-scale landscape phenomena as cultural changescapes relevant to our understanding of shifting cultural conditions today. The **eighth chapter**, 'Smart nature? Views from the cyborg tree' by Natalie Gulsrud, considers the changing cultural conditions of the present's effect on landscape architecture from a different perspective: the increasingly pervasive presence of digital communication technologies, which also impacts on what is commonly known as 'nature management'. She considers the possibilities and pitfalls involved when it is possible to take the 'human' completely out of decision-making processes regarding nature management, processes which then mediate between algorithms and the 'stuff' of nature itself—all non-human agents. Christophe Girot, in the **ninth chapter** of the volume, '"Cloudism": Towards a new culture of making landscapes', considers the concrete possibilities of actively using digital tools and methods in landscape analysis and design, and how they may reveal new knowledge about large-scale spatial relations, particularly in the use of the point cloud as a representation and working tool for landscape architects. Indeed, this tool's potential impact on design ought not to be underestimated, and Girot outlines the construction of this technology as well as its different uses. At the same time, he offers a way of beginning to ask questions about the place of old buzzwords that grew out of Romanticism, such as 'creativity' and 'genius', when certain knowledge-creating processes are delegated to digital technologies. In the **tenth chapter**, 'The Marnas digital archive: Exploring practice, theory, and place in space and time', Anne Whiston Spirn further considers the knowledge-creating and storing capacities that digital technologies make possible and how these allow us to reconsider our relationship to concrete spaces, not least by including different sets of temporalities. Her focus is on a smaller scale and on the potential for knowledge involved in affective mnemonic responses. She does this by considering a specific garden developed over many years by the late Swedish landscape architect Sven-Ingvar Andersson. Spirn also explores how photography's remediation in the digital archive can bring new understandings of such a garden, and the particular—often non-linear—temporalities this entails, thus ending this part where it started: with the knowledge potential embodied in the technique of photography, a tool often employed in and with landscape architecture.

The **third part** is titled 'Urban stories from a green planet: Green stories from an urban planet'. The chapters in this part deal with cultural, philosophical, ecological and design-related consequences of the perceived nature/culture divide, how that divide developed within—or against—a post-Romantic vocabulary, and how we might nuance or even overcome such dualistic thinking. This touches on contemporary debates concerning the Anthropocene, and also on deeper issues regarding the capacity of landscape architecture to address questions of temporality, commonality, etc. All of the chapters in this part deal with landscape architecture in relation to the urban context, not only concerning the relations between landscape architecture and urban life, infrastructures, buildings and institutions, but also in places where 'city' and 'nature' intermingle in unexpected ways. What is the position of landscape architectural thinking in situations where the relationship between the city and what has traditionally been called nature seem difficult to capture with well-known dualistic tropes? How can telling stories about these diverse urban interlinkages confront such dualisms in new ways? Can landscape architecture, in this situation, add to the city's capacity to form a common ground?

The part begins with a contribution by Gareth Doherty, constituting the **eleventh chapter** of this volume, 'The vertical and the horizontal: Combining ethnographic and geographic methods in understanding landscape'. He investigates relationships between people and land, using an autoethnography research methodology to do so. Combining geographical and ethnographic methods, an important methodological element of his project was extensive walking in Bahrain—a practice that one may not associate with the wildly automobile-dependent form of urban environment by which Bahrain often is portrayed—as a way of describing and understanding nuanced social and cultural interrelations with the landscape. Hearing and telling stories along these walking routes subsequently became a research practice in its own right, constituting a way to mark out the particular urbanism of landscape in this territory—the extreme setting of a prosperous Gulf desert state—as 'green scenery beheld vertically'. In the volume's **twelfth chapter**, by Tao DuFour, we move away from Bahrain and across the planet to a (from a Western urban point of view) equally bewildering setting: the surprisingly urban fabric of the Brazilian Amazon. This chapter, 'Toward a somatology of landscape: Anthropological multinaturalism and the "natural" world', works closely with philosophical and theoretical themes that confront nature/culture binaries, drawing on contemporary Brazilian anthropologist Eduardo Viveiros de Castro's concept of multinaturalism and placing it in relation to the phenomenology of nature developed by Czech philosopher Jan Patočka in the 1950s. The **thirteenth chapter**, 'Designing landscapes of entanglement' by Martin Prominski, moves down related theoretical avenues, but emphasizes significant contemporary European thinkers such as the French anthropologists and philosophers Philippe Descola and Bruno Latour. Proposing a starting point for a philosophy of design that marks out relationships of entanglement, Prominski discusses prominent examples of recent landscape architecture projects in light of theoretical considerations particularly inspired by actor-network-theory. The **fourteenth chapter**, 'Enlarging the urban orchestra: Rethinking current approaches to landscape architecture', records a conversation between Matthew Gandy and Henriette Steiner, in which a challenge to modernity's instrumental thinking concerning nature leads to the suggestion that there is a need to enlarge the urban orchestra, so to speak—to bring different sets of both human and non-human voices into the debate. In this contribution, particular aspects of how this may become a way of rethinking current approaches to landscape architecture and landscape architecture research are discussed. The final chapter in this part, and the **fifteenth chapter** of the volume, is 'City, nature, infrastructure: A brief lexicon' by Jane Wolff. Wolff takes San Francisco Bay as a case study to work out a nuanced view of hybrid ecologies as characteristic of the contemporary urban fabric in light of what is commonly discussed as the Anthropocene. It is along this line of thought that Wolff develops a lexicon for telling the stories of these crucial relationships, where the bay comes to reflect what the city is and vice versa yet emerging into independent hybridities. This concludes the part's different stories from a planet that is both green and urban and displays a variety of intermingling forms of the two.

The **fourth part** is titled 'Designing with the past in the future: Politics, heritage and sustainability'. The chapters in this part simultaneously look forwards and backwards in time, and consider the role of landscape architecture in current debates concerning urban development, culture, politics and design, in which questions of sustainability raise concerns about the future that touch on deeply seated ethical questions. What role might landscape architecture have in articulating and working through these issues, and in making them visible? All of this affects the way we conceive of how both the past and the future manifest themselves in our present, and it has significant cultural and political implications for how landscape architects confront, represent and preserve landscape, from the perspective of both theory and practice, as well as for the challenges we face when dealing with sites and when designing.

The first chapter of the part, 'Urgent interventions needed at the territorial scale—now more than ever', is the **sixteenth chapter** of the volume. In it, Kelly Shannon discusses how the historical time of writing her contribution is perceived by many as a time of crisis. Discussing this crisis rather as a time of plural crises covering political, economic, social and climate-related issues, Shannon begins to zoom in on how landscape architecture, because of its ability to work at very large scales and with dynamic and changing materials, is a position from which this situation may be confronted and even alleviated—not least as a way of making meaning and thereby offering qualitative measures or interventions. In the volume's **seventeenth chapter**, 'Landscape architecture and social sustainability in an age of uncertainty: The need for an ethical debate', Shelley Egoz further bolsters this self-understanding of the role of the landscape architect in the face of the current societal and economic crisis, focusing her discussion on the inherent qualities landscape settings have to offer. Landscape may, perhaps unexpectedly, play a key role in what may be called relations of social sustainability, despite the general context of uncertainty. What is key, however, Egoz argues, is the need for ethical debate, and the chapters in this part are attempts to contribute to that debate. The **eighteenth chapter**, 'Coupling environmental and sociocultural sustainability for better design: A case study of Emirati neighbourhoods and landscape' by Sneha Mandhan and Alan M. Berger, further discusses the crucial importance of joining questions of environmental and sociocultural sustainability. Using the example of Emirati residential neighbourhoods in Abu Dhabi, the authors expand on the concept of sustainability and propose ways of rethinking conventional landscape practices for a more comprehensive and contextualized consideration of social and cultural factors in environmental thinking. The importance of sociocultural factors as the crux of heritage research and practice in relation to landscape architecture is also explored in the volume's **nineteenth chapter**, 'Planning with heritage: A critical debate across landscape architecture practice and heritage theory' by Svava Riesto and Anne Tietjen. The authors take as a starting point the fact that landscape architects are increasingly involved in transforming existing urban landscapes and therefore inevitably involved in heritage work. In this way, Riesto and Tietjen help to theorize the role of heritage-led spatial practice, considering three recent landscape projects in locations in Denmark, Switzerland and Germany. This also provides a more self-reflexive understanding of the different temporalities that heritage-led spatial practice requires practitioners to consider, with the awareness that this always involves taking contested cultural terrains into account. In the **twentieth chapter**, 'The case to save socialist space: Soviet residential landscapes under threat of extinction', Christina Crawford considers the difficulty involved in even preserving the landscapes of some of the large-scale housing estates of the Soviet Union. From the point of view of a period when the capitalist economy seems to be completely pervasive, the undoubted qualities of these spaces, and the threat they face of complete annihilation by development, make it urgent to consider whether and how some of those qualities might be preserved and even understood, as it were, for current and future generations. With this view from the past into the future, the part concludes with a further consideration, in line with the previous chapters, of the capacity of landscape architecture to contribute to practices of preservation, memory and cultural heritage.

The **fifth and final part** is titled 'Philosophy of landscape architecture: Knowledge, practice and education'. The chapters in this part consider how landscape architecture and landscape architects are shaped by the heterogeneous and composite values and knowledge forms involved in landscape architecture as a profession. As a quasi-global endeavour, landscape architecture also reflects the different contexts in which its practices are conducted: the landscapes that landscape architecture constructs are not limited to greenery in traditional ways but extend to urban contexts at large, and to concrete historical and institutional settings. In the volume's **twenty-first chapter**, 'Imaginaries in landscape architecture', Ian H. Thompson points out the existence of deeply seated leitmotifs or landscape imaginaries, stating that we cannot operate without them—whether they be pastoral or the currently pervasive Edgelands imaginary. He therefore urges designers to maintain a critical awareness, since working either implicitly or explicitly with such imaginaries can also leave one open to deceptions and lures. Moreover, the imaginaries we create, foster and sustain constitute an ethical imperative that we need to take into account as well as to reflect on. The **twenty-second chapter**, 'Whose city is it? Public space as agent of change in marginalized settlements in Buenos Aires' by Flavio Janches, focuses on slum areas. Janches uses socio-territorial values as a vehicle for landscape architecture to strengthen the socialization networks, systems of daily life and cultural significations of the communities within these marginalized urban areas. Being both the means and the goal, these transformation strategies are not based on an ideal image; rather they outline a process for the articulation, adaptation and promotion of public places as spaces where the wishes and desires of society are or could be materialized, demonstrating that such imaginaries are pluralistic, correspond to concrete contexts and are inescapable, even in highly socially and culturally contested situations. The concept of the good city, seen from the citizens' perspective, and the role of landscape architecture is also core to the **twenty-third chapter**, '*Khorographos*: Space-scripting' by Michael Tawa. Tawa considers how the city can provide not only the physical amenities but also the psychological contexts that are fundamental to well-being, to a human life that connects person to community and place. He then proposes an idea of the city as choreographing and scripting the conditions for a counteractive civic space of resistance, turning to the philosophers Deleuze, Guattari, Agamben, Heidegger, Derrida and Plato for answers. Kathryn Moore also looks to philosophy, but her aim is to examine how and why design research remains—and perhaps ought to remain—highly contested, that is, in her words, weighed down with confusion and controversy, and fraught with misunderstandings and misconceptions. In the **twenty-fourth chapter**, 'Towards new research methodologies in design: Shifting inquiry away from the unequivocal towards the ambiguous', Moore targets the relationship between the senses and intelligence, which has far-reaching consequences for our understanding of language, intelligence, meaning, the senses and subjectivity. Moreover, she outlines the need to examine and reconceptualize the epistemology, pedagogy and function of design, and hence to re-evaluate some of the assumptions underlying practice-based research and research through design. These fundamental relationships—our understanding of what design, design research and design education are in the case of landscape architecture education—are closely linked both to one another and to questions of the knowledge we as landscape architecture educators consider the most important. In the final and **twenty-fifth chapter**, 'A conversation on education', we invited ten significant educators from different university institutions around Europe and North America: Samantha L. Martin-McAuliffe, Anne Bordeleau, Torben Dam, Lilli Lička, Alan Tate, Tom Nielsen, Inge Bobbink, David Grahame Shane, Cathy Dee, Catharina Dyrssen and Maggie Roe. As most university-based researchers also contribute to landscape architecture education, this conversation seeks to lay bare their intrinsic overlaps and interdependencies. We asked the contributors to reflect on the fundamentals of landscape architecture education such as the different forms of embedded knowledge related to the discipline, on what we see as the recurrent paradox of a globalized profession working in distinctive local contexts and knowledge forms, and on the tensions between the intrinsic aesthetics and ethics of the discipline.

All in all, the twenty-five chapters, in various ways and from various positions, touch upon, reflect, challenge and contribute to recurrent interpretations of the changing cultural motifs of 'nature', and we hope that readers will find that the volume is able to stimulate dialogue and discussion.

We would like to thank the many people who made this volume possible. First and foremost, we thank the authors of the different chapters, who collectively, and with incredible enthusiasm and patience, took the volume's overall aim much further that we could possibly have hoped and made collating their contributions a joyful and rewarding exercise for us. Thanks to Routledge for carrying on with the idea of a volume birthed by Ashgate, and to the line of editors who assisted us in this trajectory. Thanks to the incomparable Merl Storr, for her meticulous help with language editing. Moreover, we would like to thank doctoral fellow Kristen Van Haeren, who multi-assisted us in the final phases of the production, and our home institution, the Section for Landscape Architecture and Planning at the University of Copenhagen, for supporting our efforts in making this volume happen. We also would like to thank Dreyer Foundation for financial support for the book's production.

Notes

1 These shifts are taking place across a variety of disciplines, policies and public debates. We would like to cite a fractured and by no means comprehensive list of recent works that have informed our context: Philippe Descola, *Beyond Culture and Nature* (Chicago: University of Chicago Press, 2013); David Harvey, *The Ways of the World* (Oxford: University of Oxford Press, 2016); N. Katherine Hayles, *Unthought* (Chicago: University of Chicago Press, 2017); Tim Ingold, *The Perception of the Environment* (London: Routledge, 2011); Timothy Morton, *Hyperobjects* (Minneapolis: University of Minnesota Press, 2013); Anna Tsing, *The Mushroom at the End of the World: On the Possibility of Life in Capitalist Ruins* (Princeton: Princeton University Press, 2017).

2 'Le Jardin Planétaire', Gilles Clément website, accessed March 30, 2018, http://www.gillesclement.com/cat-jardinplanetaire-tit-Le-Jardin-Planetaire.

3 See the discussion of the idea of climate optimization in the original argument on the Anthropocene put forward by Paul Crutzen in the famous 2002 *Nature* article that coined the term: Henriette Steiner, 'Nature Created? Or, the Gentle Touch of Artificial Snow', *Montreal Architectural Review* 4 (2018): 5–18. See also James Graham et al., eds, *Climates: Architecture and the Planetary Imaginary* (Bern: Lars Müller Publishers, 2016).

4 For the idea of epistemes active within landscape architecture see Svava Riesto and Ellen Braae: "Seeing, Thinking, Designing Urban Landscapes with Epistemes", Theories and Methods in Landscape Architecture Course 2018, University of Copenhagen

Bibliography

Clément, Gilles. 'Le Jardin Planétaire'. Accessed March 30, 2018. http://www.gillesclement.com/cat-jardinplanetaire-tit-Le-Jardin-Planetaire.

Crutzen, Paul. 'Geology of Mankind'. *Nature* 415 (2002): 23.
Descola, Philippe. *Beyond Culture and Nature*. Chicago: University of Chicago Press, 2013.
Graham, James et al., eds. *Climates: Architecture and the Planetary Imaginary*. Bern: Lars Müller Publishers, 2016.
Harvey, David. *The Ways of the World*. Oxford: University of Oxford Press, 2016.
Hayles, N. Katherine: *Unthought*. Chicago: University of Chicago Press, 2017.
Ingold, Tim. *The Perception of the Environment*. London: Routledge, 2011.
Morton, Timothy. *Hyperobjects*. Minneapolis: University of Minnesota Press, 2013.
Steiner, Henriette. 'Nature Created? Or, the Gentle Touch of Artificial Snow'. *Montreal Architectural Review* 4 (2018): 5–18.
Tsing, Anna. *The Mushroom at the End of the World: On the Possibility of Life in Capitalist Ruins*. Princeton: Princeton University Press, 2017.

PART I

Landscape in the rear-view mirror

Historicizing the field

1

CULTURE, NATURE, A PUNKT IN SPICE

Peter Carl

> Yes, the newspapers were right: snow was general all over Ireland. It was falling on every part of the dark central plain, on the treeless hills, falling softly upon the Bog of Allen and, farther westward, softly falling into the dark mutinous Shannon waves. It was falling, too, upon every part of the lonely churchyard on the hill where Michael Furey lay buried. It lay thickly drifted on the crooked crosses and headstones, on the spears of the little gate, on the barren thorns. His soul swooned slowly as he heard the snow falling faintly through the universe and faintly falling, like the descent of their last end, upon all the living and the dead.[1]

Even in a tale full of conversation, music and song, it is unusual for Gabriel Conroy to hear snow ('falling faintly [...] faintly falling'). The famous conclusion to James Joyce's short story 'The Dead' is, as with Gretta Conroy listening to her past at the top of the stairs, 'like a symbol of something'. Snowflakes are invisible until they pass by a darker background (day) or as shadows falling through light (night). Conversely, they make the normally invisible air rhythmically apparent in a way that rain, fog or dust do not. Snowflakes' supposedly infinite hexagonal variation—all different in their sameness—are like people or their thoughts, their words, histories, memories or their souls. The capacity for snow to recreate things is captured by Mario Soldati:

> Whilst having lunch, he watched the snow falling against the dark russet background of the Palazzo Carignano. The snow, resting on the baroque cornices, on the mouldings of the windows, on the play of recesses and decoration, repeated—fatally and exactly rediscovered—the light traces of the pen of Guarini, when, in his first rapid sketches, he had imagined the façade of the palazzo. [...] The snow simplified everything, like a great designer.[2]

Conversely, for Wallace Stevens, also listening, it is the imagination, or human culture, which 'beholds' something where in fact there may be nothing but implacable natural processes:

> For the listener, who listens in the snow,
> And, nothing himself, beholds
> Nothing that is not there and the nothing that is.[3]

And then of course snow melts, in a pathos of slush and meltwater, to become next year's snow. Accordingly it is plausible that Gabriel Conroy heard lightly falling what most people call 'time'. However, because time as such and in the singular occurs only in philosophy or modern physics, 'temporalities' is preferred here. It is always temporality-of-something-somewhere. Moreover, as regards our own participation in temporalities, it is in fact the ephemeral present that is most obscure; we necessarily use our recollections or experience or memories or customs or traditions to anticipate or fabricate possible futures.

Stephen Dedalus' remark that 'history is a nightmare from which I'm trying to awake'[4] must be set within the cyclic history of Vico which permeates *Finnegans Wake*. The contest between linear and cyclical temporalities in *Finnegans Wake* finds linear sentences striving to complete themselves, like streams constantly encountering obstacles of portmanteau words, archaic or imported languages, elaborate puns, raw sounds, battles, drunken rants, prayers, allusions, people with multiple identities from history or myth or just publicans in Dublin, and so on. Joyce's lilting Dublin accent reading the Anna Livia Plurabelle chapter of *Finnegans Wake* (which concludes with the metempsychosis of two washerwomen into a tree and a stone[5]) is like natural music: the transformation of the Liffey into language becomes the turbulent bubbles, splashes, currents of innumerable temporalities. The mytho-philosophic water of Thales or the river of Heraclitus is recovered, suggesting that riverlanguage mediates between earth and world.

It is in this context that one appreciates Lefebvre's effort to situate rhythm with respect to linear and cyclic temporalities, to observe the mid-century turn to concrete particulars and the embodying conditions, to deploy music as the leading metaphor in his rendering of the phenomena of rhythm, and to speculate—after rejecting a laboratory of rhythmanalysis—that the rhythmanalyst ought to be a poet.[6] Indeed, after remarking the ubiquity of rhythm—from electrons and cell processes to the juices and viscera of plants, animals, people, to the seasons and celestial cycles—he speaks of 'a *garland* of rhythms [...] as if the artist nature had foreseen beauty [...] that results from all its history'[7] echoing Soldati's 'great designer'. With the dual life of Gaston Bachelard in mind—both interpreter of scientific epistemology and, through the rubric of *reverie*, author of reinterpretations of the four elements—Lefebvre precedes his advocacy of poetry with the desire to make of rhythmanalysis a 'science' (presumably intending the more general sense of '*scientia*', before experimental science). His own interpretative method appears to take the form of philosophically informed description of, for example, the polyrhythmy he observes in Parisian streets and squares.

Perhaps because Lefebvre is troubled by the contemporary mediatized superfluity of fragmentary information, references and messages, he devotes only a few sentences to ritual contexts, where rhythm dominates. Against Eliade's over-insistence upon a radical distinction between the sacred and the profane, we find that rituals are rooted in what people do anyway (*praxis*), but attuned to rhythm: voices and words move towards song or poetry; noise is clarified into the music of drums, flutes, horns or strings; bodily movements become procession, dance or significant postures; temples acquire colonnades or buttresses scrupulous with their measure, as well as images, statuary and ornament suffused with rhythm. All this is for the sake of establishing conditions propitious for communicating with deities and their claims from the deep context (animals, plants, earth, weather, heavens); and the places where this happens are generally lavish in their size, materials, ornament, overall orderliness, staffing, languages and comportment by comparison with everywhere else. We should imagine a stratification of depth of meaning descending from ritual (temple) to ceremony (palace) to drama or festival to conventions, norms or customs to habits. The role of plants and animals as manifestations—claims—of the divine, not to mention the mythic topographies of earth and heavens, teach us that the nature-culture continuum was then quite distributed. To the entropic arrow of history and the rhythmic or cyclic temporalities that permeate the continuum, we must add the time-out-of-time of ritual, ceremony, drama or festival, whose capacity for re-enactment (memory) allows recovery of original conditions in the context of history. Where, indeed, does nature leave off and language begin; should the continuum be called culture or nature; are thought or technology the opposite of nature; is there anything that is not nature; is rhythm a principal attribute of what the philosopher's term Being (if so, are the vibrations of an electron, the subdivisions of cells, the articulation of a façade, the metre of a poem, the typicality of human situations all variations on the one thing, rhythm) (Figure 1.1)?

This phenomenon is not confined to Bronze Age urban or tribal cultures, but manifests itself as well in contemporary empirical science. Although Latour continues to explore how matters of fact, from science, might be reconciled with matters of value, from culture, his early ethnography with Steve Woolgar[8] treats the Salk Institute for Biological Studies as an institution for producing statements. Through testing in the laboratory and debate among researchers, these statements move from a condition of contest, even scorn, to a condition of out-there-and-unchangeable, whereupon the statement generally loses its drama and personalities and is simply accepted as a 'fact'. Explicitly adducing Bachelard ('phenomenotechnique', laboratory apparatus as 'reified theory') and—by observing the derivation of 'fact' from Latin *facere*, 'to make'—indirectly adducing Vico's principle of *verum esse ipsum factum* ('the truth is made'), the combination of laboratory craft, debate, writing and politics by which facts are constructed and validated leads Latour and Woolgar to embed scientific research practice in wider social customs and behaviours not dissimilar to the primarily civic concerns of Vico. Despite the objections of

Figure 1.1 Evenki Shaman tent and platform, Siberia, with fish-souls and animal supports, using live and harvested trees, 1907. © Makarenko, A.A.

the scientists that they were discovering, not constructing, facts, this early version of what later became Actor-Network-Theory strove to understand the deep background to our procedures for understanding. It did not, however, go so far as to see the Salk, or their book, as places where nature understands itself; nor did it venture as far as Lefebvre's poet or Bachelard's *reverie*, although Latour and Woolgar speculate on the degree to which their ethnography might be 'fiction'. Nor, finally, did the authors imagine they were writing myth, often wrongly seen to be a preliminary form of the 'explanation' provided by experimental science (via, for example, the gnostic texts of post-sixteenth-century alchemy). The analogical, narrative character of myth, embodied in rites, stories, urban distinctions, etc., has always been more congenial for ethical ontologies than has Enlightenment science; and it is easier to speak of a continuum when not beholden to claims from certainty (distinct from 'truth', which is vaguer and more profound, holding an ethical claim, always requiring reinterpretation). However, the ethnography of *Laboratory Life* is sufficiently rich to be able to grasp why, for example, 'ecology' has increasingly become the working name for the continuum, potentially replacing 'space' in architecture and urban planning.[9] To be sure, 'ecology' is laden with guilt and veers between green redemption and the metabolism of cells, but its most important function is to decentre our—mostly Western—fascination with instrumental generalizations and concepts for the sake of acknowledging the claims upon us of the processes of the deep background, even if many landscape architects and urban designers presently attempt to absorb ecology into 'space' by favouring systems, techniques, aesthetics and cartography—projectivity, in several senses—leaving obscure how the obvious ethical concern arises from these conceptions of nature-culture.

Finitude, history, common-to-all

Electrons, molecules, cells, objects obeying trajectories, etc., only appear in isolation in textbook illustrations, replicating the isolation of the laboratory. Indeed, this capacity for isolation is a particularly human trait, bearing analogies with the ritual sites mentioned earlier (intensification of the conditions), though now dominated by the research protocols of empirical science (exclusion of all but a few conditions). Between the extremes of ritual and laboratory research lie other bounded settings, such as parks and nature reserves (reserved from urban civilization), cinemas, cloisters, not to mention the belief in an 'inner life'[10] and indeed our experience of interiority, which allows us to 'forget' the context. Robots, artificial intelligence (AI), synthetic food or similar manipulations of nature are inevitably trying to secure small, manageable increments of this continuum, thereby obscuring the degree to which one is obliged to all of 'nature'. The principle was established in the transformation over millennia of measure from practical affairs and justice to the apodictic geometry of Euclid,[11] where geometric figures became the vehicle of logical 'proofs' (*apodeixies*), to which every other geometric meaning or symbolic possibility was subsequently judged to be an arbitrary supplement (despite the fifth-century Neoplatonist Proclus' attempt to provide a three-level geometric ontology[12]). Heidegger called this moment of isolation the *Ge-stell* [en-framing], allied it with 'the forgetting of Being', and set his inquiry within that condition. Moreover, the Romanticism that so influenced Heidegger saw nature as the milieu in which the inspired artist or poet, empowered as *natura naturans*,[13] encountered the soul of the earth in a Sublime manifestation of what came to be known as 'the' sacred—that is, as a concept emancipated from the obligations of religious praxis, requiring only contemplation. As science progressively transformed nature into ever more refined fragments governed by systems, a new form of isolation was created out of the persistent translation of concrete phenomena into concepts as instrumental and disaggregated generalizations, including those spawned in philosophies seeking to avoid nature/culture, object/subject dualities, such as Deleuze and

Guattari's rhizome, territory, plateau, assemblage, etc., and my own use of 'continuum'. This two-century contest comes down to distinguishing the general—a 'horizontal' concept, pertaining to many cases, such as all triangles—from the universal—a 'vertical' structure of analogy which orients judgements and seeks to preserve the claims of the fundamental conditions, such as the Trinitarian equilateral triangle at the apex of Borromini's church of San Carlo alle Quattro Fontane, which requires the architecture, the city and its customs, a millennium and a half of theological exegesis and sacramental practice, light symbolism and so forth.

When not isolated in textbooks, electrons, molecules, cells etc., are constantly involved in processes. Nature is thus seen to be always on the job or at work across a continuum between the unimaginably small and the unimaginatively vast (biological cells, let alone toothpaste or a tea ceremony, ultimately require the structure in which resides the galactic supercluster Laniakea[14]). Sympathy with nature-as-*ergon* apparently leaves little scope for analogy: a chain of causation rather than a chain of Being. For the rigorous empirical scientist, analogical interpretation embarrasses itself with thoughts of a great designer, divine or human, as if reverting to 'primitive' modes of thought in the face of the uncanny precision of most natural processes. Positive science cannot allow itself metaphor, except in its speculative phases, which are written out of the final statements of a 'fact'; and this methodological principle has been tacitly accepted as the character of 'nature' itself. This gives rise to the perception of nature as a system of systems (the 'laws of nature'), which constitutes its 'necessity'. The primacy of univocal notation demotes language to second position,[15] encouraging cultivation of the absurd as an expression of human freedom—Dada, Artaud, Roussel, Dubuffet, etc.—as well as perception of tribal ontologies as exotic or 'poetic', closer to 'the' unconscious.

Whether or not God is a mathematician, let us agree that the utility principle serves its purpose, that the focus on 'facts' promotes a discourse that enables international collaboration (evident in all recent scientific publishing), and even that the search for General AI will eventually produce credible simulations of our involvements with the natural conditions. Do we then take Heisenberg's observation that the physicist does not love according to the equations of physics to mean that everything that is written out of the statement of scientific fact actually comprises the full scope of the natural-cultural world? That is, the statement of fact (like the concept of 'system') *depends upon* this cultural-natural world or context, or more properly is one of its modalities, like a rhythmic eddy in the Joycean rivermusic. If that can be accepted, what humans call good and evil—justice—are also part of this account (as Anaximander and Plato believed); and perhaps 'music' (or snow) is too harmonious or homogeneous a metaphor—despite innovations after Schoenberg—to characterize the continuum (or 'discontinuum').

In other words, the generosity of what I am provisionally calling the continuum is not matched by current methodological preferences. Positivism (but not every positivist) is either methodologically deaf to the music or asks us to be patient: eventually all possible music will be framed as statements of fact. The humanities, already adopting scientific methods by the time of von Humboldt, range from the ancient polis-in-the-cosmos rites, philosophies and arts ('obligation-to' the deities or 'freedom-for', commitment to the anonymous whole) to the present much more open speculations on subjectivity, agency, identity, etc., within a contest between versions of 'community' and statistical distributions of individual subjects ('freedom-from', liberty). An obvious question: nature as a relentless material process reminds us how narrow a band human culture occupies in the spectrum between leptons and Laniakea. That narrow band is for humans the matrix for life in the way that fish have the complex ecologies of streams, rivers, lakes or oceans; and perhaps metaphor is relevant *only* to humans. Nonetheless, if the situations of 'culture' depend upon their embodying conditions (and their fourteen billion years of—irreversible—evolution), how is it that some cultures experience an autonomy from nature?

The experience is quite archaic: *The Epic of Gilgamesh* positions the city—fellowship with each other and with gods, administration of justice, fecundity—against a nature whose wildness (Enkidu, Humbaba) is redeemed through sexual relations, whilst *The Book of Genesis*, roughly a millennium and a half later, takes the opposite view: Eden is an oasis of togetherness with YHWH until ruined by sexual knowledge, and Cain's less-favoured offering (sedentary crops versus Abel's nomadic sheep) precipitates a fratricidal city-founding. The suggestion that the transition to urban life is one source of the experience of cultural autonomy from nature resurfaced in the Romantic reaction to the cosmopolitan, industrial metropolis; but even Heidegger never developed his early insight that the city gives a 'direction' to nature.[16]

Although many tribal cultures seem at home in nature's death and regeneration, its modalities of violent power and fruitfulness, its multiple identities, Western culture appears to grant humans the capacity to imagine something outside nature red in tooth and claw (its 'necessity'), something with the potential to universalize nature by making it an eschatological symbol. This nature is domesticated, embodied in Eden or the Virgilian Arcadia, in cloistered and fruitful gardens, fecund fields, herds of food-giving animals, or garlands, all often associated with sacrifice and meant to persist even after death (e.g., Elysium, Heavenly Jerusalem), at least until the seventeenth-century drift towards ruins, to the *Et in Arcadia Ego* theme, or to Dutch floral paintings with their fallen petals, overripe fruit and insects. Implied in the springtime metaphors was a condition that was glorious, unique or at least generous, dignified and fair, notwithstanding humanity's evident capacity for evil. The emphasis here is moral, and is experienced as both freedom and responsibility. Is, then, Western culture named by the convergence of arrogance/hope and guilt/fear, by doubt and choice (Descartes without the epistemology)? Unlike animals or plants or stones, which are, as it were, trapped in their capacities, the world-openness of human physiology and intelligence[17] confirms Aristotle's primacy of *mythos* (narrative, plot) to Greek tragedy, since *mythos* is dependent upon *prohairesis* (choice). Under these conditions, tragedy is characteristic of being-human: the creature for whom, blessed with a memory with the capacity for re-enactment, temporality offers scope for speculating on the overcoming of death and finitude.

Although it is a post-Enlightenment habit to overvalue change and progress ('history') at the expense of the most of life that remains the same—face-to-face discourse, sleeping/waking, hunger/satiety, love/hate, gravity, seasonal change, celestial movements, etc.—it is certainly the case that we presently find ourselves operating in a substantial intermediate milieu that is part-natural and part-technological,[18] although technology is always ultimately natural (technology distinguishes itself from nature mostly by a) the inability to integrate its wastes with natural cycles—it is both a drain on resources and a pollutant—and b) the inability to reconcile any particular technological practice with the others or with the natural conditions, except at the level of the market). This appears in e-comms (more sophisticated in the delivery than the use, which is mostly reading, writing and looking at images), the engineering of food (always accompanied by appeals for natural ingredients), warcraft (ever hoping to mechanize and to distance its conflicts), architecture (with a tendency to speak of construction as 'technology', actually more like the artful assembly of industrially produced components) and cities (deemed compensation for urbanity's 'evils', the greenery is regulated according to criteria of ease of maintenance and health and safety). However, medicine is probably the most comprehensive and sensitive example of this intermediate milieu, in which consignment to laboratory conditions for a cure is simultaneously alienating and reassuring, and for which the research ethics are most rigorous (extended to animals, but not to insects or plants). None of these is wholly robotic or systematic; there are always significant components of what Latour and Woolgar termed 'laboratory craft', and subtle moments of care and judgement. Gadamer rightly declares that the claims of technology, bureaucracy, etc. (further developed into 'the scary new networks

[...] called the informatics of domination' by Donna Haraway[19]) elevate 'our adaptive capacities to privileged status'.[20] The project of technologizing life, stimulated by the ready market for progressively greater ease of accomplishment, seeks to overcome rather than to celebrate human finitude. The resulting cultural attenuation is a consequence not only of the 'flattening' character of these styles of management (complexity rather than richness, the general rather than the universal), but also of tacitly taking for granted the fundamental conditions always-already-there in our fascination with a progressive human history.

Whether or not the World Wide Web will eventually compel everyone to speak weblish or webinese, the great sea of fragments promotes the realization that, in global terms, the practice of 'local' now superimposes upon spatial proximity another based upon shared topics (aggravated by search engines which feed our 'preferences'). We do not all have Shakespeare, Al-Ghazali, Muso Soseki, totemism, eucalyptus trees or information theory in common; it is more likely that we can commonly recognize love, hate, death, an artefact for sale, the difference between an animation and the algorithms which drive it, and so forth. Indeed, the professional discourses of the arts, humanities and sciences make mutual understanding difficult, but communication at the more basic level is comparatively straightforward (so-called plain language). It is instructive to recognize that the rich social practices of primates, dolphins and insect-colonies precede measures of intelligence like tool-use or numeracy, and that these require forms of language.

Figure 1.2 Akrotiri (Thera), Minoan house fresco, before 1627 BCE. © Cima, Caroline.

This is preserved in the general phenomenon of typicality—potentially banal repetition (empty rhythm), but also pointing to what is most common-to-all, most profound, the basis for any possible communication or understanding (Figure 1.2).

Stratification of agonic communication

The depth metaphor here—'profound'—is intentional. Again, the understandings of language that are determined by linguistic or semiotic categories, such as 'code' or 'sign and signifier' (as if language operated independently of content), *depend upon* the open, dynamic, metamorphic structure that enables communication of several kinds across levels or strata: from the level of gesture to that of concepts or symbols, from the levels of nature's orders to those of thought or whatever might lie beyond that. Language is only the most articulate level of the communicative matrix in which our choices are made, and it has the capacity to adapt extremely locally (the argot of a gang) whilst always retaining quite archaic material. This open structure does not operate like a geared or cybernetic system but rather implicitly, across gaps, not unlike human conversation (or synapses, or energy levels within an atom, or the bee delivering pollen). Indeed, the different characters of the gaps—not neutral 'space', they are always charged with content—are determinative of what is possible to communicate, under different conditions. The scheme of objects<relations>field that arises from perception-of (perspectivism) occludes the much deeper structure of involvements-with or claims-upon that, in biological life, is embedded in the billions of years of evolution from the Universal Common Ancestor. The 'biosemiotics' of Jacob Uexküll's *Umwelt*[21] in which each animal (or plant or stone) has (or affords) a differently interpreted 'environment'—a proto-language—based upon its needs (or Wallace Stevens' 'beholding'), gives a useful approximation of how a culture or an ecological niche exhibit both coherence and richness (compare Lefebvre's Parisian street). Heidegger coined a term, *Ereignis*, to designate the event of appropriation by the deep context—an insight, if not the term, which deserves to be extended beyond the human situations he intended.

These communicating gaps do not arise in general, but only in particular situations. Even for humans, the world-openness happens mostly in peripheral vision; it is possible to concentrate effectively only upon one thing at a time (obvious in visual perception, where, for example, one can actually read only one title at a time on shelves of books—it is not a question of optics), although of course attention may shift rapidly. The transfer of molecules in cell metabolism, or the fox chasing the hare, for example, charge their intervals or gaps with quite specific exchanges or goals, whereas the world-openness of human dialogue endows the gap with the quality of a topic, with generous scope for errancy, confusion, wit or deep insight. Indeed, this common ground of difference is central to Socratic dialogue (Socrates depends upon his interlocutor to argue well the opposing view—the insights arise collaboratively, 'between' the disputants), based in turn on the agonic principle (the institutionalization of conflict) found in all significant settings of the *polis* and enshrined in the pro and contra sections, the *agones*, of rhetorical oratory. On this basis, we are encouraged to imagine the overall order of the nature-culture continuum as a texture of myriads of these particular agonic involvements exhibiting a stratification from geobiological processes (and temporalities) to human doubt/insight (philosophy being among the most fragile of our possibilities). This stratification is also a stratification of dependencies—the more sophisticated or articulate layers depend upon the more 'primitive' layers, the possibilities depend upon the conditions, much as a newborn child passes from a biological to a cultural being, needing to be able to coordinate and orient its whole body to particular situations before it can begin to participate in spoken language.

The grand conceptual generalizations, such as 'space', 'time', 'systems and subsystems' and 'society' obscure the infinite granularity of the rhythmic texture of myriads of particular agonic involvements. These are always agonic, because the claim is always made by an 'other': a brightly feathered bird may present an ornithological specimen, a mating display to a female of the species, a participant in an ecological niche as a potential meal or a potential predator to an insect or fruit, a potential disease or cure, a myth, a headdress ornament, fletching for an arrow, a deity, a season, etc., along with all the conditions and possibilities relevant to each situation. To a being in the continuum endowed with memory and its capacity for speculation, the granularity offers multiple schemes for larger organizations and discriminations. These range from parts of organisms—roots—to organisms—silver maple trees—to clusters or groups—a forest—to an ecological region—a piedmont—and its cultures—from Raweno (Iroquois Great Spirit, Creator) to the Tennessee Valley Authority—to attempts to name the whole. It should be emphasized that this last step is not necessary. It is not obvious that the dynamic structure of dependencies or claims can be captured in, e.g., categories or an ontology...or conversely, that 'thought' can be separated from practical involvements. We have worldhood more than we have world, which, like the good, is an ever-open demand for understanding (the 'nature' of our tragic condition). Near Eastern/Mediterranean cultures subsisted for millennia within tensional networks of analogy before the whole was named 'cosmos', reputedly by Pythagoras in the sixth century BCE and possibly anticipated by the advent of a single highest god.[22] Whether this living-inside natural-cultural continuity is sufficient to qualify as an 'ontology', or that term requires the experience and naming of the whole, Descola[23] exhaustively observes, probably overusing the term 'system', that the recorded ontological possibilities can be divided between the anthropocentrism of animism and—mostly Western—naturalism on the one hand, and, on the other, the cosmocentrism of Aboriginal totemism (in which human relations are derivative of the larger natural order) and the socio-cosmic orders of analogism in which, for example, a Dogon individual (here depending on the now-queried account of Griaule[24]) is endowed with several separable identities that can change according to circumstance, even during a day, and can move back and forth across a variety of souls, clavicle seeds, family ponds, foods, ancestors, organized according tribes, castes, functions and so forth. Shorn of the particular (collective) metaphors, we recognize here the variety of personalities an individual may adopt.

As salutary as is Descola's corrective, as part of the ontological turn in anthropology, to the overconfidence of Western culture's division between nature and culture, the assembly of ontologies in a book[25] implies that it is possible to choose between them or even select parts. Under these conditions, each individual culture loses its orienting power and acquires the character of a hypothesis (the material becomes part of knowledge-customs not native to the original cultures). Moreover, Descola fails to identify an important stratification with regard to Western culture, which is both historical and ontological. He rightly observes that pre-Enlightenment modalities of Western culture adhered to analogical styles of thought such as 'the great chain of being', the 'macrocosm-microcosm' theme and, particularly, the Empedoclean–Aristotelian four-temperament oppositions and reciprocities of the natural-cultural order. To this could be added Aristotle's applicability of the four causes to both nature and culture, as well as the constellation of meanings contained in the word 'culture' itself (from the Latin *colere*, to attend to, cultivate, respect—as in 'cult'). Similarly, we still speak of the nature of something or someone to indicate their essential qualities or properties (again from the Latin, *natura*, which has a root in *nasci/natus*, to be born, but is philosophically dependent upon Aristotle's *physei onta*—things which are by nature—inseparable from their *ti estin*, what-it-is). Much of this is the subject of an attempted recovery in Romanticism, but largely in conceptual terms—e.g., 'the' sacred, historicism, pantheistic motifs—and with that the advent of what earlier I termed 'the fascination with history'. The character of this history—effectively, death and replacement—corresponds to the temporalities of technology. If, however, we expand the historical scope of our involvement with nature through making, we can detect at least four levels, whose periods overlap. From most recent to most ancient:

IV. Embodied Information (crowds, flows, computation, economy, complexity theory, emergence, techno-bureaucracy, dispersed identity)—from mid-twentieth century
III. Instrumental concepts (management of fragments, form, space, styles, historicism, aesthetics, industrial capitalism)—from mid-seventeenth century
II. World-as-Picture (perspectival theatre, divided between *natura naturata* (nature as given, 'landscape') and *natura naturans* (the forces of generative nature, which in the sixteenth century migrate from God to the artist/inventor, reprised four centuries later in Schelling's vision of art as a higher nature))—first appears in Hellenistic culture, is then resurrected in the European Renaissance, and develops eventually into the *ego cogitans* and the *res extensa*
I. Practical Life (identity shared with contexts, metaphoric making/craft in the context of rites, myth; role of *techne* with respect to truth and moral goodness in Plato and Aristotle)—archaic

The accuracy and details of this schema are beyond present concern; more relevant is the stratification, which may be regarded as levels of emancipation from the claims of the fundamental conditions. For contemporary making, all levels are always present, and the earlier, more embodying levels provide (materio-historical) conditions for the more articulate or sophisticated later levels. The conceptual and mathematical emphasis of the later two levels have the habit of concealing their dependency upon the earlier levels; and the turn to the embodying conditions begun with phenomenology and recently incorporated into anthropology as well as human geography, AI and robotics is mirrored in the constant tension in twentieth-century visual representation between techno-futurism and 'primitivism'.[26] This in turn is behind the cultural attenuation mentioned earlier: whilst it is imaginable that the nature-culture continuum could be represented as embodied information, the vast 'flat' complexity of such a representation (the 'abstract machine') contrasts with the compact depth (richness) of the more embodied styles of involvement and particularly their more congenial support for ethical orientation (which appears difficult to derive from present mathematical or algorithmic depictions of relations[27]). Involvement with nature has always been a much richer encounter than subject-object or mind-matter. Malafouris follows Pickering's metaphor of a 'machine' in which 'emerges' the content or meaning of the interaction of human and non-human agency, despite the emphasis upon 'mind' in his title.[28] If the lesson of embodiment is that 'thought' depends upon the claims of the less articulate orders, it is the continuum which matters. 'Intelligence' is distributed, residing in the trajectories, directions or affordances invoked by the modes of involvement with—claims of—people, a stone, a lathe, the weather, a computer, etc., each of which comes with its deep background of constituencies and material conditions, and having the quality of a cultural institution.

Ethics and the analogical field

Ethics lie at the heart of the problem of culture's possible continuity with nature, since an individual ethics is a contradiction in terms; it is always a matter of a proper involvement with contexts and their deep backgrounds, even in the violation (not necessarily evil). Although the recurring theme of every agonic involvement, this is never available conceptually or in general, but only in the moral judgements of particular practical circumstances. The appeal to the embodying conditions is an effort to recover universality in the face of the limited and flattening character of the conceptual generalizations. It is clear that human culture cannot find orientation simply through intelligent bodily practice, but must address its tragic dimension, the tension between freedom and responsibility with respect to the fundamental natural conditions, which appears only in the time-out-of-time of celebration or reflection. The impossibility to frame all the music as facts is a mark of the impossibility to step outside of—to objectify—Being. When Heidegger refers to the 'strife' (*agon*) of earth and world,[29] of conditions and possibilities, in which we strive to participate, he rethinks *physis/nomos*: he argues that we humans can never find world outside the claim of earth; conversely, earth is, as it were, always-already architecture.

Let us take seriously the prevailing structures of reference—the claims—in which we find ourselves, as the basis for beginning to understand the nature of our tragic conditions and possibilities. A compelling example is this short extract (omitting some marginal remarks) from *Finnegans Wake*, which rehearses most of what has been said here:

> Pastimes are past times. Now let bygones be bei Gunne's. Saaleddies in this warken werde, mine boerne, and it vild need olderwise since primal made alter in garden of Idem. The tasks above are as the flasks below, saith the emerald canticle of Hermes and all's loth and pleasestir are we told, on excellent inkbottle authority, solarsystematised, seriolcosmically, in a more and more almighty expanding universe under one, there is rhymeless reason to believe, original sun.
>
> Footnote 2: We dont hear the booming cursowarries, we wont fear the fletches of fightning, we float the meditarenias and come back to the isle we love in spice. Punt.[30]

There are five stories superimposed here, two journeys and three transformations, within the time-out-of-time of pastimes (or rites or drama), suspended between historical and cyclical temporalities. A rough literal rendering of the passage reads: Play is re-enactment. Let the origins of pasts be present ('Now') in Gunne's theatre (contra Bacon's *Idola Theatri*). It's like this in this beautiful world, my children; and, wild before (*older*-wise), it would need ancient wisdom (older-*wise* and other-*wise*) since first (God) made second (or other, Latin) or age or older (German, i.e. temporality/change) or an altar in the Garden of Eden/the Same (the same 'beautiful world', the divine Same against which historical change is measured and the Incarnation). The rhyming tasks/flasks paraphrase the second dictum of the Emerald Tablet of Hermes Trismegistus, indicating that heaven can be influenced through alchemy's love/loathing (*coincidentiae oppositorum*) and pleasure/please stir (a stirred pot is used in a later chapter to join embalming with the copulation of ALP and HCE), as well as through lath and plaster, basic architecture. However, heaven is now a solarsytematised universe that is serio-comically[31] expanding from an almighty (usually said of God) big bang, [32] the original sun/Son/sin, according to rhymeless reason (mathematics, calculation), issuing from an inkbottle-flask. Meanwhile, in the footnote, Pharaoh Hatshepsut's cursing warriors/cassowaries sail inland (Latin, *mediterraneus*—'mid-land', but echoing the Mediterranean—and *in teneris*, in childhood) through flashes of lightning and arrows of fighting to the land of Punt for spices used in cosmetics and embalming, re-enacted as a trip to the 'Spice Islands', Irish slang for a back-garden privy. The journey successfully concludes with 'Punt!', the sound of a turd hitting the ground (one of the many modalities of the alchemical *prima materia*), creating a full stop at the end of a sentence and a punkt/point in spice/space. The similitudes may be charted as in Figure 1.4.

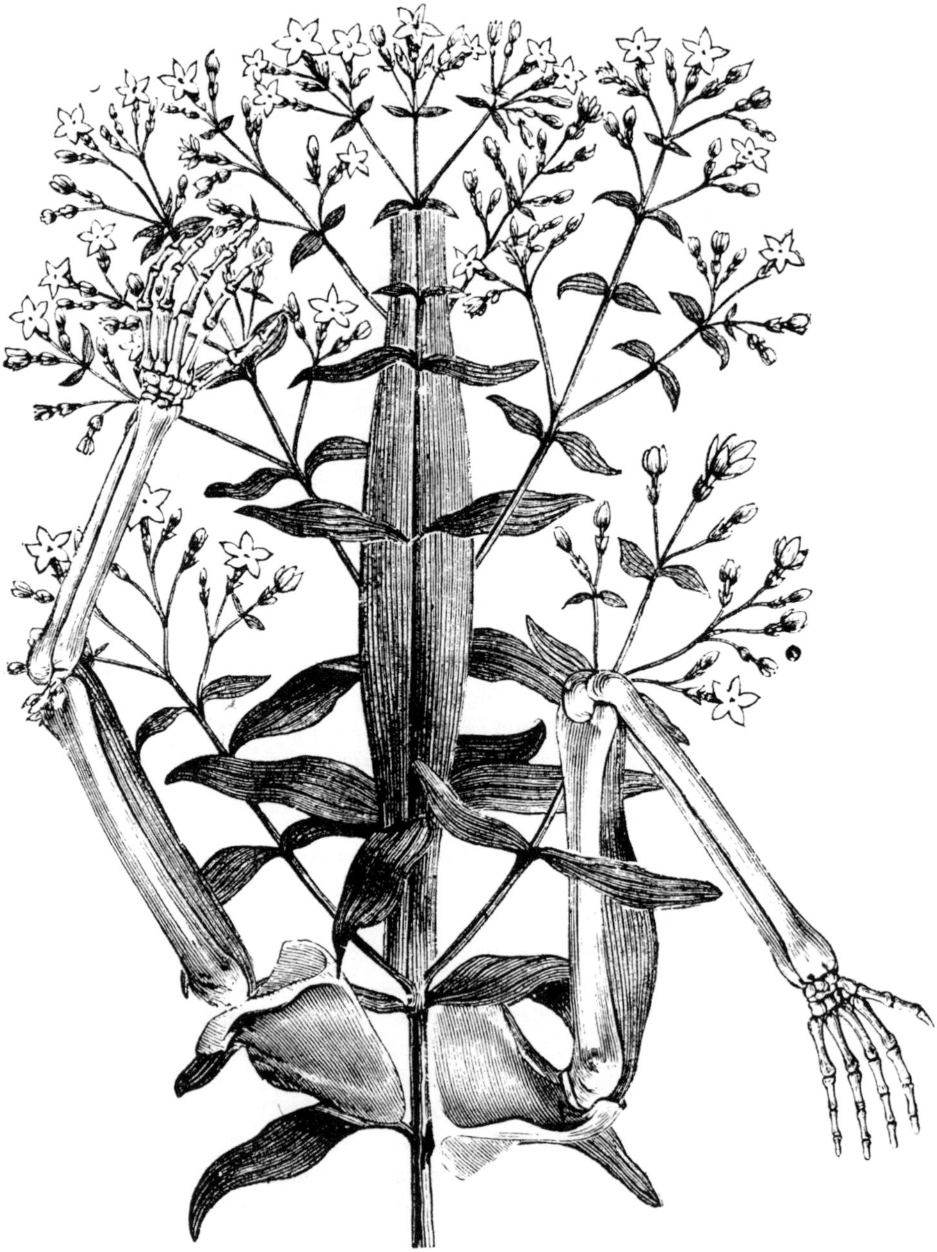

Figure 1.3 Human–plant fragments imagined to be at the disposal of culture. © Ernst, Max. *Une Semaine de Bonté*, editions Jeanne Bucher, Paris. 1934. © ADAGP, Paris and DACS, London 2018.

Unlike the examples from Descola, which attract the term 'ontology' because all of local natural-cultural life is implicated, *Finnegans Wake* assembles verbal fragments from the European tradition and its imports. We may envy the wisdom of Aboriginal cosmocentrism; and it would seem that the Western tradition represents the most attenuated and emancipated version of

TOPIC	OUTHOUSE	PUNT	CREATION EDEN	ALCHEMY	CREATION UNIVERSE
SITE	garden/privy body	sea/vessel tomb	world garden	lab-oratory flasks	galactic space
PLOT	journey	journey	transformation	transformation	transformation
PROCESS	walking peristalsis	trade voyage battle embalming	*creatio ex nihilo* naming	symbolic chemistry	big bang
OUTCOME	turd (punkt in spice)	judgement + rebirth as Osiris	judgement + salvation	transfiguration – of materials – of adept	distribution of matter + energy
		Original Sin Original Son			Original Sun

Figure 1.4 Let Bygones Be Begun. © Carl, Peter.

natural-cultural possibilities, the one most threatening to bioclimatic equilibrium, and therefore the home most in need of looking-after, the least clear about its ethical obligations. This home presents itself as a natural-cultural dilemma, which basically comes down to finding a form of practical wisdom able to reconcile ethics with the 'nature' of the deep background, the context for judgements.

Finnegans Wake might be considered a preliminary exposure of the analogical field in which that dilemma exists and might find orientation; indeed, possible ontologies are put in question. Lacking the universality that naturally arises from contesting topics in particular situations and places, the prose refuses to be simply transparent, and has its own iconicity or concreteness. Enmeshed in a language that veers between babbling sound and philosophical or theological insights, the reader can no longer assume the stability or clarity of written language—Western culture's primary vehicle of understanding, expression, orientation, influence, manipulation and confidence. Higher levels of abstraction—e.g., matrix mathematics, algorithms or theories of signification—are of limited use here; rather one must revert to one's more 'primitive', embodied levels of understanding, following the Kierkegaardian insight that wit and tragedy call upon our deepest assumptions about life's orderliness or otherwise.[33] Moreover, reading *Finnegans Wake* requires entering into collaboration and conflict with many other lexicographers, interpreters and scholars ... not all of whom (least of all me) can claim thorough understanding. Joyce's text is the site of the time-out-of-time of reflection, rites, drama, play, in which the conditions and possibilities of tragic understanding are stirred and stewed—even digested. It is not, however, an arbitrary or capricious lexico-grammatical scramble; the simplicity of its main 'plot' elements—a family, a pub in Chapelizod village on the Liffey, a wake, children learning their lessons, etc.—conforms to the basic level of communication mentioned earlier. We readily recognize in the extract, for example, that just as a stage is capable of being a living room or battlefield (or both), or an Egyptian temple a field of reeds, the sites share common grounds of difference (e.g., back garden, Eden, sea, universe—imagine Heideggerian 'clearings' in a forest of Derridean *différance*) for the *agon* which each journey or transformation has for its topic (e.g., original creation, death/rebirth or entropic dispersion). The two forms of plot or narrative (journey and transformation) re-enact the two principal forms of ritual, the procession and the sacrifice. The similitudes resonate up and down the stratification from the claims of the natural conditions to the great dramas of hopeful or fatalistic hypotheses, from child's fable to relativity.

Not all of *Finnegans Wake* exhibits so focused a meditation as this brief passage. In the face of our desire for a book or reality to be simple, concise and emotionally consistent, the work is much more accepting of absurdity, contradiction and mere coincidence than are Catholic dogma, scientific, social or political theories (and their inevitable happy endings), or even common sense. Against the numerous critics who, prompted by Joyce himself, would file *Finnegans Wake* safely under night-time dreams-and-the-unconscious, or under authorial psychological traits (e.g., glossolalia), it seems preferable to acknowledge in *Finnegans Wake* a basic honesty to the confluence of doubt and hope, of confusion and insight, of cruelty and generosity, of annoyance and inspiration that marks human culture's tragic suspension between freedom-from and obligation-to the fundamental natural conditions.

As our scientific understanding of nature has broadened, our capacity to understand human culture as part of nature has diminished; it is as if human cultural life is supplementary to nature-as-system. Historically, we have moved from deification of nature to its exploitation, even abuse, and from an ethics situated in the natural conditions to a morality oriented about individual rights and emancipation from those conditions (here distinguishing 'ethics', reserved for the order of the whole, from 'morals', referring to the judgements made in concrete practical circumstances). Science necessarily contributes to the practical concerns of 'sustainability'; but the ontological problem remains, since it is impossible to derive an ethics from logico-empirical statements of fact.

However we might formulate the continuum, it is more likely to be oriented around the rhythms of agonic involvements than conceptual dualities such as stasis-kinesis, identity-difference or psycho-geography. Similarly, the insights will come from the involvements, therefore not from texts or mathematical relationships alone. On the principle that the embodying conditions provide orientation for the more sophisticated discourses, we might look to architecture. Here Schiller's famous remark—that the temples still command respect, even though the gods have been declared ridiculous[34]—remains true, but obscurely so, since even these, like the great museums, are marked by the historicist sadness Kracauer observed of an archaicizing translation of the Bible.[35] We might accept that the collection of utopian experiments, the temples, churches or ashrams, the charities, the local communities devoted to the arts or to growing/making/recycling things, and the universities comprise an appropriately 'bottom-up' and disaggregated replacement for the more 'top-down' great rooms (e.g., Great Mosque, Córdoba) or temple topographies (e.g., Tiruvannamalai) with their rites, institutions and capacity for celebration of profound ethical reflection. Conversely, it is also possible that the claim of the ethical conditions might eventually predominate over that of individual freedom and we will find new topographies and their institutions, perhaps after the fashion of the late medieval northern Italian city-states that required four centuries to discover civic Humanism out of civil war.

Notes

1 James Joyce, 'The Dead', in *Dubliners* (London: Grant Richards, 1914), 223.

2 Mario Soldati, 'I passi sulla neve', in *Italian Short Stories: Racconti in Italiano*, volume 1, ed. Raleigh Trevelyan (Harmondsworth: Penguin Parallel Text Series, 1973): 149–174.

3 Wallace Stevens, 'The Snow Man', in *Poetry* magazine, Chicago, 1921. Cf. Johannes Kepler, *The Gift, or Hexagonal Snow* (now available as *The Six-Cornered Snowflake*, trans. Colin Hardie, Oxford, Clarendon Press, 1966, originally 1611); Wallace Stevens, *Wallace Stevens, Poems* (Poemhunter.com – The World's Poetry Archive, 2004): 91, https://www.poemhunter.com/i/ebooks/pdf/wallace_stevens_2004_9.pdf.

4 James Joyce, *Ulysses* (Paris: Sylvia Beach, 1922), 34.
5 See A.W. Forte, 'Speech from Tree and Rock: Recovery of a Bronze Age Metaphor', *American Journal of Philology*, 136, no. 1 (Spring 2015).
6 Henri Lefebvre, *Rhythmanalysis* (London and New York: Continuum, 2004).
7 Ibid., 20.
8 Bruno Latour and Steve Woolgar, *Laboratory Life: The Construction of Scientific Facts* (Princeton: Princeton University Press, 1986).
9 Charles Waldheim, *Landscape as Urbanism* (Princeton: Princeton University Press, 2016).
10 'We dream of journeys through the cosmos; but is the cosmos not in us?' Novalis, 'Blütenstaub Fragments', *Atheneum* 1, 1798).
11 Peter Carl, 'Architecture, Justice, Conflict, Measure', in *Architecture and Justice: Judicial Meanings in the Public Realm*, eds. Renée Tobe, Jonathan Simson and Nicolas Temple (Farnham: Ashgate, 2013); Reviel Netz, *The Shaping of Deduction in Greek Mathematics: A Study in Cognitive History* (Cambridge: Cambridge University Press, 1999).
12 Proclus, *A Commentary on the First Book of Euclid's Elements* (Princeton: Princeton University Press, 1970), 50.
13 'you must master the essence, the *natura naturans*', in Samuel T. Coleridge: *On Poesy or Art* (paragraph 8 of the lecture, 1818, now in *Lectures, 1808–1819: on Literature*). Bollingen Series 75 (Princeton: Princeton University Press, 1987).
14 Noam I. Libeskind and R. Brent Tully, 'Our Place in the Universe', *Scientific American* (July 2016).
15 Hans Georg Gadamer, 'Text and Interpretation', in *Dialogue and Deconstruction*, eds. Diane P. Michfelder and Richard Palmer (Albany: SUNY Press, 1989).
16 Martin Heidegger, *Being and Time* (Oxford: Blackwell, 1962), 100.
17 Arnold Gehlen, *Man, His Nature and Place in the World* (New York: Columbia University Press, 1988).
18 Occasionally termed second or third nature. See William Cronon, *Uncommon Ground: Rethinking the Human Place in Nature* (New York: W.W. Norton and Company, 1996) and Gilbert Simondon, *The Mode of Existence of Technical Objects* (Minneapolis: Univocal Publishing, 2017).
19 Donna Haraway, *Simians, Cyborgs, and Women: The Reinvention of Nature* (New York: Routledge, 1991).
20 Hans Georg Gadamer, *Reason in the Age of Science* (Cambridge, MA: MIT Press, 1981), 73.
21 Jacob von Uexküll, *A Foray into the Worlds of Animals and Humans* (Minneapolis: University of Minnesota Press, 2010) and James J. Gibson, 'The Theory of Affordances', chapter 8 of *The Ecological Approach to Visual Perception* (Boston: Houghton Mifflin, 1979).
22 Jan Assmann, *Of God and Gods: Egypt, Israel, and the Rise of Monotheism* (London: University of Wisconsin Press, 2008).
23 Philipe Descola, *Beyond Nature and Culture* (Chicago: University of Chicago Press, 2013.)
24 See W.E.A. van Beek, 'Dogon Restudied', *Current Anthropology* 32, no. 2 (April 1991): 139–167.
25 Whose Structuralism is shared by Mircea Eliade, *Patterns in Comparative Religion* (London and New York: Sheed and Ward, 1958).
26 On this tension, see Dalibor Vesely, *Architecture in the Age of Divided Representation: The Question of Creativity in the Shadow of Production* (Cambridge, MA: MIT Press, 2001).
27 Contra Max Tegmark, who professes a purely materialist version of the continuum and who would see the goal-oriented quality of current AI to be the basis for a generally accepted ethical framework in *Life 3.0: Being Human in the Age of Artificial Intelligence* (London: Allen Lane, 2017), chapter 7.
28 Lambros Malafouris, *How Things Shape the Mind: A Theory of Material Engagement* (Cambridge, MA: MIT Press, 2013); Andrew Pickering, *The Mangle of Practice: Time, Agency, and Science* (Chicago and London: University of Chicago Press, 1995).
29 Martin Heidegger, 'Origin of the Work of Art', in Martin Heidegger, *Basic Writings,* ed. David Farrell Krell, revised edition (Routledge, 1993).
30 James Joyce, *Finnegans Wake* (London: Faber and Faber, 1939), 263.
31 Also serial-cosmically, after the temporal theories of the Irishman J.W. Dunne, *The Serial Universe* (London: Faber and Faber, 1934).
32 Called the 'Cosmic Egg' by the Jesuit Lemaître, discoverer of cosmic expansion in 1927.
33 Søren Kierkegaard, *The Concept of Irony with Continual Reference to Socrates* (Princeton: Princeton University Press, 1989).
34 Friedrich Schiller, *Letters on Aesthetic Education of Man, IX* (in his journal, *Horen*, Tübingen, 1795).
35 Siegfried Kracauer, 'The Bible in German', in *The Mass Ornament: Weimar Essays* (Cambridge, MA: Harvard University Press, 1995).

Bibliography

Assmann, Jan. *Of God and Gods: Egypt, Israel, and the Rise of Monotheism*. London: University of Wisconsin Press, 2008.

Carl, Peter. 'Architecture, Justice, Conflict, Measure'. In *Architecture and Justice: Judicial Meanings in the Public Realm*, edited by Renée Tobe, Jonathan Simson and Nicolas Temple, 189-202. Farnham: Ashgate, 2013.

Coleridge, Samuel T. *On Poesy or Art* (paragraph 8 of the lecture, 1818, now in *Lectures, 1808–1819, on Literature*). Bollingen Series 75. Princeton: Princeton University Press, 1987.

Cronon, William. *Uncommon Ground: Rethinking the Human Place in Nature*. New York: W.W. Norton & Company, 1996.

Descola, Philippe. *Beyond Nature and Culture*. Chicago: University of Chicago Press, 2013.

Dunne, J.W. *The Serial Universe*. London: Faber and Faber, 1934.

Eliade, Mircea. *Patterns in Comparative Religion*. London and New York: Sheed and Ward, 1958.

Forte, Alexander S.W. 'Speech from Tree and Rock: Recovery of a Bronze Age Metaphor'. *American Journal of Philology*, 136, no. 1 (Spring 2015): 1–35. https://muse.jhu.edu/article/579063.

Gadamer, Hans Georg. *Reason in the Age of Science*. Cambridge, MA: MIT Press, 1981.

Gadamer, Hans Georg. 'Text and Interpretation'. In *Dialogue and Deconstruction: The Gadamer-Derrida Encounter*, edited by Diane P. Michfelder and Richard Palmer, 21–51. Albany: SUNY Press, 1989.

Gehlen, Arnold. *Man, His Nature and Place in the World*. New York: Columbia University Press, 1988.

Gibson, James J. 'The Theory of Affordances'. Chapter 8 of *The Ecological Approach to Visual Perception*, 127–146. Boston: Houghton Mifflin, 1979.

Haraway, Donna. *Simians, Cyborgs, and Women: The Reinvention of Nature*. New York: Routledge, 1991.

Heidegger, Martin. *Being and Time*. Oxford: Blackwell, 1962.

Heidegger, Martin. 'Origin of the Work of Art'. In Heidegger, Martin, *Basic Writings*, ed. David Farrell Krell, revised edition, 139–212. Routledge, 1993.

Joyce, James. *Dubliners*. London: Grant Richards, 1914.

Joyce, James. *Ulysses*. Paris: Sylvia Beach, 1922.

Joyce, James. *Finnegans Wake*. London: Faber and Faber, 1939.

Kepler, Johannes. *The Gift, or Hexagonal Snow*, originally 1611, now available as *The Six-Cornered Snowflake*. Oxford: Clarendon Press, 1966.

Kierkegaard, Søren. *The Concept of Irony with Continual Reference to Socrates* Princeton: Princeton University Press, 1989.

Kracauer, Siegfried, 'The Bible in German'. In *The Mass Ornament: Weimar Essays*. Cambridge, MA: Harvard University Press, 1995.

Latour, Bruno and Steve Woolgar. *Laboratory Life: The Construction of Scientific Facts*. Princeton: Princeton University Press, 1986.

Lefebvre, Henri. *Rhythmanalysis*. London and New York: Continuum, 2004.

Libeskind, Noam I. and R. Brent Tully. 'Our Place in the Universe'. *Scientific American* (July 2016): 33–39.

Malafouris, Lambros. *How Things Shape the Mind: A Theory of Material Engagement*. Cambridge, MA: MIT Press, 2013.

Netz, Reviel. *The Shaping of Deduction in Greek Mathematics: A Study in Cognitive History*. Cambridge: Cambridge University Press, 1999.

Novalis. 'Blütenstaub Fragments'. *Atheneum*, 1 (1798).

Pickering, Andrew. *The Mangle of Practice: Time, Agency, and Science*. Chicago and London: University of Chicago Press, 1995.

Proclus. *A Commentary on the First Book of Euclid's Elements*. Princeton: Princeton University Press, 1970.

Schiller, Friedrich. *Letters on Aesthetic Education of Man, IX*, in his journal, *Horen*, Tübingen, 1795.

Simondon, Gilbert. *The Mode of Existence of Technical Objects*. Minneapolis: Univocal Publishing, 2017.

Soldati, Mario. 'I passi sulla neve'. In Italian Short Stories: Racconti In Italiano, volume 1, ed. Raleigh Trevelvan, 149–174. Harmondsworth: Penguin Parallel Text Series, 1973.

Stevens, Wallace. 'The Snow Man'. *Poetry* magazine, 1921, available at: https://www.poemhunter.com/i/ebooks/pdf/wallace_stevens_2004_9.pdf.

Tegmark, Max. *Life 3.0: Being Human in the Age of Artificial Intelligence*. London, Allen Lane, 2017.

Uexküll, Jacob von. *A Foray into the Worlds of Animals and Humans*. Minneapolis: University of Minnesota Press, 2010.
van Beek, W.E.A. 'Dogon Restudied'. *Current Anthropology* 32, no. 2 (April 1991): 139–167.
Vesely, Dalibor. *Architecture in the Age of Divided Representation: The Question of Creativity in the Shadow of Production*. Cambridge, MA: MIT Press, 2004.
Waldheim, Charles. *Landscape as Urbanism*. Princeton: Princeton University Press, 2016.

2

RENAISSANCE GARDENS

Topicality and the scene of nature

Clare E.L. Guest

Gardens are real and imaginative places, images of collective aspiration or personal sanctuaries. This essay on Italian (predominantly Roman) Renaissance gardens considers how the garden as theme and metaphor worked with gardens as planned topographies, in a period when the garden 'contained' defined arguments. These gardens are more than heritage showpieces; they reflect how we 'disclose' nature by the meanings with which we endow it. Their symbolic character, where human fulfilment is figured in the perfection of place, appears alongside conceptual tendencies which use design instrumentally. They address questions whose contemporary guise appears in phenomenological and ecological concerns with the experiential character of place, or relations between nature and design; as early cases or anticipations of privately owned 'public' spaces they presage issues about communality. If they require historical scholarship, they transcend historical significance as architectural embodiments of the mental apparatus which determines our 'direct' encounter with nature.

Renaissance gardens were rooted in a teleological vision of nature as a whole working to preordained ends, permitting allegorical interpretation within a providential narrative concerning conditions of being. Nature was the book of nature, written in symbols, where everything had moral and figurative as much as physical properties, and where contemplation of creation enabled the ascent from the visible to the invisible things of God (Romans 1:20). Once that teleological and analogical understanding of nature faded, gardens lost much of their metaphoric resonance, and this affected their role as political symbols, as examined in this chapter.

The garden as topos (place, image, argument) was a pervasive spiritual, poetic and political metaphor extending beyond literary conventions of the *locus amoenus* or amorous 'pleasance'.[1] Irrespective of context, the garden as theme implies flowering, regeneration and the seasonable, accommodation or ordering of variety, and protective enclosure; this makes it central to the syncretism and anachronism of Renaissance Humanism.[2] The meaning of Renaissance gardens is not just a 'programme' imposed through a monumental itinerary but a play between a context and its topoi, with their organizing and transformative force.

As topos, the garden gives a situation which is not only rhetorical but phenomenological; it opens possibilities and promises fulfilment in the relation between person and place.[3] It is inseparable from the city as locus of striving, as reflected in paradise as political and spiritual theme. Renaissance gardens attempted to synthesize the garden as organizing topos with the garden as image of civic and urban renewal. This synthesis became increasingly reliant on conceptual tools such as perspective which abstract the relationship of person and place. By the late Renaissance, gardens such as Boboli are 'theatres of nature' whose typified design would carry certain themes to any place. The resulting generalization of the spectator is a significant legacy of Renaissance gardening, resulting from thematic handling of the garden as well as its design. Formulations such as *rus in urbe*, coined by Martial ironically (*Epigrams* 12.57), ultimately weaken the city as paradise into the city as park—symbolic depth relinquished for spatial extension in the city as urbanistic entity.

One achievement of Italian Renaissance gardening is *natura artificiosa*—the artifice of grottoes and waterworks, which 'discovers' nature in activity so that landscaping appears to generate figurative meaning. Italian Renaissance gardens concerned process, attested by ruins or antiquities, the hylomorphic drama of the grotto or the interpretative puzzles set for visitors. Their problematic character lay in the replacement of topical associations with conceptual instruments to relate place and person—usually conceived as a visitor, a transitory alien, rather than the seeker after paradise as the end of travails and place where the self is found.[4]

Topos

Already in Homer's *Odyssey*, gardens are political images: the tended orchard of Alcinous has an antitype in the pleasance of a predatory female (Calypso, Circe) which entraps and thwarts heroic activity. The Persian *paradeisos* as image of ethical action leading to a final time of happiness develops in Xenophon's descriptions into an image for the well-governed state; in Ovid's *Fasti* V. 295–330, Flora's gardens wither when the Roman senate neglects her rites.[5] The political image endured in English landscape gardens like Stowe, and survives in contemporary examples such as Little Sparta.

The garden can be a metaphor for good government, a metonym for its effects, or a situation in civic life, as in the classical *negotium-otium* pairing used by Alberti to articulate architectural decorum. *Otium* is not just leisure but reflective distance used to recreate oneself for deepened civic engagement; the mutually defining character of *otium* and *negotium* appears in the interdependence of garden and city in ancient Roman villas as rural complexes incorporating urban features (libraries and gymnasia). The Roman villa is a privatized equivalent of ancient philosophical schools whose names derived from leisurely settings—Academic grove, stoa or Epicurean garden, which Pliny regarded as originating *rus in urbe*.[6]

The garden's retirement and interiority is a mode of involvement with the world, where the theme of amorous leisure initiates experience of the numinous (e.g., Plato's *Phaedrus*) and political reflection, as in Virgil's *Eclogues*, the canonical work of pastoral literature and traditional entry text for Latin studies.[7] The garden as topos implies ethical and spiritual involvement in the city or city of God; there is continuity between ancient philosophical schools, the Roman villa and the monastery as paradise, developed at Subiaco from the ruins of Nero's villa.[8]

The fourteenth-century Tuscan Humanists Petrarch and Boccaccio, who cement the *locus amoenus* as *the* locus of poetry, view the garden's political character in terms of cultural renovation.[9] Petrarch's 1341 coronation as poet laureate on the Capitoline Hill concerned Roman civic renewal; the fame won by poetry paralleled the eternity of Roman civilization (*civitas*) and contrasted with the decayed *urbs*. Boccaccio's interest in the poetic-political garden was influenced by his 1327–40 stay in Naples, heir to a southern Italian-Sicilian tradition of gardening and antique display by twelfth-century Norman rulers.[10]

Fifteenth-century Neapolitan Humanists Pontano and Sannazaro, who delineate the poetics of the antiquarian garden, develop the Neapolitan orange grove as symbol of benevolent rule into a larger Hesperidean and Arcadian theme.[11] If they describe the citrus grove as *paradeisos* of the Neo-Latin Muses, they link the garden or Arcadia with death, loss and memory, themes suitable to the antiquarian garden as a site of *memorie*—memory and memorials, or ruins.[12]

Hortus conclusus

Christian reflection on the garden concerns creation of the species—plants, creatures and heavenly bodies—called in commentary on Genesis (hexaemeral commentary) the *ornatus mundi* or *exornatio mundi*.[13] In the exornation of paradise, potentiality blossoms into the beauty of form in the variety of the species and the individuation of each thing. Paradise is the locus of nourishment and cultivation, which concerns the soul as well as husbandry. The garden as habitat is the model for human interventions in nature; in *De natura deorum II* Cicero includes cities as instances of *ornatus mundi*.

From its inception, hexaemeral literature provided an armature for reconciling scripture with classical cosmology, notably Plato's *Timaeus*; thus Philo (d. 50CE) compared the chorus of the Muses to the planting of Paradise.[14] This syncretism developed into the cosmogonic allegories of the twelfth-century Chartres school, such as the *Cosmographia* of Bernardus Silvestris, where the 'dim and baffling' *chōra* (material substrate and place as a condition for individuation) of *Timaeus* becomes the parturition of the womb of matter.

The cosmogonic womb which blossoms into paradise in demiurgic fables had a prototype in Mary's conception of Christ; Mary's womb is frequently called *chōra* in Byzantine discussions.[15] The metaphor of Mary as enclosed garden (*hortus conclusus*), celebrated in the luscious imagery of the Song of Solomon, concerns Mary's virginity; Irenaeus compares it to the paradisal earth from which Adam was fashioned.[16]

Theologically, Mary's uncorrupted womb as paradise denotes her character as *Theotokos*, bearer of God; this concerns Christ's divine and human natures in the Incarnation, debated intensely in Byzantine Christological controversies. Thus invocations of Mary as unfading flower and uncorrupted fruit appear in the Greek Church, such as the Byzantine Akathistos hymn, a recitation of Mary's attributes influential on hymns which celebrate her endlessly as *amoena*, flower, spring, fruit, vine, cypress, garden, grove, earth or protecting wall.[17]

These Marian metaphors continued through the preaching and devotions of mendicant orders, notably rosary devotions, which gained widespread popularity in the fifteenth century. The metaphors were both popular and theologically precise: expounded in theology and homilies, diffused in songs and prayers, given ritual form in liturgy (processional hymns and sequences) and visual expression in artworks and architectural settings. They consolidated the sacred symbolism of the garden as an inalienable element—and anchored it to a female custodian or nymph as an orienting, organizing figure.

Nympha loci

The nymph is a point of convergence for understandings of what is sought in the garden and what we become within it. She illuminates the garden topically as axial planning focuses it visually, as at the Medici villa at Castello, where Giambologna's statue ofVenus-Fiorenza stood at the heart of a labyrinth at the villa's core.

The term nymph, meaning nubile girl or bride, is used of Mary and brides of Christ – virgin saints, nuns, the soul or the Church – as well as pagan motifs like the maenad.[18] It forms the refrain of the Akathistos ('Hail, nymph unwedded [*anympheute*]') and echoes through hymnody alongside its classical identifications with local deities, elemental spirits (daemons), poetic conventions and antique sculptures.[19] There was thus a spiritual tradition for invocations of Mary as nymph in the Neo-Latin poetry of Christian Humanists, such as Carmelite vicar-general Mantovano (1498)[20] or the *Coryciana* epigrams (1524). The latter commemorated Sansovino's statue of St Anne, Mary and Christ, commissioned by the papal protonotary Goritz for the church of Sant'Agostino and celebrated in Goritz's poetry festival, held in his gardens near Trajan's forum on the feast of St Anne.[21] The poems, affixed to citrus trees and statues, adapted conventions from ancient epigrams to celebrate the living presence of God in the effigy; they showed the role of gardens as setting and symbol for a deeper synthesis of ancient and Christian, presented anachronistically as a renovation.

The nymph illuminates the complex relations of classical revival and Christian controversy in Roman antiquarianism, where gardens pertained to urban renovation via archaeology, political-cultural restoration of the *civitas* through Humanism and spiritual renewal which placed poetic and antiquarian activities, as well as nature itself, within conditions of salvation.[22] These complex relations are illustrated by the interests of Sixtus IV (1471–84), who 'donated' Roman monuments to the Capitol, established the Feast of the Immaculate Conception (1476) concerning Mary's 'paradisal' freedom from Original Sin, and encouraged study of Philo and Greek Patristics and reformation of the classical Pomponian academy.[23]

In this syncretic context the nymph represented tensions as well as continuity. Christian correctives to the antiquarian garden, such as Erasmus' *Convivium religiosum* (1522), on an ideal Christian Humanist villa and Louis Richeome's *La Peinture spirituelle* (1611), describing the *hortus* of the Jesuit novitiate on the Quirinal, an area rich in villas, return to the garden as allegorical book of nature, albeit displayed through Humanist monumental epigraphy. The shift from enclosed *hortus* to antiquarian display ground appears in the *Hypnerotomachia Poliphili* (1499), the illustrated romance concerning an erotic quest for the nymph Polia through antiquities, pleasure buildings and gardens.

The *Hypnerotomachia's* nymph theme, heavily inspired by Boccaccio, exploits the literary topic as a locus where arguments are 'found'; its minute architectural descriptions reflect the loci of mnemonic techniques. Its accommodation of *ninfale*, Marian *hortus* and antiquarianism is a profanation which parodies Marian iconography and challenges classicizing ideals of the garden as scene for Christian Humanist *otium*.[24] Unlike the hexaemeral garden as locus of self-knowledge, the antiquarian landscape of tombs and fragments is a place of loss and evanescence whose *nympha loci* dissolves in a delusive dream from which the narrator finally awakens.

With its extended monumental topography devoid of civic values, the *Hypnerotomachia* subverts the (Roman) theme of urban renovation as the fashioning of an earthly paradise. It signals the presence of fantasy at the heart of antiquarianism and its engagement with *memoria*, shown in the association of garden, nymph and tomb shared with contemporary epigrams such as Pontano's *Tumuli*.[25] Unlike the species as encyclopaedic allegories in hexaemeral gardens, the sculpture garden prizes fragmentary epigraphy whereby a randomly damaged object 'addresses' an unknown audience. In place of the perfection of the hexaemeral garden, the ruin, subject to variation and rearrangement, elicits completion in the imagination as promising a potential wholeness greater than the integrity of the complete object.[26] The sketches made of Roman gardens by artists such as Marteen van Heemskerck (1530s) suggest the diverse readings to which 'casually' displayed antique fragments were amenable In such sketches, 'random' views were orchestrated perspectives including fabricated juxtapositions; antiquarian landscapes are scenes, brought to unity first by Humanist theatricals and subsequently by planning.[27]

The wastelands where antiquities were recovered and studied promoted interest in abandoned settings, reproduced in artificial grottoes which initially represented 'rough and broken antiquity'.[28] The ruination which deforms and reveals appears contemporaneously in Vesalius' anatomical plates (1543), where ruins form the locus of the damaged figure, mortal or monumental.[29]

Juxtaposition of antiquities and opened bodies proclaims the analogy between archaeology and scrutiny of nature's hidden structures. It also exhibits with pathos the passivity of the figure to its observer notable in Roman garden poetry on antique statues.

If antiquarian gardens precluded completion by their provisory assemblages and their artefacts' imperfect state (prompting numerous reflections on mutability and loss), nature in these garden concerns not only the perfection of the species but processes of change, i.e. *meteorological* processes concerning particularly water in transformation and atmospheric effects (e.g., precipitation) displayed by tufa.[30] The grotto develops from niche or artificial cave to become centrepiece of *natura artificiosa*, a theatre of metamorphosis which ingeniously 'discloses' the emergence of form from the darkness of matter and the protean transformation of the elements.[31] Later Renaissance gardens such the Villa D'Este at Tivoli, famous for its tufa, or Pratolino (shaped by Aristotle's *Meteorologia*) were dominated by waterworks and hydraulic marvels, especially automata.[32]

The relation of antiquarianism and meteorology touches the nymph fountains which appear in Rome from the 1460s, featuring a recumbent antique or pseudo-antique statue inscribed with a version of Giannantonio Campano's epigram in which the *nympha loci* begs visitors to her sacred spring to leave her sleeping to the murmuring waters.[33] The epigram concerns animation through poetry and through the vital yet soporific 'lymph'; it recalls the Neoplatonist Ficino's contemporary discussions of the nymph as quickening water spirit and 'local daemon' who determines the differentials and distinct character of a place.[34]

The configuration of fountain, grotto and sleep suggests Orphic initiations on the twin underworld springs of Memory and Lethe, whose waters bring recollection and oblivion: teachings known to Renaissance Humanism through Plato's Myth of Er (*Republic* 10), Plutarch's association of Lethean waters with Dionysus, and Pausanias' account of the mysteries at Trophonius, whose springs brought memory and oblivion.[35]

As *memoria* and symbol of oblivion, the sleeping nymph reflects Humanist ambivalence about antiquities exhumed from 'Hades' and their potentially 'daemonic' reanimation.[36] The nymph is rendered innocuous by sleep, her voice channelled by the poets she 'inspires' (often to panegyrics about her new owner).[37] There is continuity between the nymph as figure for latent vitality controlled and the grotto, where nature's processes are revealed and exploited into spectacular waterworks and hydraulic mechanisms that display the vivifying lymph in transformation. The Roman sculpture garden 'enlivened' by poetry and inscriptions develops from the 1530s into concerns with *natura artifiosa*, centred on grottoes where nature's hidden operations are disclosed.

The speaking, sleeping nymph descends from the Hellenistic epitaphs of the *Greek Anthology*, where fictional tombs address wayfarers in wastelands; she signals a shift from the dialogue of lover and beloved in the *locus amoenus* to an epigrammatic convention involving a passive object and a generalized, anonymous audience. The nymph's address to 'whoever' (*quisquis*, *quicunque*) and her injunction to be left undisturbed reappear in the *lex hortorum*—the Roman garden inscriptions designating it as 'public' or semi-public and stipulating behaviour within it.[38] The *lex* had enduring vitality: public closure and proposed development of Villa Borghese in 1885 was blocked by recourse to the *lex hortorum* stated in a seventeenth-century inscription.[39]

This case shows the *lex hortorum* interpreted retrospectively, with the garden conceived as a public amenity in the modern sense. The epigraphic generalization of 'whoever you are' denotes an impersonal garden public comprising transitory 'users'. The stipulations of the *lex hortorum* in its lengthier versions, as at Villa Giulia (1551–5), show the antique epigram transforming into an anticipation of the bye-laws of a municipal park.[40] This is the forerunner of the 'people's garden', developed for education and surveillance, which becomes finally the leisure facility where recreation 'no longer appealed to the visitor's need for freedom, his virtue or morality, but merely restored his capacity for work'.[41]

Field and theatre

The antiquarian garden is an extensive field whose features are coordinated topically; its associative character is more pervasive than an iconographic programme, and transcends distinctions between formal and informal planning. Formal gardens, reproducible in maps and vistas, dominate Renaissance garden history, but innovative parks (Gennazzano, Bomarzo) and important antique displays (Cesi gardens, Carpi *vigna*) were partially or wholly informal.[42] Van Cleve's c. 1550s painting of the Cesi gardens (inconsistent with written descriptions) shows courts and formal gardens alongside *vigne* with informally arranged antiquities where remains (and buildings in construction) mediate between fabric and nature[43] (Figure 2.1). The image idealizes how sculpture gardens 'contained and corrected' the landscape of ruins into a *rus in urbe* within the field of the garden-city, unified topically yet exhibiting various emphases—episodic and eclogic.[44]

The non-axial garden as field appears clearly at Bomarzo, where material, stylistic and imaginative consistency is exemplified by sculptures carved from living rock—literal embodiments of the identity of place and theme. When Francesco Sansovino identified Bomarzo with Sannazaro's *Arcadia* he was not offering an iconographic key but a topical context, whereby the park's poetic coherence corresponds to the narrative space of Sannazaro's elegiac romance, with its deepening shadows of chaos and mourning.[45]

When perspective orders the field, it acquires a conceptual character which is concretized as theatre. The authoritative example is Bramante's Belvedere Court, where permutations of the monumental Roman arch are deployed in a gigantic work of topographic enclosure. The Court develops Hellenistic-inspired Roman transformations of a hillside into a monumental complex (e.g., Temple of Fortune at Palestrina) into a courtyard extended to contain a hill. Topographic particularities are expunged to create a site whose salient quality is extension and whose characteristics are determined by architectural manipulation. Within the arcades of a palace, the hill

Figure 2.1 Hendrik van Cleeve, Cesi Palace and Gardens c. 1584. Prague: National Gallery. Prisma archivio/Alamy Stock Photo.

is stratified physically into terraces and contained conceptually as theatre, culminating in the cavea-like exedra of the uppermost terrace.

The Court demonstrates the abstraction and artificiality obtained by perspectival control of a topography. The resulting theatricality is typified in the monumental arch which frames antiquities, focuses vistas and relates disparate elements, effecting convergence of axial planning and antiquarian topography. As nymphaeum, revived by Bramante at the Belvedere, the arch is also a fountain-grotto which frames theatrically the disclosure of nature within the garden.

Bramante's Belvedere influenced conceptualized models of progression through a site, exhibited in works by (or attributed to) Vignola—Villa Giulia, Caprarola, Horti Farnesiani—which control vision and movement, prescribing an itinerary through a scene which manufactures themes.[46] Instead of the 'inspirational' meandering promoted by episodic arrangement, at Villa Giulia and Horti Farnesiani the visitor rehearses a pre-ordered experience, simultaneously actor and dazzled spectator, as calculated obscurities (baffled movement, concealed or sunken features) are succeeded dramatically by revealed vistas.[47] At Horti Farnesiani, the ascending path forced the visitor to keep turning outwards from features recessed in the hillside (semi-circular 'theatre' and grotto or cryptoporticus) towards the vista of the forum and the Basilica of Maxentius, framed by the upper arch of the entrance gate and the cryptoporticus.[48] The garden is a theatre and frames the city as a theatre.

The Horti Farnesiani, in a site of unique archaeological and mythological significance on the Palatine hill, shows the topical articulation which made objects and places holders and generators of meaning translated into a scenographic model where things are experienced theatrically within a pre-ordered visual field. Perspective as scenic instrument worked alongside other elements, like epigrams apostrophizing visitor-actors to solve enigmas, admire artworks, contemplate nature, leave the flowers untouched, etc. The theatrical character of Renaissance gardens lies in the manner of experiencing things (statues, inscriptions, fountains, plants) as they lie within a field of predetermined vision and movement.

Rus in urbe and *sacro monte*

Successful integration of axial planning and topical field appears at Villa Lante at Bagnaia (commenced 1568), associated tenuously with Vignola[49] (Figure 2.2). Created by Cardinal Gambara, Bishop of Viterbo and Counter-Reformation Inquisitor, Bagnaia is largely expunged of statuary enacting a programme. The garden shows the progressive articulation of *natura artificiosa*, from grotto to parterre via the central 'water chain', which identifies the axis of vision and movement with the transformations of nourishing, reflective lymph.[50]

The attenuated iconography enhances the harmony of design, site and material, intensifying the pervasive sense of relatedness within the garden as a special, sequestered place. Bagnaia shows virtuoso handling of the reciprocity of containment and extension in perspective; its enclosure is strengthened through mutual definition with the townscape and distant horizon. Paired buildings, symmetrically disposed to each side of the central fountain axis, develop the pleasure complex (Bagnaia as 'reconstruction' of Varro's Aviary) into concerns with the integration of building groups in a landscape. The relation of villa to townscape appears in the topographic frescoes of the villa's principal loggia.

Villa Lante shows the villa as exemplar for urban topography. It was completed by the cardinal-nephew of Sixtus V, the pope whose urban regeneration planned traversal of the abandoned *monti* with avenues linking the basilican churches, centred on Sixtus' titular church of Santa Maria Maggiore beside his Villa Montalto.[51] This coordination of Roman pilgrimage topography, with its celebrated recycling of ancient obelisks as axial markers, cemented the theatricality

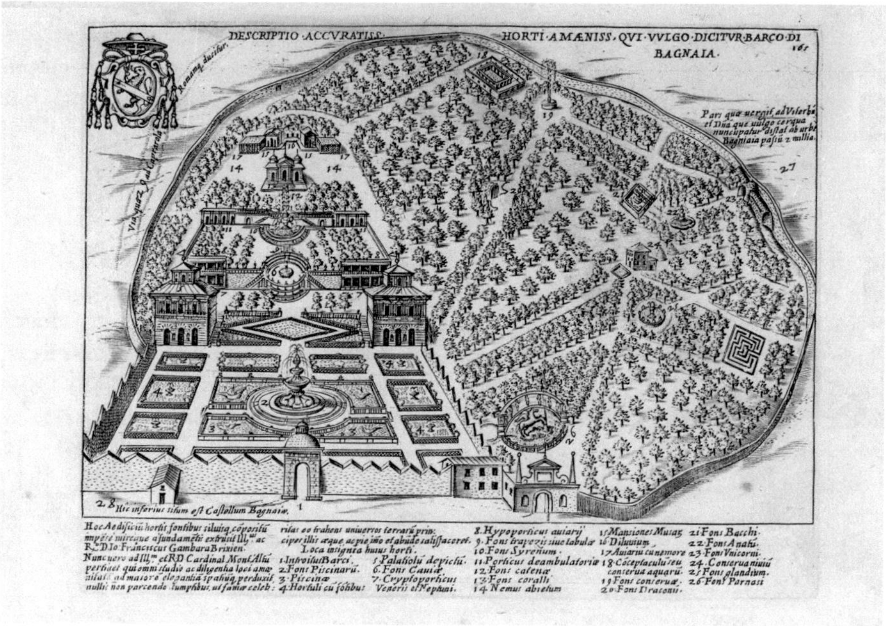

Figure 2.2 Giacomo Lauro, Villa Lante at Bagnaia, Antiquae Urbis splendour, 1641–2. Biblioteca Hertziana – Max-Planck-Institut für Kunstgeschichte, Rome.

of axial planning to ritual requirements and planned reinvigoration of active life.[52] The Salone Sistino depiction of Guerra and Nebbia (Vatican Library), like Gianfrancesco Bordini's print, shows the *rus in urbe* traversed by routes linking the pilgrimage *loca*; the star-shaped network, simultaneously sacred and instrumental, predominates over the city as topical field. It exhibits the conceptual innovation of Sistine urbanism compared with earlier Roman Christian landscapes, which developed from antiquarian vistas showing religious processions, charitable acts and eremitical wildernesses in place of triumphs, mythological scenes and ruins.[53] Sixtus' Villa Montalto, with radial pathways and intersecting avenues through uncultivated land, reflected his conceptual urbanism, focused on extension and organization of circulation.[54] Standing like a *giardino segreto* at the heart of his reorganized pilgrimage topography, it is not a symbol of civic life, but an ideal counterpart to urban design.

These instrumental tendencies appear in other pilgrimage places, such as the late Renaissance *sacri monti*, mountainside complexes which created an itinerary through chapels each housing a 'mystery' (usually an episode from the life of Christ or Mary), first developed by Franciscans as 'New Jerusalems' to substitute pilgrimage in Palestine.[55] The landscape and monuments of the *sacro monte* constitute a totality, yet it is a place of *topomimesis*, a 'transposition' of structures associated with Christ's life, simultaneously historic and miraculous.[56] The *sacro monte* replicates the spiritual value of the unique pilgrimage site; it does not imitate one monument (e.g., Holy Sepulchre), but an entire topography.[57]

This forms the setting for orchestrated, participatory contemplations—simultaneously concrete scenes in the sacred 'theatres' (whose painting and statuary show extreme verisimilitude, with real clothes and props) and imaginative exercises followed by the pilgrim, on the model of the rosary with its sequential meditations on the life of Mary, or Loyola's *Spiritual Exercises* with their composition of place. This 'theatrical imposition of the sacred' exploited the terrain, but the spiritual 'reality' of the place was in the representational overlay. Thus at Varese, dedicated to the rosary, chapels housing the Joyful, Dolorous and Glorious Mysteries lay respectively in fertile, desolate and luminous areas of the mountain, linked by a processional path. In this topo-mimetic coincidence of artifice and sacredness, place is simultaneously illuminated spiritually and instrumentalized into a panoramic backdrop for sacred theatre.

Subtle personification of the ground

The *sacri monti*, where a topography was transformed topomimetically into a holy land, stretched landscape as scene to conceptual limits. Like villas, they could be sites for urbanistic projection: Scamozzi at Villa Duodo, Monselice (1589–1607) designed a complex of villa, reliquary church and hillside chapels, modelled on the seven Roman basilicas and granted the same indulgences.[58] Our narrative of progressive scenic generalization shows perspectival planning coordinated with a movement from topos to concept, like the shift from Marian *hortus* metaphors to rosary devotions, whose instrumental potential is exploited in the *sacro monte*.

We conclude with the Villa D'Este at Tivoli, commenced in 1560 by Ippolito D'Este, which accommodates symbolic, topographic and scenic-conceptual tendencies (Figure 2.3). Like its closest

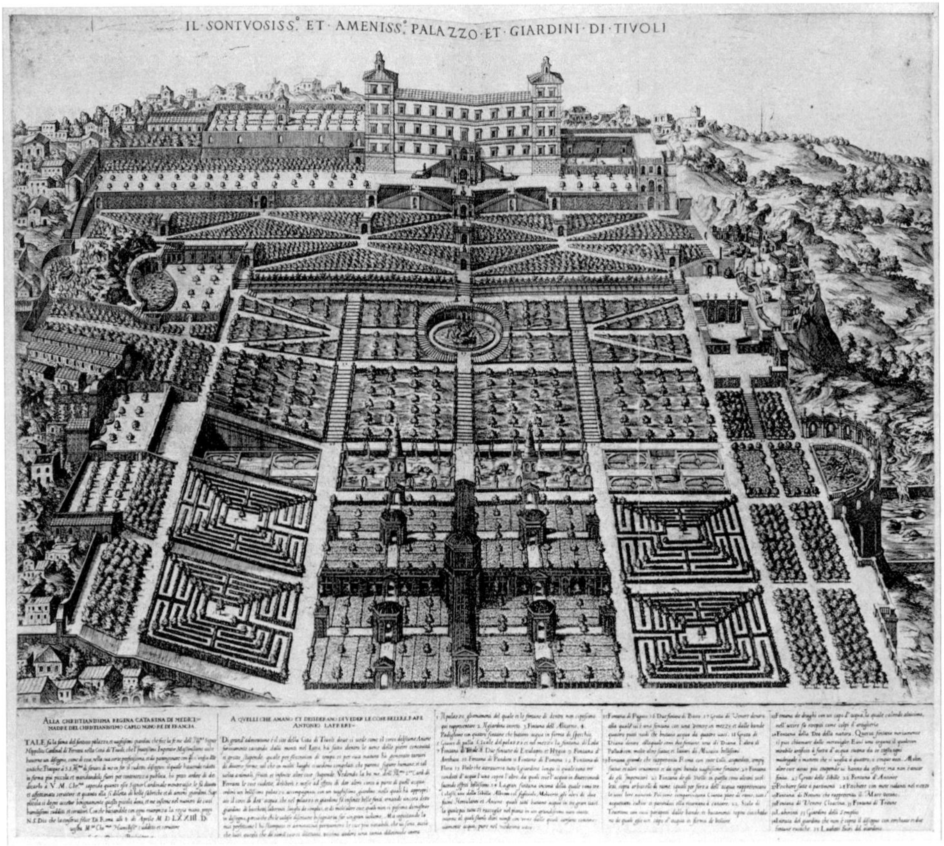

Figure 2.3 Etienne Dupérac, Villa D'Este at Tivoli, 1573. New York: Metropolitan Museum. Harris Brisbane Dick Fund, 1941. Photo: www.metmuseum.org.

model, Hadrian's Villa as *memoria* of world empire, the Villa exhibited antiquarian encyclopaedism (promulgated by the principal architect, Ligorio), which it synthesized with a traditional tripartite allegorical structure (natural-mythical-sacred) in an axial hillside garden whose paths were narrative journeys articulated by 'theatres'.[59] It progresses from nature's latent processes and oceanic cycles to meditations on myth, prophecy and history culminating in rebirth and apotheosis.[60]

The garden represents nature meteorologically, through multiple allusions to water as element of transformation, and topographically, as a 'subtle personification of the Tiburtine ground', explored through mythic and archaeological associations.[61] The *nympha loci* of Tivoli's grottoes and cascades is Albunea, who conflates Ino-Leucothea, Dionysus' aunt and nurse, drowned and metamorphosed into a marine goddess, with the Tiburtine Sibyl, whose image appeared in the whirlpool of the Aqua Albunea and who prophesied Christ's Coming to Augustus.[62] The *nympha loci* dominates the Fountain of Tivoli and slumbers in a grotto in the villa courtyard; associated in the Water Organ with the *Dea natura*, she symbolizes universal nature as well as Tivoli.[63] The nymph embodies the ascent from nature to myth to sacred; she 'activates' the analogical structure of the garden through her orientation to chthonic and chaotic elements.

The gardens form the nymph's *memorie* (memories/memorials) or, more dramatically, her awakening: the dormant nymph is vitalized. The raging torrents of the garden are her voice, 'recreated' at the Water Organ; the aquatic transformations and long paths are her story. The nymph's speech as water spirit and pythoness is an unleashing of forces half-hidden or half-understood; her waters are veils of metaphor.[64] The garden finds cohesion in her story, raised beyond a schematic programme to become an enactment in which metamorphosis is an essential structure of meaning, not a thing portrayed. Rather than the garden being a metaphor, like the *hortus conclusus*, its metamorphic character gives a structure of meaning compatible with nature as process.

The Villa takes engagement with grounding to unique lengths, dramatized in multiple journeys into darkness: labyrinths, the 'Hesperidean' orientation to sunset (shown in Muziano's 1568 Salone fresco), the dragon as symbol for ocean and the sun's journey under the earth, the subterranean course of the world's waters, the Tiburtine grottoes with their metamorphic 'sculpted' *tartaro*, the sibyl rising from the abyss and her death and transformation – like Hippolytus – in water.[65] This grounding is, however, mediated by the garden as scenic field; the Rometta fountain, which epitomizes Rome as a theatre scene, shows the complexities of representation. If nature in the garden must be disclosed within a providential framework, the Villa is a field of revelation which discovers the relations between *natura naturans*, human (ethical) choice and final, paradisal reward.

Renaissance gardening remained rooted in the allegorical tradition which encouraged structures of multiple meaning, giving figurative depth and balancing the conceptual tendencies of theatre as paradigm. The circulating, transforming waters of the Villa D'Este reflect a typological tradition of Christian allegory and a poetic practice of disclosing natural philosophy through mythology. Its planning shows the attempt to project analogical thought onto a physical site. Theatre played a dual role in this project, providing a continuum of artifice which accommodated multiple levels of representation, where metaphor could play a key role of revelation in a synthetic totality.

The problematic aspect lay in the generalized promotion of the garden as 'scene of nature' – thus vulnerable to critique at that level. The lurid exhortations to rape nature in early modern science identify exposure of nature's secrets with the stripping of analogical thought; they are the ultimate responses to the *nympha loci*.[66] As we face the ecological consequences of that aggressive attitude, Renaissance gardens show ethical and existential reflection as necessary preconditions for meaningful intervention in nature.

Notes

1 Ernst Curtius, *European Literature and the Latin Middle Ages* (London: Routledge, 1953), 183–202, notes Libanius' fourth-century codification of landscape delights (springs, plantations, gardens, soft breezes, flowers and birdsong). Hellenistic fantasy landscapes decorating porticoes were called *topia* (Vitruvius 7.5.2).
2 Denis Ribouillault, *Rome en ses jardins* (Paris: CTHS-INHA, 2013), 42–8.
3 See Terry Comito, *The Idea of the Garden in the Middle Ages and Renaissance* (Hassocks: Harvester, 1979).
4 Comito, *Idea*, 108.
5 Xenophon, *Oeconomia* 4.13–25; *Anabasis* 1.2.7, 5.3.9–12.
6 Pliny, *Natural History* 19.19.50–1.
7 Comito, *Idea*, 105; Curtius, *European Literature*, 190.
8 See Augustine, *De ordine*; Clarence J. Glacken, *Traces on the Rhodian Shore* (Berkeley: California, 1967), 213, 303, 313. The garden of the Abbey at Cassino was called 'paradise in the Roman fashion'; see Georgina Masson, *Italian Gardens* (London: Thames and Hudson, 1966), 46.
9 Petrarch, *Canzoniere*; Boccaccio, *Genealogia deorum gentium*, Book 14.
10 Comito, *Idea*, 3–11; Masson, *Italian Gardens*, 47–50 on Islamic features.
11 Serlio's description of Poggio Reale, the Neapolitan suburban palace, laments Italy as a garden destroyed by political misfortune (*Terzo Libro*, 1540). 'Hesperides' becomes a convention for pleasure gardens, such as Falda's views of Roman 'Hesperides' (c.1670).
12 Sannazaro ends *De partu virginis* 3.510–13 with the Neapolitan Muses' caves whose citrus groves recall Persian orchards; Pontano, *De hortis hesperidum* 1.315, elegizes gathering oranges with his dead wife. On Hades as landscape of loss or *locus inamoenus*, see Sannazaro, *De partu virginis* 1.364; Ovid, *Metamorphoses* 10.15.
13 Comito, *Idea*, 125–47; Clare Guest, *Understanding of Ornament in the Italian Renaissance* (Leiden: Brill, 2015), 37–48. The theme persists into Humboldt's *Cosmos* (1845–62).
14 Philo, *De plantation*, 126–31.
15 Leena M. Peltomaa, *The Image of the Virgin Mary in the Akathistos Hymn* (Leiden: Brill, 2001), 135–9. In *Cosmograpia* 2.1, Noys calls Natura 'blessed fruitfulness of my womb' (Luke 1:42), echoing the Marian salutation.
16 Irenaeus, *Adversus haereses* (c. 180). Cf. Jerome, Epistle 9 (PL 30, 132) on Mary as garden of delights.
17 Peltomaa, *Image*; for Mary as *hortus*, etc. in Latin hymnody, see the fifty-five-volume *Analecta Hymnica Medii Aevi*, e.g., Conrad of Haimburg's 'O Maria paradisus', c. 1360), *AH* 3.5, which enumerates the four rivers flowing out of Eden and the virtues represented by each plant.
18 *Greek Patristic Lexicon*, s.v. '*nymphe*'.
19 See *Analecta Hymnica*, vol. 1, *Cantiones Bohemicae* (thirteenth to fifteenth centuries).
20 Mantovano, *Adulescentia* (1498), Eclogues 7–8.
21 *Coryciana*, ed. J. Ijsewijn (Rome: Herder, 1997), 31, 149, 257, 371.
22 Nicholas V's Vatican gardens project was praised as a 'most perfect space of paradise'; see Carroll Westfall, *In This Most Perfect Paradise* (University Park: Pennsylvania Press, 1974), 149–50, 154–65.
23 Charles L. Stinger, 'Greek Patristics', in *Rome in the Renaissance*, ed. Paul A. Ramsey (Binghampton: CMERS, 1982); Kathleen W. Christian, *Empire Without End* (New Haven: Yale, 2010), 103–42, 161; *Catholic Encyclopaedia*, s.v. 'Immaculate Conception'.
24 Polia is described finally as a withered flower of antiquity, unlike Mary as unfading flower (and *regina poli*), whose tomb sprouts flowers signalling her incorruptible flesh in contemporary Assumption depictions (Botticelli, Pinturicchio, Raphael). The acrostic formed by the chapter headings, which identifies the author Colonna and his beloved Polia, echoes the acrostics of rosaries and Marian verses (e.g., Ulrich Stöcklins in *Analecta Hymnica*, 6.17, 6.23); Colonna was probably a Dominican monk in a period of Dominican promotion of rosary devotions.
25 See Pontano, *Tumuli*, I.44. Pontano's fictional graves sprout flowers, reviving the ancient topos of epigrams as flowers.
26 Leonard Barkan, *Unearthing the Past* (New Haven: Yale, 1999), 124, 207.
27 See Kathleen W. Christian, 'For the Delight of Friends, Citizens and Strangers', in *Rom zeichnen*, eds. Tatjana Bartsch and Peter Seiler (Berlin: Gebr. Mann, 2011), 129–56. Christian, 'Landscapes of Ruin and the Imagination in the Antiquarian Gardens of Renaissance Rome', in *Gardens and Imagination*, ed. Michel Conan (Washington, DC: Dumbarton Oaks, 2008), 117–37.
28 Annibale Caro, letter on the *vigna* of Giovanni Gaddi (1538), in David Coffin, *Gardens and Gardening in Papal Rome* (Princeton: Princeton University Press, 1991), 35–7.

29 Clare Guest, 'Art, Antiquarianism and Early Anatomy', *Medical Humanities* 40, no. 2 (2014), 97–104.
30 Pontano wrote a *Liber meteororum* (1505).
31 Alberti, *De re aedificatoria*, 9.4 discusses artificial grottoes and ancient use of pumice chips ('Travertine foam') for rustication; Claudio Tolomei (1543) coins *natura artificiosa* in description of grottoes, waterworks and tufa. See Elisabeth B. MacDougall, *Fons Sapientiae: Renaissance Garden Fountains* (Washington, DC: Dumbarton Oaks, 1994), 12–14; Coffin, *Gardens*, 35–7.
32 Hervé Brunon, 'Les monuments des eaux de l'univers', in *Les éléments et les metamorphoses de la Nature* (Bordeaux: Blake, 2004), 33–53.
33 MacDougall, *Fons Sapientiae*, 37–55; Christian, *Empire* , 134–40, 178–82. The *Hypnerotomachia* features a sleeping nymph statue from whose breasts hot and cold water flows, ogled by satyrs and inscribed '*Tokadi Panton*' ('to the mother [*tokas*] of all'), possibly recalling profanely Mary as *Theotokos*. Later fountain nymphs with irrigating breasts included Ammannati's meteorological allegory *Hera* (installed at Pratolino in 1580) and the gigantic satyress of Cristoforo Madruzzo's Papacqua nympheum, Soriano nel Cimino (c. 1560). Madruzzo appears in the Bomarzo inscriptions and was a neighbour of fellow cardinals Farnese (Caprarola) and Gambara (Villa Lante, Bagnaia).
34 Ficino, *Commentary on Phaedrus*, Chapter Summaries 2, 38; Summary to Plato's *Apology*.
35 Plutarch, *Moralia*, 565–6; Pausanias, *Description of Greece*, 9.39.8; Pliny, *Natural History*, 31.11.15. See William Guthrie, *Orpheus and Greek Religion* (1952; repr. Princeton: Princeton University Press, 1993), 168–78; Comito, 'Beauty Bare: Speaking Waters and Fountains in Renaissance Literature', in MacDougall, *Fons Sapientiae*, 41–2.
36 Ficino, Summary to *Apology*, discussed sculpture as daemonic, comparing the daemon's presence 'in' the statue to the face reflected in the mirror or the voice in the echo.
37 Christian, *Empire*, 135.
38 Christian, *Empire*, 196–201; MacDougall, 'The Sleeping Nymph: Origins of a Humanist Fountain Type', *Art Bulletin* 57, no. 3 (1975): 48; Coffin, *Gardens*, 244–57, 268–9.
39 Coffin, *Gardens,* 244–5.
40 Ibid., 246–7; 269.
41 Martin Warnke, *Political Landscape* (Cambridge, MA: Harvard, 1995), 82.
42 Claudia Lazzaro, *The Italian Renaissance Garden* (New Haven: Yale, 1990), 121, contrasts real or counterfeit ruins in parks as memories with the past reborn in ordered gardens.
43 Coffin, *Gardens*, 22–4.
44 Christian, 'Landscapes of Ruin', 119.
45 Sansovino, dedication to Vicino Orsini, in Sannazaro, *Arcadia* (Venice, 1578). Orsini's correspondence discusses Bomarzo as memorial to his dead wife and quasi-therapeutic expression of his melancholy.
46 Vignola is thought to have provided a general design for Horti Farnesiani with detailing attributed to Giacomo del Duca. See Coffin, *Gardens*, 69–75, 255; Giuseppe Morganti, *Gli Orti Farnesiani sul Palatino* (Rome: French Academy, 1990).
47 Co-architect Ammannati's 1555 letter discusses Villa Giulia as a theatre, with the sunken nymphaeum as 'heart of the perspective'; see Giacomo Balestra, *La fontana pubblica di Giulio III* (Rome: Battarelli, 1911), 69, 74.
48 Coffin, *Gardening*, 73–4.
49 Lazzaro, *Italian Renaissance Garden*, 243–69; Carla Benocci, *Villa Lante* (Viterbo: Ghaleb, 2008). The hydraulics expert Ghinucci was supervisory architect.
50 Coffin, *Gardening*, 93–5.
51 The Salone Sistino fresco states that Sixtus' basilican roads opened a way (*via*) to the stars, invoking Mary as *stella maris* and *stella matutina*.
52 Sixtus' practical schemes centred on the Esquiline Città Felice, to which he extended the Acqua Felice (commemorated in the Moses fountain).
53 See the 1580s decorations of the Vatican Tower of the Winds by Tempesta and Matthijs Bril, discussed in Ribouillault, *Rome*, 133–212. At the Casino of Pius IV, Santi di Tito depicted the parable of the vineyard in Pius' villas (1563).
54 Villa Montalto is a garden park, a type associated with parvenu papal families as a suburban equivalent to rural aristocratic estates; the largest, Villa Pamphili, extended to 240 acres in the 1670s (see Ribouillault, *Rome*, 160–1). Coffin, *Gardening*, 139–58, notes the Villa Giulia estate as a precedent.
55 Early sites were Varallo, north of Milan (1491, successively redeveloped with involvement of Carlo Borromeo) and San Vivaldo, Montaione (Tuscany, 1500); most developed from the 1580s in the Alpine foothills.

56 Luigi Zanzi, *Atlante dei sacri monti prealpini* (Milan: Skira, 2002), 61–5.

57 At Graglia, Andrea Velotti (d. 1624) projected an area with 100 chapels illustrating scenes from Genesis to the Ascension and local hills renamed after biblical mountains.

58 Adriana Augusti, Guido Beltramini, Sandra Vendramin, 'Casa Duodo', in *Vincezo Scamozzi 1548–1616*, eds. F. Barbieri and G. Beltramini (Venice: Marsilio. Exhibition catologue, 2003), 301–19. Varallo, initiated as a reconstruction of Palestinian pilgrimage sites, was reinterpreted in Galeazzo Alessi's proposals (1565–9) as an ideal city.

59 See Guest, *Understanding of Ornament*, 552–67, on Ligorio's antiquarian use of threefold allegory.

60 Ippolito's humanist Muret composed epigrams dedicating the gardens as Hesperides to Hercules and Hippolytus, drowned and resurrected as Virbius, priest to Diana at Nemi.

61 U. Foglietta, *Tiburtinum Hippolyti Estii* (Tivoli: Tiburis Artistica, 2003), 5. On the allusions to Tiburtine archaeology, see Denis Ribouillault, 'Le ville dipinte del Cardinale Ippolito D'Este a Tivoli', in *Delizie Estensi*, eds. Francesco Ceccarelli and Marco Folin (Florence: Olschki, 2009), 341–71; David Dernie, *The Villa D'Este at Tivoli* (London: Academy Editions, 1996), 19–25. Creation of the Villa involved ruthless destruction of local property.

62 Cesare Nebbia's frescoes in the second Tiburtine room of the Villa depict Ino's story; the adjoining chambers are dedicated to Noah and Moses.

63 Dernie, *Villa D'Este*, 64–9; the Water Organ formerly held statues of the sibyls.

64 Renaissance Platonism linked poetry and prophecy as forms of 'divine frenzy'; see Ficino, *De divino furore* (1457). Pegasus appears behind the Sibyl at the Fountain of Tivoli.

65 Ligorio describes the dragon of the Hesperides as a figure for the ocean (*Libro dell'antichità*, Turin, State Archive, a.II.J.23, vol. 44); the dragon signifies temporal cycles, rivers, exhalations, the 'virtue' which germinates seeds and the sun's journey under the earth. On water as 'foundation' of the earth, see *Ligorio, Libro dell'antichità*, National Library, Naples, XIII B9, 114r; see ibid., 35r, on the 'infernal' Tiburtine cascades. Dupérac's map notes the metamorphic Tiburtine *tartaro*.

66 Mary Midgley, 'The Remarkable Masculine Birth of Time' (the title quotes Bacon), in *The Essential Mary Midgley* (London: Routledge, 2005).

3

THE BIRTH OF LANDSCAPE FROM THE SPIRIT OF THEORY

Alexander von Humboldt's artistic and scientific *American Travel Journals*

Ottmar Ette

His life beginning on 14 September 1926 and ending just before he reached the age of 90 in 2016, French author Michel Butor is the creator of *Paysages planétaires* (*Planetary Landscapes*), a work that is as cryptic as it is humorous: landscapes unfold in which the most diverse parts of our planet are connected with one another and are perpetually in motion. *Planetary Landscapes* begins, for example, with a hybrid composition in a poem entitled 'ALASKAMAZONIE',[1] which is filled with life everywhere:

> Planetary Landscapes
> *Les cimes des conifères*
> *le royaume des corbeaux*
> *la petite et la grande Ourse*
> *les aurores boréales*
> *les restes des chercheurs d'or*
> *les traîneaux sur la toundra*
> *les mâts généalogiques*
> *le cuivre et les dents de morse*
>
> La mer, houles et replis, avec les cris des mouettes, grand large et marées, avec les chants des baleines au loin. Par les fenêtres du navire, nous voyons défiler fjords et glaciers. Soudain des blocs se détachent et tendent dans les chenaux en éclaboussant. Voici des chasseurs qui rentrent avec viandes et fourrures.
>
> *L'empire des colibris*
> *les aurores boréales*
> *cyclones dévasteurs*
> *les traîneaux sur la toundra*
> *les radeaux sur les grands fleuves*
> *le cuivre et les dents de morse*
> *les auréoles de plumes*
> *le royaume des corbeaux*[2]

Flora and fauna at high and low latitudes, the tundra and the tropics, land and sea, heat and cold permeate one another without fusing together in movements of global dimensions that converge in the American continent; other parts of *Paysages planétaires* see different continents become part of a process of reciprocal exchange. Oceans and lakes illuminate one another through the connecting, life-giving element of water. As in his epochal experimental text *Mobile*, here too everything is connected via diverse forms of transportation and subject to a reciprocal transformation process that affects humans, animals, plants, rocks, the wind and water. Nothing on this planet stands alone. Michel Butor shows us a world in which everything is interrelated.

In rapid succession, Michel Butor's readers may traverse the 'ETATS ZUNI'[3] or the 'VIETNAMIBIE',[4] the 'OCEAN PAPOUINDIEN'[5] or the 'CASPERTZIENNE ANTILLAISE',[6] the planes of the 'PACIFIC SANDWICH'[7] or the 'MONGOLIE TROPICALE'.[8] No area exists as a lone entity. This is not a planetary idyll, however. In 'ANDES AFRONIPPONES',[9] we are given warnings that threaten to shake the earth:

> D'un horizon à l'autre les trompes se répondent pour avertir de l'imminence du danger. Serait-ce le cataclysme annoncé? Toute la province est menacée, toute la nation, le continent même. Ne résistent que quelques îlots d'humidité.[10]

It would thus be wrong to view Butor's planetary landscapes as a playground for harmless exercises, as the point of intersection for an innocent game in which everything is connected: the playful movements in Butor's world are aware of catastrophe, they are aware of cataclysm. They not only depict its natural beauty; they are also signs of an imminent downfall, since everything is connected to everything else through landscapes—planetary strata, in which the local, regional or national always invoke the transareal and planetary.

Alexander von Humboldt also designed planetary landscapes of this kind in his scholarly and literary mapping. He, too, was born on 14 September (albeit in 1769) and also died shortly before his 90th birthday. The writer and scholar sketched numerous landscapes in his *American Travel Journals*; these, too, consisted of far-reaching global strata and were often characterized by huge transformations, or indeed immense catastrophes. For example, in the section 'Geognosy of America' in *Journal I*, he offers a geological-literary vision. In this meandering passage that begins in the north of Venezuela, a place with which he was familiar, it is fascinating to observe how Alexander von Humboldt describes global movements as if in a time-lapse, presenting them as a moving image of coastlines and inland areas that stretch from the northern parts of South America across the Caribbean islands, directly to the north up to Hudson Bay, then south towards the Amazon, but also taking in the Strait of Magellan, in order to expand this hemispheric construction of America across the planet: Humboldt's vision reaches across the Atlantic to Africa and Tibet in the east, and across the Pacific towards South-East Asia and the stretch of islands to the west of the coast. In this moving image, everything is connected to everything else: nothing stands alone in this planetary landscape.

Humboldt's vision of a 'Geognosy of America' reveals a planetary landscape in which nothing stands alone and in which nothing is static. Nothing stays put and nothing is spared from immense transformations that fuse together and separate islands and continents, and which transform islands into continents and continents into islands. In his creation, Humboldt draws on his own observations, on available cartographies, and occasionally on indigenous sources. There is continuous activity in the form of earthquakes and other natural disasters that change the earth's surface; that cause mountains to cave in and submerged sandbanks to re-emerge; that cause channels to form between continents and continents to drift; and which cause islands to separate from land and create new basins through flooding—such as the Caribbean or the

Mediterranean, the Black Sea and the Baltic Sea. These Humboldtian images 'jump' from continent to continent and from lake to sea, offering a comparative perspective, primarily in a *transareal*[11] manner as a *history of movement* (as opposed to a history of space): for the author of *Kosmos*, spaces are always spaces of movement.

This is certainly the case in the first volume of Alexander von Humboldt's *Personal Narrative of a Journey to the Equinoctial Regions of the New Continent*, in his 'Essay on the Geography of Plants'. The geography of plants is described not as a territorializing history of plant distribution on the earth's surface, but rather as a history of plant migration. For Humboldt, then, exploring the geognosy of America is a science that takes in the entire world: it is one in which coastlines ceaselessly change, in which continents are transforming constantly and islands emerge, move, and become submerged again, and are subject to no less 'rupture' and drift than continental masses themselves. Humboldt did not yet have access to a theory of tectonic plates, yet he sometimes comes remarkably close to this in his images of planetary landscapes.

In geological time frames, continental masses are astonishingly mobile. Alexander von Humboldt never doubts the fact that America and Africa were connected to one another at some point. He writes about the outline of the continents in his sketches time and time again, describing the Atlantic as the 'Atlantic longitudinal valley'.[12] In his 'Geognosy of America' sketch, he includes not only volcanic activity, but also the rotation of the earth and its profound influence on the entire planet. Indeed, with his ideas about constant changes and evolution, Alexander von Humboldt must be considered a significant trailblazer for the evolutionary theory of Charles Darwin, who was unsurprisingly an admirer of Humboldt when he was younger. The long-held yet misleading idea of the epistemological division between *Humboldtian science* and Darwinian evolutionary theory has long been debunked, and Humboldt's *American Travel Journals*—which should be understood as the true origins of Humboldtian science—are evidence of this; even at this early stage, they contain the basic principles of Alfred Wegener's much-contested thesis of the 'continental drift', which Wegener developed from 1912 onwards. Although he has erroneously been labelled a polymath, Humboldt thus does not stand for the residue of a traditional concept of science, but for a transdisciplinary science that was one of many influences on nineteenth-century historical scholarship.

Indeed, in his oft-cited reflections, Alexander von Humboldt drew on mythic constructions of flood from *The Epic of Gilgamesh* or the *Bible*—and ideas about catastrophes and spring tides that derived from these—as well as on concepts relating to the 'accuracy of fit' of the African and American coastlines that were crystallizing during the early modern period as a result of increasingly precise cartographic materials. It was not in vain that Humboldt measured the angles of these two continental masses in the parts where they extend and recede, and thereby identified key indicators that suggested the migration or drifting of the continents. In *Journal I*, he connected these angles with further indicators, such as common plants in America and Africa. Here, however, he not only addressed the geography of various plants, but also offered cultural and historical reflections about the differing development of humans in the south and in the colder, northern part of the globe.[13]

In fact, Humboldt drew conclusions about this cultural landscape that were based on all kinds of plant, animal and human migration across large distances, and he describes these movements in a number of other texts too. From today's perspective, however, his conclusions are sometimes problematic. For example, as a result of the diversity of the constantly available food in tropical areas, he claims that people 'are freed from the need to dry fruits or to orientate their culture towards survival. This explains the slow development of an intellectual culture, that is, the eternal childhood of farming in the hot zone. This need creates the arts.'[14] However, these kinds of conclusions do not in any way compromise the dynamic, *history-of-movement* focus of Humboldtian theory.

Let us return to the planetary landscape. In the early decades of the twentieth century, Alfred Wegener arduously pushed his concept of 'mobilism'—the shifting of individual continents—and spoke out against the powerful and influential followers of 'fixism'. Yet with his scientific ideas about movements of all kinds, Alexander von Humboldt came very close to this kind of mobility concept even earlier. For in addition to a theory of catastrophe and an increasing cartographical 'accuracy of fit', his thinking was also driven by ideas about the earth's rotation; as a result of his studies of volcanoes, he also put forward theories about internal earth forces that would go on to become tectonic plate theories in the twentieth century. Migration was the epistemological basis for Humboldt's thinking—and it opened up for him new, promising perspectives about future landscapes of theory that would only become part of common knowledge in the twentieth and twenty-first centuries.

At the core of the Humboldtian epistemology, which was in no small part developed in the infinite measurement sequences and rich store of field research described in the *American Travel Journals*, is a focus on the mobility and relationality of all objects on our planet. For Humboldt, everything on our rotating globe is subject to constant motion and transformation: land, water, air, mountains, high plateaus and lowlands, continents, islands and archipelagos, as well as plants, animals and, last but not least, humans with their perpetually changing cultures. Humans must continue to adjust to the new paths and movements—and thereby to the changing conditions of life.

It is on the basis of this theory—this Humboldtian epistemology of life and movement, of nomadic knowledge and transdisciplinary mobility—that Alexander von Humboldt's landscapes emerge. They are, in an excellent way, *landscapes of theory*. For Humboldt, landscape was born out of the spirit of theory. His empirical field research and precise 'measuring' of nature (which Friedrich Schiller once famously called 'shameless'[15]) are certainly of critical importance; however, it is notable that this theory first produced the Humboldtian landscapes as well as his landscapes of theory. Two questions arise, then: what is a landscape of theory? And in what contexts did these kinds of landscapes develop for Alexander von Humboldt?

Landscapes of theory

It should first be iterated that the term 'landscape' is defined in relevant geography manuals in general terms as a 'section of the earth's surface [characterized by] its external appearance [...] or its geographical position'.[16] More specifically, the term relates to 'the combined effect of the respective components and geofactors', whereby the corresponding landscape is not the sum of these geofactors 'but rather their integration into a geographical set of conditions or geosystem'.[17] 'Landscape' corresponds terminologically to the terms 'material' and 'life', in that it relates to the 'connections between the different [phenomena] that are unified in a landscape', as well as to their interrelation.[18] Making a distinction between a natural landscape and cultivated landscape appears to be problematic; the former tends to be avoided in more recent research histories of geography.[19] From a geographical perspective, the term 'landscape' refers to a complex system of interrelation and synergy between different factors that cannot be reduced to a particular physiognomy. Consequently, the term 'landscape' may in the first instance be understood as the designation of a complex interdependent geosystem.

In the second instance, it is necessary to determine the aesthetic dimension of the term 'landscape'. A narrow definition of the terms 'landscape' and 'theory' can be found in Joachim Ritter's remarks on Petrarch's 'turn to nature as a landscape'.[20] Here, in relation to the question of the function of the aesthetic—while contemplating nature as landscape—Ritter locates theory in the Aristotelian 'sphere of the festival and festive play'.[21] Taking this line of argument further, 'theory' is then connected with a modern idea of freedom whose origins lie predominantly with Friedrich Schiller. Landscape thereby holds potential as a place—or perhaps even more as an area of play—for a theory that represents a space (*Freiraum*) outside of a direct instrumental rationality, in whatever form this may take. This is how theoretical thinking and aesthetic form converge within the scope of art.

Ritter's discussions are undoubtedly more deeply rooted in Alexander von Humboldt's understanding of nature than is generally claimed; Humboldt's *Kosmos* is repeatedly referenced in Ritter's work. And Joachim Ritter is thereby significantly closer to Carl Ritter, founding father of the modern discipline of geography together with Humboldt, than is acknowledged in his influential essay on 'landscape', which addresses the construction of the individual and subjectivity. Yet Joachim Ritter clearly sought to distance himself from predominant geographical definitions of landscape, as he is keen to emphasize in a footnote of his essay. Within this conceptual framework, to separate geographical from philosophical/aesthetic thought is as impossible as a separation of the Apollonian from the Dionysian in Nietzsche's philosophy. It is, however, precisely this area of intersection that the philosophical thought and scientific practice of Alexander von Humboldt sought to negotiate.

Indeed, Ritter found it necessary to free Carl Troll's understanding of landscape from the 'connection between the "subjective" and the "aesthetic" that was so important to him'[22] when it came to natural sciences. Troll's approach may be geographical, but it is certainly compatible with other methodological approaches. He defines landscape as a part of the earth's surface 'that establishes a spatial unit of a particular character, based on its external appearance, the interplay between its phenomena, as well as its internal and external positional relations, and which merges into landscapes of another character at geographically natural borders'.[23] Is it possible that what Carl Troll calls the 'interplay' of phenomena opens up precisely this area of play for theory that is no longer necessarily tied to the idea of subjectivity, or even associated with the kind of central perspective that is anchored in the modern subject? In other words: might it be possible to conceptualize a force of the aesthetic as an aesthetic force,[24] without linking the idea of landscape to the central perspective of the subject?

Even today, Humboldt's theorem of *multiple connectedness* is present in geographical defini tions of landscape (it is a term that caught on at the time, but which will be returned to in the context of the present article). The idea of the landscape refers to a geoecologically construable, complex system of interrelation and *interplay*, which makes the term 'landscape' applicable in very diverse ways—and aligns it with the notion of an open, polylogical system.

In this complex relationality of areas of tension and force fields, an increasing number of further aspects and dimensions of the term 'landscape' are being developed in cultural studies and cultural-theoretical contexts, and some of these are of significance in the present study of Alexander von Humboldt's landscapes of theory and writing. For the art theoretician W.J.T. Mitchell, a landscape emerges from a triangular force field that is established through the relationship between *place* and *space*.[25] Drawing on canonical studies by Michel de Certeau[26] and Henri Lefebvre,[27] the US art historian employs his terminological triangulation in order to harness relationships between places and spaces into a triangular force field that takes into account places of locality and spaces of mobility. His understanding of landscape thus only emerges in its actual configuration—and in fact in its true form—on the basis of this relationship.

A moment of movement is thus fundamentally inherent in this idea of landscape that is necessarily *vectorially* defined by—and charged by—the interplay between places and spaces. Landscape is unthinkable without the inclusion of vectoricity: as a term of movement, vectoricity refers to a poetics of movement[28] that is understood as advancement.

Of these abovementioned perspectives, Mitchell's mostly unconvincing thesis that posits an intimate relationship between landscape painting and imperialism is largely unimportant in terms of his historical findings. However, his suggestion that landscape can only evolve out of movement, and that it is inextricably linked to the exertion of power—the power of thinking in particular—is more significant (an example of which is of course Petrarch's famous *Ascent of Mont Ventoux* from 26 April 1336). It is safe to assume that Alexander von Humboldt was familiar with this theoretical and writing tradition, and that he referred to this implicitly when he artfully set out his diverse literary representations of mountain ascents and gave these philosophical as well as epistemological emphases. In this way, Petrarch's *Mont Ventoux* is still present in the American Andes or other mountains of the 'new world' in those literary stagings in which Humboldt's travel writing features an aestheticization of mountain landscapes.

For example, in his *American Travel Journals*, Humboldt describes the adverse conditions of an ascent of the Silla de Caracas saddle long before other mountaineering accomplishments in the High Andes of the Quito province:

> Fantasies, intuitions are keeping me busy at this altitude, and since it always happens to geognosts at this altitude, this folly eases the pain of the discomfort I am overcoming. A rationally thinking man is exhilarated at the mountain's summit, and he perceives his condition, as compared to that of the mob accompanying him, as an intellectual stimulation that gives him more energy than any food or drink. Fantasy works as a soothing balm full of miraculous curative powers that nature bestows upon the suffering man as a constant companion, and it heals the wounds of the physical organism, such as those deep wounds that are struck by one's own and others' faculty of reason.[29]

The transcendental qualities of the landscape are transformed into the transcendence of the thinking man in the face of adversity (Humboldt and Bonpland had long been deserted by the accompanying men, who had taken all of their food and drink back to Caracas). Thus nature deploys an aesthetic force over the landscape, a force that imaginatively and creatively inspires not only thinking and writing, but also the healing of physical and spiritual ailments. This includes ailments that are caused by one's own or others' (Apollonian) faculty of reason: Humboldt takes delight in a distancing from this reason through the Dionysian experience of a rugged mountainous landscape that is characterized by extremes, just like the one that Petrarch once espoused as a space of thought and perception.

If we take this idea as a point of reference for understanding landscapes as legible texts,[30] and to explore their legibility from the perspective of a variety of disciplines, from geometry and geography to art history and visual culture, then it is worth also rendering legible a landscape that is heavily charged with vectors—including *paysage littéraire*, for example, to which French-speaking scholars are increasingly turning their attention[31]—in a narrower sense as *landscapes of theory*.

To claim that Alexander von Humboldt was one of the early masters of this kind of vector-charging and semanticization of legible landscapes of theory is no exaggeration. In his landscapes, he combines the Apollonian with the Dionysian in a form of science that is exhilarating and joyous—and he does so not only at mountain summits.

Precisely which parts of these landscapes does Humboldt render legible and vividly imaginable? The landscapes are sketched in the form of a literary text or a painting, but also in phonotextually interpretable sound patterns, synaesthetic compositions and technical drawings; these sketches reveal a model—a model that lays bare a landscape's complex theory and epistemology in a self-reflexive and meta-reflexive motion, all in a single gaze. Theory is rendered visible in artistic form, and—in the best case—becomes quasi-graphically clear, like a quasi-simultaneously perceivable total impression.[32]

Regardless of the artistic medium in which the landscapes of theory are sketched, and whether they depict deserted stretches of sand or highly populated archipelagos, desolate mountainous regions or inundated riverside landscapes, they always stage and embody a model of the movement of life forms and cultural norms, registering historical developments and contemporary refractions in a mobile network of coordinates. These landscapes of theory serve to choreograph the hermeneutic exercises in understanding that they seek to achieve in a sensuously comprehensible way. Landscapes are moving images of imagining and thinking, of writing and of living: an experience of fantasies of one's own and fantasies that have been encountered. Not just from a geographical or art-historical perspective, either, but from a philological perspective, too—landscapes are full of life and presuppose movement, in the sense of both motion and emotion. They are landscapes of emotion in the poetic and poetological senses.

In the first instance, this involves a process of spatialization that enables the theoretical foundations of an artistic, scientific or technical sketch to become visible. The so-called *spatial turn* has had the effect of silently making movement—the vectoricity that is inscribed in the spaces and stimulated by movement—disappear behind the spaces that have been (re)constructed in the sketch.

This visible-making, in the sense of both visualization *and* rendering visible, primarily affects vectorization in the form of a landscape of theory insofar as *places* of movement and *spaces* of movement appear as highly mobile choreographies. The choreographies have (animated) vectors that take in both historically accumulated movements and movements that are prospectively anticipated in the future. Landscape is always oriented towards the future; thanks to its high movement coefficients, it explores that which horizons in motion are professing—or rather, what they are uncovering. For Humboldt, landscape alone is not simply a witness to a historical having-become, but rather an area of play for future sketches as well as sketches of the future. After all, in the future a piece of this freedom will be scientifically and artistically consolidated in the landscape.

It is precisely the aesthetic dimension and the specific liberty of artworks that mean that they do not only open up a past of represented things and undoubtedly important memory functions through their presence and presentation. More significantly, the horizon of prospective representation—a kind of *prospection*—is simultaneously brought into the image and into the field of vision via the artwork. This process of convergence within the artwork means that the landscapes of theory release perspectives about those parts of a theory or epistemology that are already present vectorially, but which are not yet really *thought out* (or indeed formulated). And it is precisely here that the aesthetic power of artistic landscape forms is located.

Beyond spatialization, landscapes of theory therefore always also have a future-oriented dimension and a prospective aesthetic power in which a future thinking-out already takes form, and in which the gaze is prepared for immanent forms of composition. It is therefore not simply a matter of consolidating that which has already been expatiated or that which can be expatiated, but also of a consolidating pre-emption of future developments pertaining to complexity and multiple connectedness. In this way, landscapes are moving images of that which stood before and that which is to come, the future: this is clearly evident in the future volcanic ascents that are shown in the drawings of the ascent of Pico del Teide on the island of Tenerife, and even in the sketch of Humboldt's 'Painting of Nature in the Tropics'. *Un volcan peut en cacher un autre*: behind one mountain is hidden another mountain; behind one landscape is hidden another. In the landscape of the present, 'behind' or 'beneath' the landscape of the present one, we prospectively read the landscapes of the future, through which we begin to roam.

These considerations may productively be linked to an anthropological and, in the broadest sense, cultural-theoretical approach that draws on the social sciences. This approach uses the imagery of landscape to bring together the phenomena of modernity and globalization—phenomena that were of such critical importance for Alexander von Humboldt's project about a different modernity.[33] In his 1996 book *Modernity at Large: Cultural Dimensions of Globalization*,[34] Arjun Appadurai, an Indian scholar working in the US, begins by sketching his own path—'born into the ruling classes of the new nations'[35]—and thereby produces a kind of vectoricity that is illuminating not only in terms of his autobiography. Indeed, it seems to run through all of his theoretical constructions, for these are highly influenced by notions of mobility.

In one of the most influential passages of the book, Appadurai notes the importance of considering the new cultural order on a global scale as an order that no longer allows us to think 'in terms of existing center-periphery models (even those that might account for multiple centers and peripheries)'[36]—vocabulary that in many ways reminds us of Alexander von Humboldt's sketches that span the entire world, sketches that put forward a future understanding of transareal connections that is balanced and based on an outward-facing world consciousness.

In contrast to common models of globalization and development, and even in contrast to the canonical ideas of Wallerstein[37] or Wolf,[38] Appadurai offers his own suggestion for five dimensions that play the role of landscapes in his text: *ethnoscapes*, *mediascapes*, *technoscapes*, *financescapes* and *ideoscapes*.[39] He adds that in this division of different landscapes, the common '*-scape*' suffix is intended to point to the fact 'that these are not objectively given relations that look the same from every angle of vision but, rather, that they are deeply perspectival constructs, inflected by the historical, linguistic, and political situatedness of different sorts of actors'.[40]

This is how different kinds of landscapes emerge, constructed according to their chosen perspectives and the actors of the respective 'scapes'; these are, however, also landscapes that, on account of their multiple connectedness, draw attention to their networked nature while retaining their respective 'views'. In the interplay between *situatedness* and *perspectival set*, landscapes are formed which cannot be reflected without the vectoricity of changing perspectives. However, are these theoretical landscapes of *ethnoscapes*, *mediascapes*, *technoscapes*, *financescapes* and *ideoscapes* to be understood as landscapes of theory or the like?

At best, these 'scapes' may only be considered to be landscapes of theory to a limited extent. Appadurai's model is more interested in the general imagery of scapes than in the thinking and unfolding of concrete landscapes. This also applies to Appadurai's intelligent idea of culture, which is sceptical towards any form of essentialism.[41] Accordingly, the relatively open idea of culture that Appadurai develops in the first part of his volume defines 'culture as the process of naturalizing a subset of differences that have been mobilized to articulate group identity'.[42] This may be tied into a mobile concept of a (cultural) landscape that is oriented towards movements and flows of all kinds, whereby all of the landscape's processes of naturalization are subjected to a critical revision. Was it not Humboldtian scholarship that developed possibilities of thinking opposed to a clear division between 'nature' and 'culture', and thereby offered us possibilities for future thinking that we ought to apply in a more targeted manner in years to come?[43] Humboldt's landscapes of theory have not yet been properly investigated, and they open up even more diverse prospects for future paths of scientific exploration.

Landscapes of multiple connectedness

Alexander von Humboldt's *American Travel Journals* give us a vivid image of the slow production and emergence of ideas, and the basic principles of Humboldtian science, through the process of writing itself. Time and again, his sketches of basic theorems are literally interrupted as certain ideas appear to force their words upon him, indeed seem to harass him, to the point where he must write them down quickly or insert them into his work.

This applies in particular to the theorem that becomes a kind of Humboldtian axiom over the course of his scientific and literary work. Midway through a sketch of his Mexican travel journey in French, Humboldt noted down an idea in German in the moment that it became clear to him; revealingly, this then appears as an insertion in his climatological–geoecological reflections:

> L'évaporation, causée par la chaleur, produit le manque d'eau et de rivières, et le manque d'évaporation (source principale du froid atmosphérique) augmente la chaleur. Alles ist Wechselwirkung. Tout le plateau depuis Oaxaca à Chiuaua est de plus triste monotonie de construction. D`immenses plaines, des bassins à sol uni de 30–40 lieux quarrés, généralement le triple plus long que large, dirigé le diamètre plus long du nord au sud, entourés par des collines ou hauteurs à contours uniformes et ondoyants et élevés à peine de 150–200 t. au-dessus des plaines voisines.[44]

That Alexander von Humboldt was an outstanding scholar who joins the ranks of a long list of scientific observers of what we now call geoecology or geoecological systems is a fact that Humboldt research has made increasingly clear, especially over the past two decades.[45] The fascinating aspect of Humboldt's outlook in the *American Travel Journals* is that we can see this founding principle of his thought grow more mature in a range of contexts and connections as well as in diverse disciplines and subject areas. Even if it may seem that formulations 'reveal themselves' to him, this is not a case of sudden inspiration.

Indeed, the fundamental realization that everything is connected to everything (if not always at the same time and place) permeates Humboldt's thinking in every area: in his analysis of ecosystems as much as in his investigation of early American codices; in the area of geology and volcanology as much as in the history of the 'discovery' and conquest of the American hemisphere; in plant geography and climatology as much as in anthropology, the history of languages and zoology. As he writes his manuscript, an investigation of individual subjects is continually transformed into a pattern of relations and interdependencies that he recognizes with an increasing degree of precision. Over the course of his travels through the American tropics, Humboldt's gaze is increasingly drawn not to individual phenomena, but to their complex multiple connectedness. Alexander von Humboldt became a researcher and thinker of a relationality that sought to apprehend everything it encountered in its multiple connectedness as well.

This is illustrated in a particularly impressive section of the abovementioned investigations of the volcanoes of the High Andes of the Quito province. His analyses draw as much on philological studies of historical reports on volcanic eruptions as they do on his own extensive field research, and they lead Humboldt to draw the conclusion that his investigation is less about individual volcanoes to which certain phenomena can be attributed, but rather about a volcanic landscape that is related to a multitude of communications, and that the interconnections between these communications must be researched. He emphasizes:

> Je crois qu'au fond toute la province depuis Pichincha, Cotopaxi à Tungurahua, Carihuairazo, et le Sangay n'est qu'un seul Volcan, un assemblage de concavités dans lequel fermente la matière acidifiable. Toute la partie élevée de la prov. peut être regardée comme une seule montagne; ce que nous nommons Pichincha, Cotopaxi ne sont que plusieurs cimes qui couronnent cet immense dos. Ce grand Volcan a plusieurs bouches, tantôt il dirige ses matières vers Tungurahua tantôt vers Cotopaxi. [...] C'est pour cela que la terre s'ouvre où l'on l'espère le moins où extérieurement rien ne l'indique.[46]

Humboldt's focus is decidedly on the connections, the 'communications'[47] between individual volcanoes and volcanic eruptions that have continued to shake the province at different places through all historical eras, and especially in the 1790s, not long before his own stay in the province in 1802. Rivers would occasionally get closer to one another without converging ('sans se communiquer');[48] however, at times even geographically distant volcanoes might interact and directly communicate with one another via their flowing lava. Larger distances were not of importance in this process: 'Cela n'est rien géologiquement et l'on peut considérer la province de Los Pastos et le district de la ville comme appartenant à ce même grand Volcan que j'ai décrit.'[49]

This is how, at the tip of Humboldt's quill and before our eyes, a complex and multiply connected volcanic landscape is born, one that is in constant movement and in which everything appears to be connected with everything—if not always simultaneously. In this context, it is unsurprising that Humboldt very soon began to take significant interest in exploring the question of communication and relations between the most diverse parts of the earth's surface—which is constantly in motion—by identifying certain phases of especially strong volcanic activity and positioning the activity in relation to the rest of the world.

Looking at Humboldt's planetary landscapes—with their moving continents, torn-off islands or flood catastrophes on a biblical scale—it is already evident how, in the area of geology and especially in the field of so-called plate tectonics, Humboldt translates very diverse individual phenomena into landscapes of multiple connectedness. It is also clear that Humboldt believed that his theories lent themselves to the field of volcanological research. For here, too, was his axiom valid: everything is interrelated—especially in planetary interplay.

It is frequently possible to observe a phenomenon in the *American Travel Journals* that we could call (travel-literary) superposition. That is, often beneath the present journey is an earlier journey, beneath one town is a previously visited town, beneath one landscape is a landscape that has been traversed earlier (or indeed will be traversed later), which Humboldt stitches into the present movement as if it is a collage.

This also applies to individual stages of his journey through the equinoctial areas of the new continent itself. Here, the traces of landscapes through which he has travelled before (and sometimes through which he is yet to travel) appear beneath the region that is being depicted, and a complex portrait of landscape superpositions arises. *Un paysage peut en cacher un autre.* On the way through the Andes, the travelling group crosses the 'Inca palace' in Guanani, a place that Humboldt and Bonpland investigated and sketched a number of times. Here, a completely different landscape suddenly opens up in a supplementary commentary:

> Malgré le froid qu'il fait à Guanani (nous eûmes 7 ½ ° de R.) la position de ce palais est bien belle, pittoresque. Il se trouve au sommet des Andes et on y joui[t] d'une vue immense sur les plaine[s] de Puira et Lambayeque, bordé par l'horizont de la mer pacifique. Lorsque nous passâmes, ces plaines étaient couvertes d'une brume épaisse de laquelle sortaient en forme d'isle les pointes des rochers situés au Sudouest et on devinait plus qu'on distinguait l'horizont de la mer. Ces groupes de rochers isolés ressemblaient à la vue des Canaries que nous eûmes du haut du Pic de Teyde.[50]

If there is a landscape that continues to influence all others, then it is the archipelagic landscape of the Canary Islands as Humboldt saw it from the peak of Pico del Teide for the first time. This was the first (and perhaps for this reason, a very memorable) extra-European landscape that the Prussian traveller encountered.

Indeed, the relationality of the archipelago is different to that of other places or phenomena that stand in direct connection with one another, such as the volcanoes in the Quito highlands.

For the island landscape of the Canaries is a specific model of space and movement: its archipelagic structures reveal the most varied forms of landscape expression, construction and discontinuity, for example in the separation of islands from one another by water. This Humboldtian model landscape was the first extra-European landscape that he encountered, and represented an archetypal original structure (*Ur-Struktur*) for him. This archipelagic structuring appears in diverse texts, contexts, places and locations in the *American Travel Journals*. It forms the foundational landscape of theory in which Humboldt is continually interested in his writing.

Aside from the sequence of Humboldt's travels, the reason for the ubiquity of the Teide and the Canaries' archipelagic structure can be found in the fact that the islands of the archipelago reveal a basic figure of relationality and multiple connectedness, in that each island has its own logic, form and function, yet is simultaneously relationally connected to all the other islands of the archipelago.

In addition, this open relational structuring of an isolated *island world* and a multiply connected *world of islands* always points beyond the individual archipelago; it connects the different islands with islands in other archipelagos, too, via trans-archipelagic connections. Humboldt describes this concept by referring to the Canaries alongside the Caribbean archipelagos or island worlds in the *American Travel Journals*. For Humboldt, archipelagos are figures of multiple connectedness and therefore emblems of a relationality that is of great epistemological relevance for Humboldtian science.

For this reason, they are not only ubiquitous in his entire oeuvre, but also create a basic pattern of Humboldtian writing, a pattern that almost obsessively recurs exactly where one least expects it: in the expansive steppes of Siberia, for example, where Pico del Teide returns as seen from above, together with the Canaries' variously shaped islands.[51] It is in his model landscape of the archipelago that Humboldt's relational, multiply connected thinking is most incisively landscape-focused. The multiple connectedness of archipelagic structuring produced in Humboldt a mode of thinking that is expressed in his cultural landscapes of the 'Inca palaces', with their scattering across wide disparate parts of the Inca kingdom, and in the different volcanoes in the High Andes of Quito. It was out of this structuring that his landscape of theory was born, together with its embodiment of a relational epistemology. Everything is interrelated, after all.

With their transareal connections and links, Humboldtian landscape pictures give rise to discontinuously interconnected and overlapping landscapes in which the sensory and tangibly perceptible world opens up dimensions that cannot be accessed through physical movements alone. In the view from above, Humboldt's 'thinking man' becomes aware of his limits, but also of his potential for a life with broader connections that are continually expanding. He is developing a world consciousness that longs for more distant worlds while focusing on a concrete place. Thus it is not in *Kosmos* that landscapes extending beyond the planetary emerge for the first time.

For there is no single place in which the entire world in its great diversity is accessible to the traveller, writer and scientist at once. Nowhere is there a place in which everything exists. As a result, everything is always characterized by something missing, a lack that longs for transareal landscape images with their superpositions.

If there is no place of wholeness, if there is always something missing, then the vectorization of one's own life is the necessary consequence. This is brought about through an interplay between physical and emotional movement at first, but there are also those landscapes that emerge out of Apollonian intelligence and Dionysian passion, landscapes that may still be considered landscapes of theory. These are landscapes of a theory of the world, landscapes of a theory of humankind: landscapes of a life that does not wish to be satisfied with the limitations of the planetary.

Translated from German by Leila Mukhida.

Notes

1 Michel Butor, 'Paysages planétaires,' in *Oeuvres complètes, Vol. XII: Poésies 3 (2003–2009)*, ed. Mireille Calle-Gruber (Paris: La Différence, 2010), 738. I would like to thank Patrick Suter (University of Bern, Switzerland) for kindly drawing my attention to the text.

2 Ibid. See also two works by Patrick Suter: 'Butor et le livre-installation: montage de textes, oeuvre plurielle, transits entre univers culturels', in *Les graphies du regard/Die Graphien des Blicks: Michel Butor und die Künste*, ed. Christof Weiand (Heidelberg: Universitätsverlag, 2013); 'Butor transaréal', in *Dix-huit lustres*, eds. Amir Biglari and Henri Desoubeaux (Paris: Classiques Garnier, 2016).

3 Butor, *Paysages planétaires*, 740.

4 Ibid., 746.

5 Ibid., 747.

6 Ibid., 748.

7 Ibid., 750.

8 Ibid., 751.

9 Ibid., 759.

10 Ibid.

11 For more about the epistemology of this movement concept, see Ottmar Ette, *TransArea: Eine literarische Globalisierungsgeschichte* (Berlin and Boston: Walter de Gruyter, 2012).

12 See also the keynote I gave at the conference organized by Julio Ortega, entitled 'El Valle Longitudinal: Alejandro de Humboldt y las relaciones transatlánticas' (VII International Conference on Transatlantic Studies, 'After Transitions/Global Humanities/Transatlantic XXI Century', Brown University, Providence, RI, 22 April 2015).

13 Alexander von Humboldt, *Amerikanische Reisetagebücher*, Tagebuch I, 47r–53r.

14 Ibid., 50r.

15 See the oft-cited letter from Friedrich Schiller to Christian Gottfried Körner, dated 6 August 1797 from Jena. Vgl. *Schillers Briefwechsel mit Körner: von 1784 bis zum Tode Schillers*. Zweite, wohlfeile Ausgabe. Vierter Teil, 1797–1805 (Leipzig: Verlag von Veit & Comp. 1859), 47.

16 Ernst Neef, *Das Gesicht der Erde*, 3rd ed. (Zürich and Frankfurt am Main: Verlag Harri Deutsch, 1974), 700. For an extended discussion of the term 'landscape of theory', see Ottmar Ette, *Viellogische Philologie: Die Literaturen der Welt und das Beispiel einer transarealen peruanischen Literatur* (Berlin: Verlag Walter Frey and Edition Tranvía, 2013), 36–46. This is the further development of ideas from chapters 1, 2, and 11 of Ottmar Ette, *Literatur in Bewegung: Raum und Dynamik grenzüberschreitenden Schreibens in Europa und Amerika* (Weilerswist: Velbrück Wissenschaft, 2001).

17 Neef, *Gesicht der Erde*, 700.

18 Ibid.

19 Ibid.

20 Joachim Ritter, 'Landschaft: Zur Funktion des Ästhetischen in der modernen Gesellschaft', in *Subjektivität. Sechs Aufsätze*, ed. Joachim Ritter (Frankfurt am Main: Suhrkamp, 1974), 142.

21 Ibid., 144.

22 Ibid., 179.

23 Carl Troll, 'Die geographische Landschaft und ihre Erforschung', in *Studium Generale 3* (Hamburg: Springer, 1950), 165.

24 See Christoph Menke, *Kraft: Ein Grundbegriff ästhetischer Anthropologie* (Frankfurt am Main: Suhrkamp, 2008).

25 See W.J.T. Mitchell, 'Preface', in *Landscape and Power*, 2nd ed., ed. W.J.T. Mitchell (Chicago and London: University of Chicago Press, 2002).

26 See Michel de Certeau, 'Pratiques d'espace', in *L'invention du quotidian, Vol 1: Arts de faire*, ed. Michel de Certeau (Paris: Gallimard, 1990).

27 See Henri Lefebvre, *La production de l'espace* (Paris: Anthropos, 1974).

28 See Ottmar Ette, 'ZwischenWelten der Literatur(wissenschaft): Auf dem Weg zu einer Poetik der Bewegung im Kontext der TransArea Studies', in *Cultures à la derive—cultures entre les rives/Grenzgänge zwischen Kulturen, Medien und Gattungen: Festschrift für Ursula Mathis-Moser zum 60. Geburtstag*, ed. Doris Eibl, Gerhild Fuchs and Birgit Mertz-Baumgartner (Würzburg: Königshausen & Neumann, 2010).

29 Humboldt, *Amerikanische Reisetagebücher*, Tagebuch III, 39r–39v, 1.

30 See W.J.T. Mitchell, 'Imperial Landscape', in Mitchell, *Landscape and Power*, 5, as well as Mitchell's introduction to his edited volume.

31 The now quite substantial body of critical literature tends to espouse more static conceptions of the landscape. To name a few examples: Simon Schama, *Paysage et mémoire* (Paris: Seuil, 1999); Marc Desportes, *Paysages en movement: Transports et perception de l'espace, XVIIIe–XXe siècles* (Paris: Gallimard, 2005).
32 On the significance of the total impression (for both Humboldt brothers) from a stylistic as well as scientific-historical perspective, see Jürgen Trabant, 'Der Totaleindruck: Stil der Texte und Charakter der Sprache', in *Stil: Geschichte und Funktionen eines kulturwissenschaftlichen Diskurselements,* eds. Hans Ulrich Gumbrecht and K. Ludwig Pfeiffer (Frankfurt am Main: Suhrkamp, 1986); Gerhard Hard, 'Der Totalcharakter der Landschaft: Re-Interpretation einiger Textstellen bei Alexander von Humboldt', in *Alexander von Humboldt: Eigene und neue Wertungen der Reisen, Arbeit und Gedankenwelt,* ed. Gerhard Engelmann (Wiesbaden: Steiner, 1970); Birgit Schneider, 'Der "Totaleindruck einer Gegend": Alexander von Humboldts synoptische Visualisierung des Klimas', in *Horizonte der Humboldt-Forschung: Natur, Kultur, Schreiben*, eds. Ottmar Ette and Julian Drews (Hildesheim, Zürich and New York: Georg Olms Verlag, 2016).
33 See Ottmar Ette, *Weltbewußtsein: Alexander von Humboldt und das unvollendete Projekt einer anderen Moderne* (Weilerswist: Velbrück Wissenschaft, 2002).
34 Arjun Appadurai, *Modernity at Large: Cultural Dimensions of Globalization* (Minneapolis and London: University of Minnesota Press, 1996).
35 Ibid., 10.
36 Ibid., 32.
37 See Immanuel Wallerstein, *The Modern World System*. 2 vols (New York and London: Academic Press, 1974).
38 See Eric R. Wolf, *Europe and the People Without History* (Berkeley: University of California Press, 1982).
39 Appadurai, *Modernity at Large*, 33.
40 Ibid.
41 Ibid., 12.
42 Ibid., 15.
43 In particular, see here the opening of Philippe Descola, 'Leçon inaugurale, Chaire d'Anthropologie de la nature, faite le jeudi 29 mars 2001', http://www.college-de-france.fr; Philippe Descola, *Par-delà nature et culture* (Paris: Gallimard, 2005); Ottmar Ette, 'Natur und Kultur: Lebenswissenschaftliche Perspektiven Humboldtscher Wissenschaft', in Ette and Drews, *Horizonte.*
44 Humboldt, *Amerikanische Reisetagebücher*, Tagebuch IX, 27r, 2–27v.
45 See also Ulrich Grober, 'Humboldt, Haeckel und 50 Jahre Ökologie' in Ette and Drews, *Horizonte.*
46 Humboldt, *Amerikanische Reisetagebücher*, Tagebuch VII bb u. c, 22r, 2–22v, 1.
47 Ibid., Tagebuch VII bb u. c, 22v, 1.
48 Ibid.
49 Ibid.
50 Ibid., bb u. c, 58r, 3.
51 See Ottmar Ette, 'Amerika in Asien: Alexander von Humboldts "Asie centrale" und die russisch-sibirische Forschungsreise im transarealen Kontext', in *HiN: Alexander von Humboldt im Netz: Internationale Zeitschrift für Humboldt-Studien* 8, no. 14 (2007), http://www.hin-online.de.

Bibliography

Appadurai, Arjun. *Modernity at Large: Cultural Dimensions of Globalization.* Minneapolis and London: University of Minnesota Press, 1996.

Butor, Michel. 'Paysages planétaires.' In *Oeuvres complètes, Vol. XII: Poésies 3 (2003–2009)*, ed. Mireille Calle-Gruber, 738. Paris: La Différence, 2010.

de Certeau, Michel. 'Pratiques d'espace'. In *L'invention du quotidian, Vol 1: Arts de faire*, ed. Michel de Certeau, 137–191. Paris: Gallimard, 1990.

Descola, Philippe. 'Leçon inaugurale, Chaire d'Anthropologie de la nature, faite le jeudi 29 mars 2001'. http://www.college-de-france.fr.

Descola, Philippe. *Par-delà nature et culture.* Paris: Gallimard, 2005.

Desportes, Marc. *Paysages en movement: Transports et perception de l'espace, XVIIIe–XXe siècles*. Paris: Gallimard, 2005.

Ette, Ottmar. 'Amerika in Asien: Alexander von Humboldts "Asie centrale" und die russisch-sibirische Forschungsreise im transarealen Kontext'. In *HiN: Alexander von Humboldt im Netz: Internationale Zeitschrift für Humboldt-Studien* 8, no. 14 (2007). http://www.hin-online.de.

Ette, Ottmar. 'El Valle Longitudinal: Alejandro de Humboldt y las relaciones transatlánticas.' Keynote, VII International Conference on Transatlantic Studies, 'After Transitions/Global Humanities/Transatlantic XXI Century', Brown University, Providence, RI, 22 April 2015.

Ette, Ottmar. *Literatur in Bewegung: Raum und Dynamik grenzüberschreitenden Schreibens in Europa und Amerika*. Weilerswist: Velbrück Wissenschaft, 2001.

Ette, Ottmar. 'Natur und Kultur: Lebenswissenschaftliche Perspektiven Humboldtscher Wissenschaft'. In *Horizonte der Humboldt-Forschung*, ed. Ottmar Ette and Julian Drews, 13–51. Hildesheim, Zurich, New York: Georg Olms Verlag, 2016.

Ette, Ottmar. *TransArea: Eine literarische Globalisierungsgeschichte*. Berlin and Boston: Walter de Gruyter, 2012.

Ette, Ottmar. *Viellogische Philologie: Die Literaturen der Welt und das Beispiel einer transarealen peruanischen Literatur*. Berlin: Verlag Walter Frey and Edition Tranvía, 2013.

Ette, Ottmar. *Weltbewußtsein: Alexander von Humboldt und das unvollendete Projekt einer anderen Moderne*. Weilerswist: Velbrück Wissenschaft, 2002.

Ette, Ottmar. 'ZwischenWelten der Literatur(wissenschaft): Auf dem Weg zu einer Poetik der Bewegung im Kontext der TransArea Studies'. In *Cultures à la derive—cultures entre les rives/Grenzgänge zwischen Kulturen, Medien und Gattungen: Festschrift für Ursula Mathis-Moser zum 60. Geburtstag*, ed. Doris Eibl, Gerhild Fuchs and Birgit Mertz-Baumgartner, 41–57. Würzburg: Königshausen & Neumann, 2010.

Grober, Ulrich. 'Humboldt, Haeckel und 50 Jahre Ökologie'. In *Horizonte der Humboldt-Forschung*, ed. Ottmar Ette and Julian Drews, 181–190. Hildesheim, Zurich, New York: Georg Olms Verlag, 2016.

Hard, Gerhard. 'Der Totalcharakter der Landschaft: Re-Interpretation einiger Textstellen bei Alexander von Humboldt'. In *Alexander von Humboldt: Eigene und neue Wertungen der Reisen, Arbeit und Gedankenwelt*, ed. Gerhard Engelmann, 49–73. Wiesbaden: Steiner, 1970.

Lefebvre, Henri. *La production de l'espace*. Paris: Anthropos, 1974.

Menke, Christoph. *Kraft: Ein Grundbegriff ästhetischer Anthropologie*. Frankfurt am Main: Suhrkamp, 2008.

Mitchell, W.J.T. 'Preface' and 'Imperial Landscape'. In *Landscape and Power*, 2nd ed., ed. W.J.T. Mitchell, PAGE RANGES. Chicago and London: University of Chicago Press, 2002.

Neef, Ernst. *Das Gesicht der Erde*, 3rd edition. Zürich and Frankfurt am Main: Verlag Harri Deutsch, 1974.

Ritter, Joachim. 'Landschaft: Zur Funktion des Ästhetischen in der modernen Gesellschaft'. In *Subjektivität. Sechs Aufsätze*, ed. Joachim Ritter, PAGE RANGE. Frankfurt am Main: Suhrkamp, 1974.

Schama, Simon. *Paysage et mémoire*. Paris: Seuil, 1999.

Schiller, Friedrich, letter to Christian Gottfried Körner, dated 6 August 1797 from Jena. Vgl. *Schillers Briefwechsel mit Körner: von 1784 bis zum Tode Schillers*. Zweite, wohlfeile Ausgabe. Vierter Teil, 1797–1805 (Leipzig: Verlag von Veit & Comp. 1859).

Schneider, Birgit. 'Der "Totaleindruck einer Gegend": Alexander von Humboldts synoptische Visualisierung des Klimas'. In *Horizonte der Humboldt-Forschung: Natur, Kultur, Schreiben*, eds. Ottmar Ette and Julian Drews, 53–78. Hildesheim, Zürich and New York: Georg Olms Verlag, 2016.

Suter, Patrick. 'Butor et le livre-installation: montage de textes, oeuvre plurielle, transits entre univers culturels'. In *Les graphies du regard/Die Graphien des Blicks: Michel Butor und die Künste*, ed. Christof Weiand, 43–61. Heidelberg: Universitätsverlag, 2013.

Suter, Patrick. 'Butor transaréal'. In *Dix-huit lustres*, eds. Amir Biglari and Henri Desoubeaux, 423–445. Paris: Classiques Garnier, 2016.

Trabant, Jürgen. 'Der Totaleindruck: Stil der Texte und Charakter der Sprache'. In *Stil: Geschichte und Funktionen eines kulturwissenschaftlichen Diskurselements*, eds. Hans Ulrich Gumbrecht and K. Ludwig Pfeiffer, 169–188. Frankfurt am Main: Suhrkamp, 1986.

Troll, Carl. 'Die geographische Landschaft und ihre Erforschung'. In *Studium Generale 3*, 163–181. Hamburg: Springer, 1950.

von Humboldt, Alexander. *Amerikanische Reisetagebücher*, Tagebuch I, 47r–53r.

von Humboldt, Alexander. *Amerikanische Reisetagebücher*, Tagebuch III, 39r–39v, 1.

von Humboldt, Alexander. *Amerikanische Reisetagebücher*, Tagebuch IX, 27r, 2–27v.

von Humboldt, Alexander. *Amerikanische Reisetagebücher*, Tagebuch VII bb u. c, 22r, 2–22v, 1.

Wallerstein, Immanuel. *The Modern World System*. 2 vols. New York and London: Academic Press, 1974.

Wolf, Eric R. *Europe and the People Without History*. Berkeley: University of California Press, 1982.

4
FLIGHT FROM MODERNITY
Historicizing the aerial promise in landscape architecture[1]

Jeanne Haffner

In the aftermath of World War II, a novel way of conceptualizing urban space began to emerge in a variety of academic disciplines, among them sociology, anthropology, urban studies, architecture, urban planning and geography. At the centre of this new way of analysing space was a focus on the experience of urban inhabitants on the ground, that is, how local residents used and perceived the spaces in which they lived and worked. This idea, which was often encapsulated in the term 'social space' (*l'espace social*), provided an alternative to Euclidean notions of mathematical or abstract space and the functional notion of needs laid out in the Athens Charter in 1933:[2] 'social space' presumed that urban space was not pre-existing, but rather was socially 'produced' in everyday life.

One of the most famous proponents of this approach was the French Marxist urban sociologist Henri Lefebvre. Analysing the history of the social space of various societies, Lefebvre suggested that the hierarchical social structures created by capitalism were reproduced in the physical world, in the organization of urban space. The most crucial task for late twentieth-century social scientists, according to Lefebvre, was to better understand this process in order to find ways of changing it. As he wrote in the mid-1970s, 'knowledge of (social) space is now being established as a science, even though it is still in an early stage.'[3]

Examining the history of this idea reveals its connection to aerial photography—a technique of observation and representation closely tied to the French colonial state and military. Although Lefebvre and his colleagues, including the philosophers Michel Foucault, Guy Debord, and Michel de Certeau, often equated aerial photography with 'top-down' urban planning programmes that ignored the true needs of local inhabitants, the very idea of social space was actually engendered with the help of the view from above. Beginning in the 1920s and 1930s, social scientists working in the then-burgeoning fields of human geography and ethnography, as well as history, experimented with aerial photography in their work investigating the spaces of human habitation in French colonies as well as in France. Crucially, the device helped these researchers to see the connection between social organization and spatial organization.[4] After World War II, this way of analysing space inspired a new approach to landscape and urban design practice that challenged modernist architecture by drawing attention away from the rational organization of urban space from above and towards the more dynamic everyday experiences and social and psychological needs of urban residents on the ground. The view from above gave rise to the view from below.

If post-war social scientists used aerial photography to merge sociological analysis and urban planning and design, in recent years urban designers and architects have looked to this same military technique to bridge urban design and landscape architecture. The architect and urbanist Charles Waldheim coined the term 'landscape urbanism' in the 1990s to describe 'a disciplinary realignment currently underway in which landscape replaces architecture as the basic building block of contemporary urbanism'.[5] Landscape architect James Corner has called it a 'disciplinary collusion' that entails a different conception of landscape itself, from a 'bourgeois aesthetic' to a broader set of urban processes.[6] Aerial vision is therefore not only used in the discourse of landscape urbanism as a disciplinary bridge. It is also the basis for a systems-based conception of landscape, which in turn impacts concrete design projects on the ground. 'The idea of landscape', Waldheim writes, 'has shifted from scenic and pictorial imagery to a highly managed surface best viewed, arranged, and coordinated from above'.[7] Maps and plans are key here, but so too are aerial photographs'.

Yet it is not just the expanded vision of the earth offered by an airplane that draws Waldheim and others to such representations; it is the promise of aerial photography to challenge the legacy of modern architecture—and perhaps modernity itself. 'The dogma of modernist planning', wrote architects Mohsen Mostafavi and Ciro Najle in the early 2000s, 'with all its reductivism, needs to be supplanted by a whole range of new responses to the urban condition. Landscape urbanism in all its various guises constitutes one of many such approaches'.[8] Utilizing aerial reflections on the world and increasingly abstract techniques of representation, proponents of landscape urbanism thus hope to reshape the world itself. While histories of aerial vision have demonstrated that this desire is not new, a key question remains: Can the view from above still be linked to a spatially oriented critique of capitalism?

Capitalism, modernity, and ethnography from the air

In the interwar period, the aeroplane was the subject of much enthusiasm among public and politicians alike. Responding to the passion for aeroplanes and aerial views among the public, politicians on both the left and the right used aviation and aerial photography to bolster their political programmes. From above, the world appeared as a *tabula rasa* upon which a new human history could be written.[9]

In contrast, interwar French social scientists interpreted the god's eye view in a radically different way. These pioneers in ethnography, history, and human geography used the aeroplane to bring to light the history and social complexity of French colonies as well as France. One key example was the Africanist Marcel Griaule, an aviator and ethnographer who had served as an aerial spotter and navigator during World War I. In the mid-1930s, while based at the Musée de l'Homme in Paris, Griaule used aerial photographs to study the cosmology and iconography of Dogon society in West Africa. He reasoned that if aerial photographs could help him to see trenches hidden deep within the landscape, they could also aid in bringing to light the invisible aspects of cultural, social, political, and economic organization.

From the air, Griaule noticed the chequerboard pattern of Dogon agricultural fields (Figure 4.1). On the ground, he discovered that this same pattern was reproduced on the facades of sanctuaries, painted on rocks, and weaved into funerary blankets (Figure 4.2). This observation led him to argue that the Dogon had an integrated cultural system in which the various spheres of their lives were harmoniously integrated in the Dogon unconscious, and that this system was visible in the Dogon landscape. Griaule repeatedly asserted that this 'scientific discovery' would not have been possible without the aid of the aeroplane.[10]

Figure 4.1 Griaule argued that this chequerboard pattern linked various aspects of cultural and social life in the Dogon unconscious. © Fonds Marcel-Griaule, Bibliothèque Éric-de-Dampierre, LESC/CNRS, Université Paris Nanterre.

Interwar architects and urban planners also experimented with the use of aerial photography in their work. In 1935, for instance, the architect and urban planner Le Corbusier published a book titled *Aircraft* in which he used aerial photographs to critique the 'disorderly' organization of space in Western capitalist societies. He contrasted the 'natural' organization of space in non-Western areas, where people were 'happy' and connected to their surrounding landscape, with the chaos of capitalist cities in the West, where people (according to Le Corbusier) were individualistic, unsociable, and disconnected to the landscape in which they were situated.[11]

Aerial photography and 'social space' in post-war France

Ideas about the connection between the 'social' and the 'spatial' were therefore already percolating in the interwar period. But it was only in the aftermath of World War II, during an era of urban reconstruction, that the concept of 'social space' began to appear in numerous published works in the social sciences and to take hold within urban planning circles in Paris.

World War II provided the opportunity for social scientists to become involved in urban planning programmes, and the notion of 'social space' helped sociologists to differentiate themselves within interdisciplinary urban research teams. It was Paul-Henry Chombart de Lauwe, one of Griaule's students, who, drawing on pre-war anthropological theories of spatial organization, first developed this idea, with the aid of aerial photography. In his 1952 work, *Paris et l'agglomération parisienne: L'étude de l'espace social dans une grande cité*, Chombart de Lauwe argued that the 'social space' of French cities demonstrated not only the continuing spatial segregation of Paris, but also the continuing inegalitarian nature of French society more generally. Far from the ideals of the French Revolution, he suggested, France still epitomized a society divided

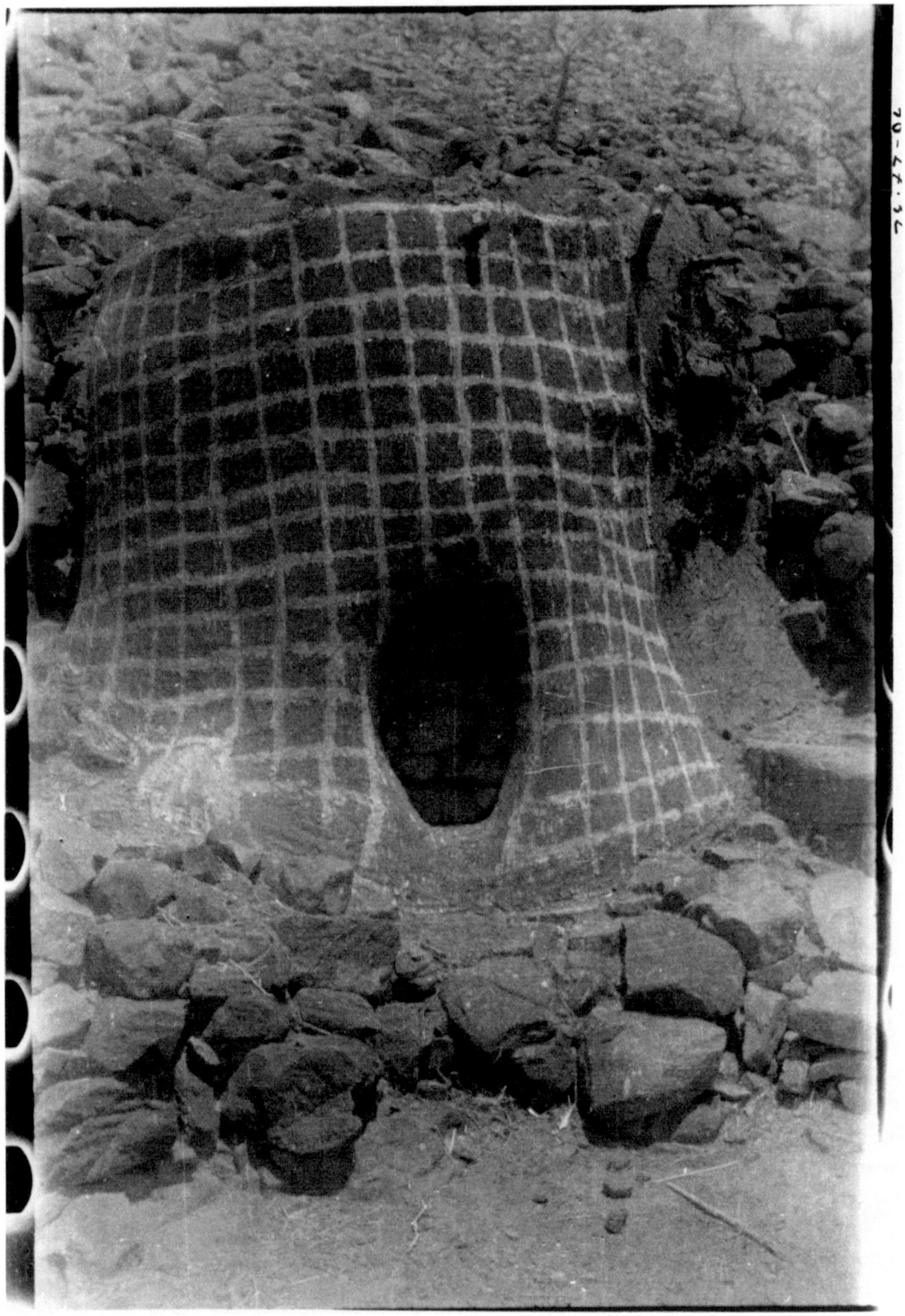

Figure 4.2 Griaule argued that this chequerboard pattern linked various aspects of cultural and social life in the Dogon unconscious. © Fonds Marcel-Griaule, Bibliothèque Éric-de-Dampierre, LESC/CNRS, Université Paris Nanterre.

along class lines. Chombart de Lauwe used the notion of social space in this work to argue for an alternative notion of 'needs' to that outlined by the Congrès Internationaux d'Architecture Moderne (CIAM), an organization founded in 1928 for the purposes, at least ostensibly, of distributing ideas about the modern movement in architecture.[12] In 1933, CIAM published the Athens Charter, a document that outlined ninety-five key points for the creation of a 'rational' city. Any such city, the charter suggested, would be divided into various zones for living, and these zones would be based on what CIAM considered the four main functions, or needs, of urban inhabitants: living, working, recreation, and circulation. In contrast to the functionalist notion of needs outlined in this document, Chombart de Lauwe argued that architects and planners should instead consider the psychological, recreational, and spiritual needs of urban residents, and design cities with these in mind.

Aerial photography provides, of course, views of the surface of the earth. But, for Chombart de Lauwe, the technique was key to gaining insight into what lay *below* this exterior. For instance, in his 1952 study, Chombart de Lauwe used aerial photographs—and graphic representations based on aerial photographs—to show that systems of social relations were intimately tied to individual actions on the ground. Aerial photographs of Paris made visible, in a new and more exciting way, a fact that anyone who had walked through the streets of the city already knew: that this environment was divided both geographically and socially. Bourgeois residents mostly lived and worked in the west, while working-class individuals largely inhabited the east.[13]

The wide and geometrical streets of the bourgeois neighbourhoods in the west were lined with classical houses that had plenty of open space around them, while the working-class sections in the east were easily identifiable by their narrow, irregular streets and closely knit buildings (Figure 4.3). Chombart de Lauwe linked these apparent differences in spatial relations with differences in habits, familial norms, and levels of sociability. Bourgeois individuals in the west, he argued, had very little contact with local merchants and neighbours in daily life, while workers located in the eastern parts of the city were found to have much deeper ties to their neighbourhood. This study served to reinvigorate old political concerns, and successfully joined urban planning and Marxism. What was novel, after all, was not Chombart de Lauwe's conclusion about the social and geographical division of Paris, but rather the methodological means by which he presented the information. Chombart de Lauwe's use of aerial photographs demonstrated, in a manner that government officials, planners, social scientists, and the public alike viewed as 'objective' and scientific, the continuing relevance of Marxist theory in the post-war era. Chombart de Lauwe argued that the discipline of ethnography had much to contribute to the endeavour to analyse and find solutions to urban problems. The analysis and solution of urban problems in France, he contended, required an outsider's perspective; this could be achieved, *literally*, with the use of an aeroplane, and furthered by ethnographic fieldwork on the ground. Both a top-down and a bottom-up perspective were necessary for obtaining a truly 'objective' look at the problems facing post-war French cities.[14]

After this publication, the contribution of sociology to urban planning, architecture, and landscape architecture—and qualitative methods more generally—started to become more accepted. This is clearly evident in the example of the planner Alexandre Burger, who drew upon Chombart de Lauwe's notion of 'social space' in his award-winning 1957 study on the use of aerial photography in urban planning, in which he used aerial photographs to compare the socio-economic characteristics of two sections of the city of Colmar in north-eastern France.[15] By placing social facts within the physical world in which they had evolved, Burger concluded, aerial photography made sociological data more comprehensible to urban planners.

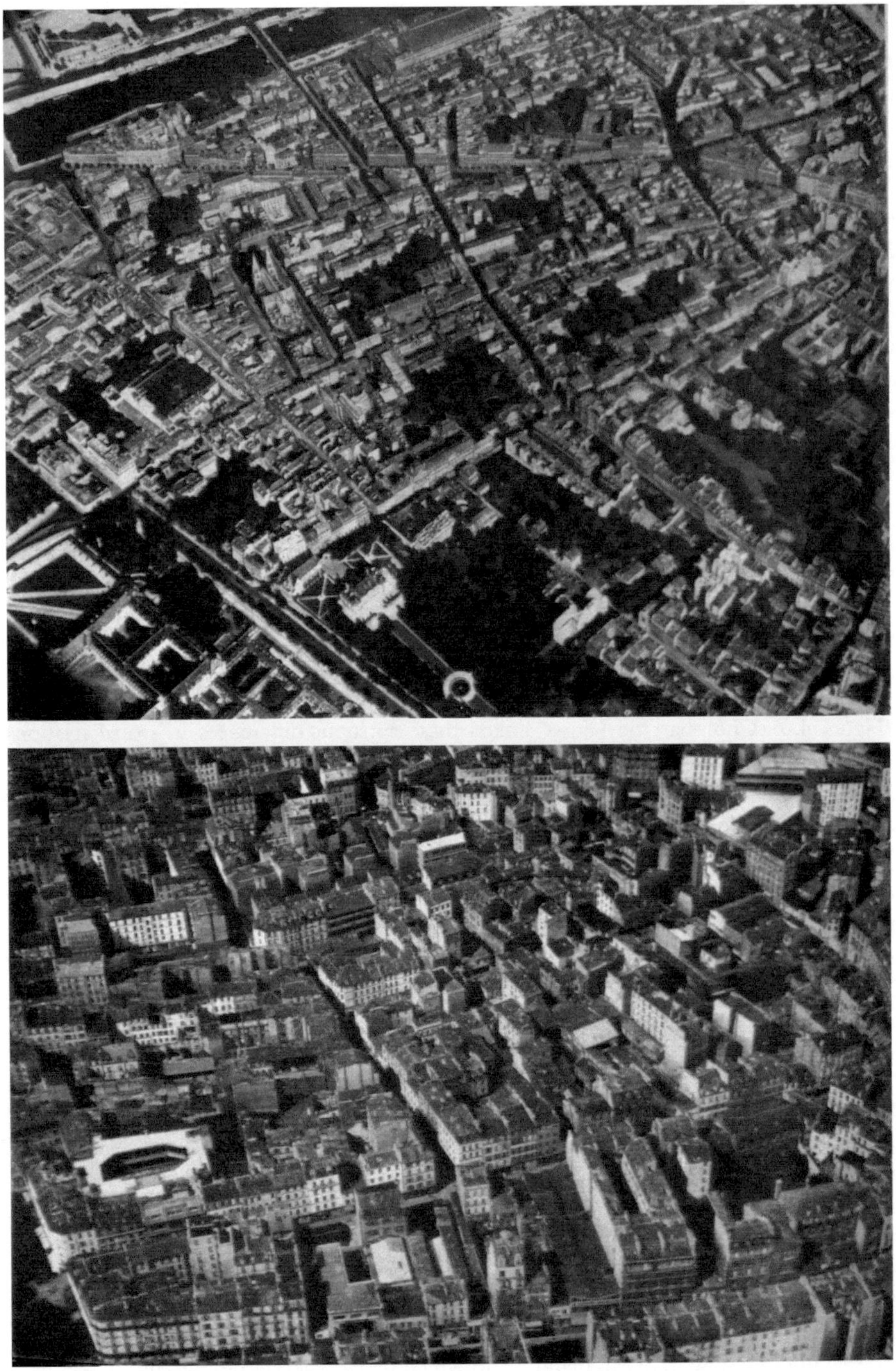

Figure 4.3 Using aerial photographs, Chombart de Lauwe claimed to discover a correlation between the urban fabric seen from above and differences in sociability and everyday life on the ground: the wide, tree-lined streets of the bourgeois sections of western Paris (top) contrasted with the narrow and chaotic streets of the working-class neighbourhoods in the eastern parts of the city (bottom). © Presses Universitaires de France.

Aerial photographs at a large scale clearly exposed the geographical contours of *quartiers* within Colmar, as well as the city's sociological character and the level of sociability among local residents. What planners were viewing on an aerial photograph of an agglomeration, he explained, was not only a city's geographical traits but also its sociological composition. Only aerial photographs provided a planner with a view into the relationship between disparate spaces within a city and how these networks (*réseaux*) had evolved over time.

As a case in point, Burger cited the example of aerial views of two different sections of Colmar. The first, a view of the Saint-Marie neighbourhood, appeared relatively homogenous on a simple topographic plan. Yet aerial photographs of this section revealed crucial details, allowing urban planners to see into its character at a glance. For instance, looking at an aerial photograph, one could clearly see the existence of two types of housing within the section: individual houses (*pavillons isolés*) and collective apartments (*grands ensembles*). Examining the photograph even more closely, a planner could see that each type of habitation was associated with a different type of garden: those attached to the *pavillons isolés* had trees, while those in the *grands ensembles* were used solely for growing food. This difference, Burger argued, pointed to different socio-economic levels, as those living in individual houses could afford to lose valuable land for cultivation, while those living in the *grands ensembles* could not. Going further, Burger compared the 'social space' of Sainte-Marie with a neighbouring section of Colmar, Château-d'Eau, which aerial photographs revealed to have a very different character. Château-d'Eau was composed entirely of spacious houses, all with decorative gardens. Each house had a garage, and the streets were harmoniously aligned and well maintained. The form of the houses, gardens, and surrounding streets revealed that the population of Château-d'Eau was wealthier than that of Sainte-Marie.[16]

Burger's analysis of the social space of Colmar, based on aerial photographs, was inspired by Chombart de Lauwe's study of the social space of Paris. It shows the role of photographic techniques and empirical observation in the development of a new approach to urban planning. By helping to quantify urban space and its sociological dimension, aerial photographs presented an alternative way of designing urban environments that was not utopian but, in the eyes of its advocates, 'scientific'.

Sociological expertise and the study of the built environment

As the idea of 'social space' developed and began to take hold in the social sciences and urban planning, so too did the perception of the application of sociological expertise to the fields of urban planning and architecture. Throughout the late 1950s and 1960s, sociologists were increasingly in demand within interdisciplinary, government-sponsored studies of urban problems.

Lefebvre was one such urban sociologist. In the late 1950s and 1960s, just as he was leaving the French Communist Party, Lefebvre became involved in government-funded studies alongside Chombart de Lauwe and other sociologists, such as Raymond Ledrut, who was based at the University of Toulouse. Like Chombart de Lauwe, Ledrut and Lefebvre used the concept of 'social space' in both their government and academic work to offer a new way of conceptualizing urban problems, e.g. of the *grands ensembles*. The main issue, they argued, was not the buildings themselves, as many at the time were suggesting. Rather, the key problem was the organization of space in French cities. In other words, at fault was not architecture but the social, political, and geographical landscape that surrounded it. With this argument, sociologists such as Lefebvre succeeded in shifting the focus of investigation with regard to the *grands ensembles* from architecture to (social) space.

Yet, unlike Chombart de Lauwe, Lefebvre and Ledrut—along with many other contemporaries, such as the philosophers Michel Foucault and Michel de Certeau—criticized exactly the technique of representation that had inspired the concept of social space in the first place.

Lefebvre's disdain for visuality was part and parcel of a more general critique of Western capitalism.[17] The emergence of this mode of production, Lefebvre argued, was accompanied by the hegemony of sight and the 'spectacle' of consumer capitalism, a sense that he equated with distance and passivity. Techniques of representation such as aerial photography only served to 'flatten' the complexity of social life, reducing its richness to an abstraction impoverished of meaning and experience. Although planners claimed that these qualitative methods were indispensable to the 'scientific' analysis of urban environments, Lefebvre warned that they actually concealed more than they revealed.

In particular, Lefebvre asserted that these abstract techniques masked the omniscience of the state in local affairs, especially in urban landscape management, under the neutral umbrella of 'science'. According to Lefebvre, the state maintained its vast bureaucracy largely by collecting data on all areas of social life. Visual methods were particularly important in government record-keeping, since officials, planners, and others considered them more objective than other methods because they were quantifiable. The quasi-colonial presence of the state in local affairs, Lefebvre argued, perpetuated the boredom, monotony, and sterility of everyday life under Western capitalism, wherein authenticity was lost to abstract representation.

Lefebvre was here expressing concerns that were typical within intellectual circles in Paris at the time. Michel Foucault, for instance, in *The Order of Things* (first published in France in 1966), likened the emergence of humankind as a subject of research in the human sciences to a king gazing upon his subjects from afar: the methods used to compare, measure, and 'order' social life reduced the complexity of humankind, Foucault argued, much like the taxonomic tables used by natural scientists to visibly order their knowledge of the natural world.[18] Ironically, a concept that had been engendered by the holistic view of the world offered by the aeroplane in Chombart de Lauwe's work years earlier now represented the distance that Ledrut, Lefebvre, and others perceived between their own sociological 'bottom-up' approach and the 'top-down', authoritative approach of government officials and planners. A literally vertical perspective was transformed into one that was so metaphorically.

Understanding this shift requires us to look deeper into the context in which it unfolded. In the wake of the atomic bomb, the Algerian War, and the first satellite photograph of the Earth, the distanced view from above was dehumanized. In contrast to World War I, in which aerial photography was utilized primarily for the purposes of surveillance, during World War II it was used as an instrument of aerial bombardment. The destruction caused by aerial bombardment left an indelible mark on the technique. By the 1960s, aerial photography epitomized governmental power over local citizens.

This was matched by a shift in the scale of urban planning programmes. The widespread construction of the *grands ensembles* in the 1960s went hand in hand with the acceleration of a long-standing programme to decentralize French cities that had begun in the 1940s under Vichy. Only through the development of provincial centres, it was thought, could France reach its full economic, social, and cultural potential.

In 1965, the new head of the District of Paris, Paul Delouvrier, initiated an ambitious programme to address the problem of growth in Paris. Developed in collaboration with intellectuals, planners, and many others, the *Schéma directeur d'aménagement et d'urbanisme de la région de Paris* proposed to deal with the rapid expansion of the suburban areas of Paris by creating five 'new towns' (*villes nouvelles*) around the city that would be connected to the centre. Roads as well as railway networks, including the suburban trains, would connect these new towns to the centre. As historian Larry Busbea explains, 'the purpose of this vast new circulation system, integrated into Paris proper and branching out across through the axes of the *Schéma Directeur*, was to bring the region into a state of equipoise, in which any location was equally within reach of the centre.'[19] In other words, much like the proponents of *l'aménagement du territoire*, Delouvrier proposed addressing the growing chaos of Paris and other French cities by implementing an urban scheme on a grander scale than ever before. The goal was to decrease dissatisfaction among urban residents by creating new centres of activity, and by providing better transport to and from the older city. As Delouvrier explained:

> Paris and its future must respond to three imperatives: to put to the best use a space that is very dense, old and small in extent; to respect the complexity of the functions and structures indispensable to the equilibrium between its territory and that of its region and, equally, the ensemble to the national territory.[20]

Aerial photographs, maps, and diagrams, meant to make the city of Paris and its surroundings more 'legible', were central to the *Schéma directeur* as proposed by Delouvrier. He even began by defining the Parisian region 'as seen from an aeroplane'.[21] Delouvrier used these visual aids to pinpoint areas of growth and develop a transport network to serve specific locations. He continually stressed the need to approach the problem of growth through a plan that encompassed the totality or 'ensemble' of the *agglomération parisienne* rather than focusing on isolated parts. Aerial photographs, maps, and diagrams were central to this holistic or 'synoptic' perspective, which was intended to be not only geographical but also social, economic, and cultural.

Creating the 'bottom-up'

If 'social space' for Chombart de Lauwe was intended to encapsulate the view from above, therefore, for Lefebvre and many others it signified only the view from below. During the 1950s and 1960s, when the French left was riddled with conflict, the new analytical category of 'social space' provided left-wing sociologists such as Lefebvre and Ledrut with a way of continuing to discuss their political concerns while moving away from the French Communist Party. In addition to the traditional Marxist categories of class, race, and gender, left-wing thinkers now had a new arena in which to explore the effects of capitalist modes of production: urban space. If 1968 did not succeed in revolutionizing spatial relations in Paris, it did, at least, provoke a new era of theorizing (social) space.

Lefebvre, working closely with Ledrut, outlined this spatially oriented left-wing programme through the creation of a new journal, *Espaces et Sociétés*, the first issue of which appeared in 1970. One of Lefebvre's first articles in *Espaces et Sociétés* was titled 'Reflections on the politics of space'.[22] In it, Lefebvre challenged the notion that 'space' in all its many forms—rural, suburban, and urban—was neutral, and that the practice of urban planning was a 'scientific' endeavour. On the contrary, he argued, 'space is political!'[23] Urban space may appear to be homogenous and shaped by scientific management, yet it was in fact produced by social relations. Modern urban planning, which was tied to the French state, effectively disguised the political nature of space in order to perpetuate its practice of social control.

Lefebvre proceeded to compare urban planning techniques in post-war France to the process of French colonization: the large-scale construction of the *grands ensembles* and new towns all over France under de Gaulle's Fifth Republic had resulted in what he referred to as the 'internal colonization' of French cities. 'There are no colonies in the old sense,' he wrote, 'but there is already a metropolitan semi-colonialism.'[24] The drastic changes occurring in cities across France, he suggested, were indicative of a major transformation in French society and culture. His call for a 'politics of space', of which the journal *Espaces et Sociétés* was to be a key vehicle, was intended to bring the significance of these spatial changes to light, not only for intellectuals, planners, architects, and government officials but for the French public more generally.

In the 1970s, Lefebvre expanded upon his theoretical reflections on the politics of space in two works, *The Urban Revolution* and *The Production of Space*.[25] In *The Urban Revolution*, Lefebvre used the terms 'urban' and 'urban society' to signify his theoretical approach. As he wrote, 'the expression "urban society" meets a theoretical need. It is more than simply a literary or pedagogical device, or even the expression of some form of acquired knowledge; it is an elaboration, a search, a conceptual formulation.'[26] The project of theorizing the urban was to serve a practical function: to help promote an 'urban revolution', by which he meant social revolution, by bringing to light the political practices embedded in the organization of space in French society.

The Production of Space of 1974 explored the connection between the spatial and the social, and the role of space in social revolution, in more depth. Lefebvre devoted an entire chapter to 'social space', in which he insisted that 'space' was not a thing but rather 'a social reality—a set of relations and forms'.[27] Social relations, and especially relations of production, he suggested, simultaneously produced space and were produced by it. Lefebvre here attempted to use the idea of 'social space' to counteract the widely held notion that space was simply an empty container in which objects and people were placed; this view of space, he argued, was not only an 'error' but an 'ideology'—it was promoted by those whose interests it served, namely the French state, urban planners, and architects.[28] On the contrary, Lefebvre contended, space was full: full of social relations, culture, and history.

In all of these works, and others, Lefebvre came back to the theme of colonization. Urban space had been colonized, he wrote, 'through the image, through publicity, through the spectacle of objects'.[29] During the urban revolution, according to Lefebvre, the political nature of space would finally be uncovered: 'space reveals its true nature as [...] a political space, the site and object of various strategies.'[30] Any move to transform politics and social relations in France and elsewhere, therefore, must include a corresponding emphasis on the organization of urban space. 'If', Lefebvre exclaimed, 'there is a connection between social relationships and space, between places and human groups, we must, if we are to establish cohesion, radically modify the structures of space.'[31]

In elaborating upon the theme of 'internal colonization', Lefebvre even referenced Griaule's 1930s work on Dogon society in West Africa, which, as we have seen, was carried out with the aid of aerial photography. Lefebvre, in a work titled *De l'État*, compared the morphological forms of habitation produced by Dogon social structures and cultural values to those produced in France by the capitalist mode of production and state dominance. In Dogon society, 'the head, limbs, male and female genital organs, and feet are represented by the grouping of huts: command huts, huts for socializing between men and women, huts for storing work tools, and so on.'[32] By contrast, in Western capitalist society—a society dominated by visual images—spaces of human habitation and urban space more generally were disconnected from the human body:

> The space of state control can also be defined as being optical and visual. The human body has disappeared into a space that is equivalent to a series of images. Perspectival space inaugurates this scotomization of the body, which was preserved, albeit in changed form, in symbolism. In modern space, the body no longer has a presence; it is only *represented*, in a spatial environment reduced to its optical components.[33]

By the 1970s, then, the anthropological critique developed in the 1930s, as we saw in the case of Griaule, had become a full-fledged attack on contemporary urbanism. With the aid of aerial photography, the notion that spatial organization and social organization were intimately bound to one another was not considered just an interesting theory; it was seen as a scientific *fact*, which could be proven with quantitative methods. Even after the technique was discarded in the 1960s and 1970s, the language of aerial photography, that is, 'social space', remained a powerful reminder of this, and of the apparent disconnect between the ideals of France as an egalitarian society (as put forth in the French Revolution) and its present reality. Any attempt at social change would therefore be incomplete without a total transformation of urban space.

Yet historicizing the view from below does not only allow us to see more clearly the genesis of Lefebvre's ideas. It also demonstrates that the distance between 'top-down' and 'bottom-up' urban planning—a division still widely used today—was not as great as Lefebvre and others would have us believe. Both emerged, in fact, from the same conceptual and institutional sources. Going still further, we can see that he and others actually constructed the notion of the 'bottom-up' during the 1960s in order to carve out their sociological terrain within interdisciplinary research teams. In the process, they crafted the very idea of the 'top-down', as well as the powerful bifurcation that remains strong in the social sciences and design disciplines even today.

This 'bottom-up' perspective—the view from below—succeeded in giving Lefebvre and others authority within government-sponsored urban research teams. It helped to distinguish the contribution of sociology to urban planning from the contributions of many others, especially geography. While the organization of space in France may never have been transformed in the way that Lefebvre and his post-war predecessors had hoped, the outside perspective offered by the aeroplane succeeded in making sociologists 'insiders' to planning practice.

Coda

Decades after the 1970s, practitioners in all design disciplines still strive to go beyond an outsider's perspective and obtain an insider's view of the ground. What they're looking for, namely, ways of integrating the local and the global, or what Chombart de Lauwe called '*normes particulières*', constitutes an ontological shift that can be traced, in part, to this post-war story. The historian of architecture and technology Antoine Picon has argued as much when he writes that, by opening up new types of information, new ways of mapping in post-war France and elsewhere in Europe led to an *epistemological* shift that remains intact today: rather than mapping monuments or architectural objects, today's designers are more concerned with 'events, simulations, and scenarios', that is, local experiences, occurrences, and potential happenings, that can be represented with ever more abstract digital technologies.[34]

This representational and ontological permutation is at the core of landscape urbanism, which has been described as 'one of the most vigorous subfields [of landscape architecture] to have emerged in the last 20 years.'[35] Landscape urbanism uses abstract representations of the human and nonhuman world to capture, analyse, and shape networks that lie at the intersection of the global and the local. 'The term landscape', writes architect Alex Wall, 'no longer refers to prospects of pastoral innocence but rather invokes the functioning matrix of connective tissue that organizes not only objects and spaces but also the dynamic processes and events that move through them.'[36] Landscape urbanists themselves often point to the Office for Metropolitan Architecture's 1982 proposal for the *Parc de la Villette* in Paris as an early example of the trend of using the landscape 'surface'—a horizontal plane that includes buildings, roads, and the spaces in between them—as an organizing principle for 'both [anticipating and accommodating] any number of changing demands and programs'.[37]

Waldheim, Bélanger, and others have argued, moreover, that representing the new 'landscape' of landscape architecture is the first step in changing not only the physical surface of the earth but also the social and economic structures that produce it. Mostafavi and Najle describe landscape urbanism as challenging the 'dogma' of modernist architecture by being more practical than utopian; more interested in process than form; and, most importantly, by including both private and public spaces within the purview of design practice. For them, expanding the scope of landscape architecture, from parks to more complex, temporal, and flexible infrastructure networks, holds political significance: it will, they write, 'set the scene [...] for democracy in action'.[38] Just as ethnographer Chombart de Lauwe used aerial photos to look at his home country of France from afar, as if it were a foreign nation, landscape urbanists use even more abstract views to see familiar settings in a different light.

While 'new' technologies such as Google Earth, Google Street View (with its famous Pegman), and many others might be novel technologically, therefore, their use and interpretation can be traced to a much longer history that designers are both descended from and still deeply embedded in. Yet if landscape urbanism shows just how relevant the tensions of the post-war era are to designers today, it also highlights how much has changed, especially with regard to the use and interpretation of visual representations. In a world in which images of the world from above, and especially urban environments, are everywhere, such representations are rarely questioned in the way that Foucault, Lefebvre, Debord, and others had hoped. Instead, visualizations of the planet have taken on a life of their own, becoming what the philosopher Jean Baudrillard called 'simulacra',[39] that is, the merging of the real and the imaginary. In this scenario, representations are perceived and treated as being more 'real' than that which they represent. Rather than searching, like Chombart de Lauwe, for what lies 'below' the surface, in landscape urbanism it is the 'surface' itself—fundamentally, a visual representation—that has become the true site of investigation and transformation.[40]

Notes

1 An earlier version of this paper was presented at 'Spaces of History/Histories of Space: Emerging Approaches to the Study of the Built Environment' (University of California, Berkeley, 30 April–1 May 2010) and published on eScholarship (https://escholarship.org/uc/item/8p97g3x9). See also Jeanne Haffner, *The View from Above: The Science of Social Space* (Cambridge, MA: MIT Press, 2013).

2 Congrès Internationaux d'Architecture Moderne (C.I.A.M.), *La Charte d'Athenes* or *The Athens Charter*, 1933 (Paris, France: The Library of the Graduate School of Design, Harvard University, 1946).

3 Henri Lefebvre, *De l'État* (Paris: Union Générale des Éditions, 1976–78). Cited in Henri Lefebvre, *State, Space, World: Selected Essays* (Minneapolis: University of Minnesota Press, 2009), 226–7.

4 I am building here on the work of the French anthropologist Marion Segaud. In *Anthropologie de l'espace: Habiter, fonder, distribuer, transformer* (Paris: Armand Colin, 2008), Segaud demonstrates how interwar French social scientists connected spatial organization and social organization, but she does not point to the crucial role of aerial photography in this development.

5 Charles Waldheim, *Landscape Urbanism Reader* (NY: Princeton University Press, 2006), 11. I would like to thank John Davis, PhD candidate at the Harvard Graduate School of Design, and John Beardsley, Director of Landscape Studies at Dumbarton Oaks, for many insightful discussions about landscape urbanism.

6 James Corner, "Terra Fluxus," in Charles Waldheim, ed., *The Landscape Urbanism Reader* (Princeton Architectural Press, 2006): 22–33, p. 23.

7 Charles Waldheim, 'Aerial Representation and the Recovery of Landscape', in *Recovering Landscape: Essays in Contemporary Landscape Architecture*, ed. James Corner (New York: Princeton Architectural Press, 1999), 121.

8 Mohsen Mostafavi and Ciro Najle, *Landscape Urbanism: A Manual for the Machinic Landscape* (London: Architectural Association, 2003), 8.

9 James Scott, *Seeing Like a State: How Certain Schemes to Improve the Human Condition Have Failed* (New Haven: Yale University Press, 1998), 89–90.

10 Marcel Griaule, 'Blasons totémiques des Dogon', *Journal de la Société des Africanistes* 7 (1937): 69–79.
11 Le Corbusier, *Aircraft* (London: The Studio, 1935).
12 It is important to note that, of course, CIAM was a very complex organization, and the history of the modern movement in architecture was much more complicated than it appears at first glance; it may not have been a 'movement' at all. See Giorgio Ciucci, 'The Invention of the Modern Movement', in *The Oppositions Reader*, ed. K. Michael Hays (New York: Princeton Architectural Press, 1998), 552–79, and Eric Mumford, *The CIAM Discourse on Urbanism, 1928–1960* (Cambridge, MA: MIT Press, 2000).
13 Paul-Henry Chombart de Lauwe, *Paris et l'agglomération parisienne: L'étude de l'espace social dans une grande cité* (Paris: PUF, 1952), 106.
14 For a discussion of the complexity of the term 'objectivity,' see Lorraine Daston and Peter Galison, *Objectivity* (New York: Zone Books, 2007).
15 Alexandre Burger, *Photographies aériennes et l'aménagement du territoire: L'interprétation des photographies aériennes appliquée aux études d'urbanisme et d'aménagement du territoire* (Paris: Dunod, 1957).
16 Ibid., 66–71.
17 See Martin Jay, *Downcast Eyes: The Denigration of Vision in Twentieth-Century French Thought* (Berkeley: University of California Press, 1993).
18 Michel Foucault, *The Order of Things: An Archaeology of the Human Sciences* (New York: Pantheon Books, 1971).
19 Larry Busbea, *Topologies: The Urban Utopia in France, 1960–1970* (Cambridge, MA: MIT Press, 2007), 121.
20 Paul Delouvrier, *Schéma directeur d'aménagement et d'urbanisme de la région de Paris* (Paris: District de la Région de Paris, 1965), 121.
21 Ibid., 47.
22 Henri Lefebvre, 'Réflexions sur la politique de l'espace', *Espaces et Sociétés* 1 (1970): 3–13.
23 Ibid., 10.
24 Ibid., 11.
25 Henri Lefebvre, *The Urban Revolution* (Minneapolis: University of Minneapolis Press, 2003); Henri Lefebvre, *The Production of Space* (Oxford: Blackwell, 1991).
26 Lefebvre, *Urban Revolution*, 5.
27 Lefebvre, *Production of Space*, 116.
28 Ibid., 94.
29 Lefebvre, *Urban Revolution*, 21.
30 Ibid., 44.
31 Ibid., 92.
32 Ibid., 230.
33 Ibid., 234.
34 Antoine Picon, 'Events, Simulations, and Scenarios', in *Digital Culture in Architecture: An Introduction for the Design Professions* (Basel: Birkhäuser, 2010), 191–207. For a rich overview of the role of aerial vision in landscape architecture and other design disciplines, see Sonja Dümpelmann, *Flights of Imagination: Aviation, Landscape, Design* (Charlottesville: University of Virginia Press, 2014).
35 Brian Davis & Thomas Oles, 'From Architecture to Landscape', *Places Journal* (October 2014). Accessed Mar 24, 2018. https://doi.org/10.22269/141013.
36 Alex Wall, 'Programming the Urban Surface', in *Recovering Landscape: Essays in Contemporary Landscape Architecture*, ed. James Corner (New York: Princeton Architectural Press, 1999), 233.
37 Alex Wall, 'Programming', 237. Quoted in Brian Davis & Thomas Oles, 'From Architecture to Landscape', *Places Journal*, October 2014. Accessed Mar 24, 2018. https://doi.org/10.22269/141013.
38 Mohsen Mostafavi and Ciro Najle, *Landscape Urbanism: A Manual for the Machinic Landscape* (London: Architectural Association, 2003), 8.
39 Jean Baudrillard, *Simulacres et Simulation* (Paris: Éditions Galilée, 1981).
40 Alex Wall, 'Programming'.

Bibliography

Baudrillard, Jean. *Simulacres et Simulation*. Paris: Éditions Galilée, 1981.

Burger, Alexandre. *Photographies aériennes et l'aménagement du territoire: L'interprétation des photographies aériennes appliquée aux études d'urbanisme et d'aménagement du territoire*. Paris: Dunod, 1957.

Chombart de Lauwe, Paul-Henry. *Paris et l'agglomération parisienne: L'étude de l'espace social dans une grande cité*. Paris: PUF, 1952.

Ciucci, Giorgio. 'The Invention of the Modern Movement.' In *The Oppositions Reader*, edited by K. Michael Hays, 552–79. New York: Princeton Architectural Press, 1998.

Congrès Internationaux d'Architecture Moderne (C.I.A.M.). *La Charte d'Athenes* or *The Athens Charter*, 1933. Paris, France: The Library of the Graduate School of Design, Harvard University, 1946.

Corner, James. 'Terra Fluxus.' In *The Landscape Architecture Reader*, edited by Charles Waldheim, 21–35. New York: Princeton Architectural Press, 2006.

Daston, Lorraine, and Peter Galison. *Objectivity.* New York: Zone Books, 2007.

Davis, Brian and Thomas Oles. 'From Architecture to Landscape.' *Places Journal* (October 2014). https://doi.org/10.22269/141013.

Delouvrier, Paul. *Schéma directeur d'aménagement et d'urbanisme de la région de Paris.* Paris: District de la Région de Paris, 1965.

Dümpelmann, Sonja. *Flights of Imagination: Aviation, Landscape, Design.* Charlottesville: University of Virginia Press, 2014.

Foucault, Michel. *The Order of Things: An Archaeology of the Human Sciences.* New York: Pantheon Books, 1971.

Griaule, Marcel. 'Blasons totémiques des Dogon'. *Journal de la Société des Africanistes* 7 (1937): 69–79.

Haffner, Jeanne. *The View from Above: The Science of Social Space.* Cambridge, MA: MIT Press, 2013.

Jay, Martin. *Downcast Eyes: The Denigration of Vision in Twentieth-Century French Thought.* Berkeley: University of California Press, 1993.

Le Corbusier. *Aircraft.* London: The Studio, 1935.

Lefebvre, Henri. 'Réflexions sur la politique de l'espace'. *Espaces et Sociétés* 1 (1970): 3–13.

Lefebvre, Henri. *De l'État.* Paris: Union Générale des Éditions, 1976–8.

Lefebvre, Henri. *The Production of Space.* Oxford: Blackwell, 1991.

Lefebvre, Henri. *The Urban Revolution.* Minneapolis: University of Minneapolis Press, 2003.

Lefebvre, Henri. *State, Space, World: Selected Essays*, edited by Neil Brenner and Stuart Elden. Minneapolis: University of Minnesota Press, 2009.

Mostafavi, Mohsen, and Ciro Najle. *Landscape Urbanism: A Manual for the Machinic Landscape.* London: Architectural Association, 2003.

Mumford, Eric. *The CIAM Discourse on Urbanism, 1928–1960.* Cambridge, MA: MIT Press, 2000.

Picon, Antoine. 'Events, Simulations, and Scenarios'. In *Digital Culture in Architecture: An Introduction for the Design Professions*, edited by Antoine Picon, 191–207. Basel: Birkhäuser, 2010.

Scott, James. Seeing *Like a State: How Certain Schemes to Improve the Human Condition Have Failed.* New Haven: Yale University Press, 1998.

Segaud, Marion. *Anthropologie de l'espace: Habiter, fonder, distribuer, transformer.* Paris: Armand Colin, 2008.

Waldheim, Charles. *Landscape as Urbanism: A General Theory.* Princeton: Princeton University Press, 2016.

Waldheim, Charles. 'Aerial Representation and the Recovery of Landscape'. In *Recovering Landscape: Essays in Contemporary Landscape Architecture*, edited by James Corner, 120–39. New York: Princeton Architectural Press, 1999.

Wall, Alex. 'Programming the Urban Surface', in *Recovering Landscape: Essays in Contemporary Landscape Architecture*, ed. James Corner, 233–49. New York: Princeton Architectural Press, 1999.

5

BEYOND INNOCENCE

The norms and forms of colonial urban landscapes

Tom Avermaete

Colonial urban landscapes are a phenomenon in their own right. Designed and realized under extreme political power regimes, they illuminate some of the most fundamental tensions within the discipline of landscape architecture.

When we discuss and criticize urban landscapes, we commonly maintain that they are designed and realized as spatial and material harbingers of positive experiential, social and economic change. They bring open space and air to the city; they introduce green and urban nature; they offer spaces for leisure, relaxation or even food production. From this perspective, landscape designs are sometimes even considered prime loci of modernization and development, as well as sites of acculturation and emancipation where newcomers to the city are confronted with urban mores and practices. This understanding of urban landscapes as sites of prospect should come as no surprise, since many twentieth-century urban landscapes were designed from a perspective of 'reform': the idea that urbanism, architecture and landscape architecture were going to transform social life by providing the counter-forms for a new way of living.[1]

Colonial urban landscapes, however, also rather explicitly reveal another set of characteristics of landscape architecture which are far less emancipatory and can more aptly be described as 'confirmative' and 'limiting'. They illustrate how landscape design can also have the capacity to firmly locate, restrict and even divide citizens. These 'other' capacities triggered American sociologist Janet Abu-Lughod to speak of colonial urban landscapes as figures of 'urban apartheid'.[2] Following Abu-Lughod's argument, I maintain that these colonial conditions bring to the surface one of the most fundamental paradoxes of the discipline: urban landscapes have the power to create new social and cultural possibilities, but they simultaneously sustain the capacity to restrict or negate social and cultural practices. This paradoxical condition requires us to rethink the very premises and perspectives from which we analyse urban landscapes, and more particularly how we conceive of their reciprocal relations to political, social and cultural practices. In this chapter, French colonial planning will offer a point of departure from which to reflect upon this paradoxical capacity of urban landscapes, and more specifically on the methods and theories we use to approach it.

The invention of a discipline between metropole and colony

In order to develop my argument, I propose to focus on the role of landscape architecture in the planning of new Moroccan cities during the colonial regime of the French Protectorate between 1912 and 1956. After the Treaty of Fez defined the French Protectorate over Morocco in 1912, the first Resident General, Marechal Hubert Lyautey, initiated a programme for the creation of new cities, *villes-paysage* (landscape cities), which would act as models and symbols of the modernity and progress that the French colonial power was bringing to the North African territory.[3] For Lyautey, the development of new cities was one of the main and urgent challenges that the new French territory in North Africa faced. As a result, and seemingly untroubled by the unfolding turmoil of the First World War, he launched a large urban planning and construction programme in which landscape architecture would play a major role.

The challenges that Lyautey faced were not only related to the Moroccan territory. After all, urbanism was a new disciplinary field in the 1910s. Hence, when the Marechal wanted to embark on his ambitious urban planning programme for the new colonial territory, he could not rely on well-established methods. As a result, he was forced to surround himself with specialists from various disciplines and to conceive of urban planning approaches as they were being implemented. Lyautey therefore invited some of the main figures of the emerging field of French town planning to Morocco on research missions.

It is very telling that Jean-Claude-Nicolas Forestier, a landscape architect who edited the influential publication *Grandes villes et systèmes de parcs* (*Large Cities and Park Systems*), was one of the first experts to be invited.[4] Forestier was called to Morocco to conduct a specific study of the 'reserves of land for the creation of parks and public gardens in the cities of the Protectorate'.[5] He was rapidly followed by the highly esteemed French architect Henri Prost, who was summoned to Morocco in 1913 for a mission of three months. Prost returned in 1914 with the lawyer Guillaume de Tarde at his side, and subsequently stayed in Morocco for ten years to plan numerous *villes nouvelles* (new towns) and transform several existing cities into new *paysages urbains* (urban landscapes). Forestier and Prost collaborated closely, and in 1911 both became founding members of the French Society of Architects, an organization whose main ambition was to define a new science of cities that would be called 'urbanism'.

The French attempt to establish the field of 'urbanism' was the fruit of a broad cultural and professional debate on the transformation of cities in the late nineteenth and early twentieth centuries. New societal issues were emerging as a result of the industrialization of cities. Countless studies were being conducted on hygiene, unhealthy housing conditions, new ways of working, and changing living conditions. A corps of experts from different fields, to which Forestier and Prost belonged, were gathering in the Musée Social.[6] This private organization had been founded in 1894 as a research centre that would investigate, record and—above all—remediate the large societal changes that were taking place at the end of the nineteenth century. In order to achieve this goal, the Musée Social gathered together various professionals, ranging from doctors, hygienists and sociologists to lawyers, engineers, architects and landscape architects. A first part of the joint venture of these various experts was to construct new urban knowledge through surveys, conferences, debates and studies on housing and the extension of cities.

However, the actions of the Musée Social did not remain limited to the analysis of urban conditions. Its members also regarded it as their task to come up with remedial proposals for problematic urban neighbourhoods in the form of new plans for urban transformation and development. The design of green urban landscapes would play a very important role in these projective ventures. As early as 1903, a leading role was taken in this respect by the well-known French architect Eugene Hénard, who made a strong plea to reintroduce green landscapes into the dense city centre of Paris. Hénard proposed to use the zone of the obsolete fortifications as the basis for the design of an extra belt of parks that, together with the existing green spaces in the city, would form a dense archipelago of green islands in the French capital. But he also made proposals on a smaller scale. For the Parisian boulevards, Hénard suggested breaking up the form of the corridor street and turning it into a *boulevard à redans* (stepped boulevard). His idea was that the combination of all the boulevards would create a sinuous figure of green and air that would transform the city into a continuous green urban landscape.

It is essential for my argument that proposals such as Hénard's were looked upon within the context of the Musée Social as amendments and improvements to existing urban conditions. The new urban landscapes were looked upon as vehicles to install more hygienic living conditions, and as propellers of new urban practices as well as new land logics. On many occasions it was emphasized that these design proposals for urban landscapes by members of the Musée Social were all geared towards the 'amelioration of the living conditions'[7] to the extent that 'questions of hygiene, art, and comfort were of prime importance for the physical development of our race and its moral and intellectual elevation.'[8] In other words, the urban landscape was regarded as a site of hygienic, social and cultural reform (see Figure 5.1).

This idea of the landscape as a site of reform also informed Prost and Forestier when they encountered the territory of the Protectorate in Morocco. Their aim was to define a manual of principles for the reform of existing settlements and towns into new green cities. Their examples were urban developments in England, Germany and the United States. Forestier was well acquainted with the work of Olmsted and his park system as it had been developed in Boston, New York and Baltimore. He also studied the creation of garden cities in England according to the theories of Ebenezer Howard, and developments in Australia and in cities such as London, Vienna and Cologne. He maintained:

> Men are not made to be heaped up in anthills, but scattered over the land which they are to cultivate. The more they congregate, the more they become corrupted. The infirmities of the body as well as the vices of the soul are the infallible effects of large numbers. [...] The remedy would be to flee the great cities and return to the country. Perhaps this will happen one day, but today social organization is needed for these enormous groups. For their dangers and disadvantages, it is possible to provide palliatives, if not a remedy, and municipalities can try to avoid, in part, consequences that are truly formidable, but it must be done methodically.[9]

And, he continued, we need the 'improvement of the living conditions of the mass of the inhabitants [...and] the increase of morality'.[10] For Forestier and Prost, the modern city was to be based on clear-cut scientific principles that would first be developed in the heartland of the metropolis and could be tested on colonial ground 'in a process of trial and error'.[11] 'Urbanism' and 'colonial urbanism' were understood as being in reciprocal relation: the former was substantiated by the latter, which was considered a testing ground for theoretical and practical principles. Prost and Forestier maintained that the colonial city was the ideal testing ground, since it posed very complex problems for the science of urbanism, not least because it was required to engage with 'the particularities of geography, climate and mores'.[12]

The landscape as site of innovation and emancipation

Just like other members of the Musée Social who were at work in Morocco, Forestier believed that urbanism was all about social reform, and that it was directly related to the embellishment of the city.[13] One of the ways to achieve this social reform was by conceiving a general plan for the development of cities. This plan rested on three key principles: the zoning of urban functions, the definition of traffic arteries as the structuring elements of the city, and the introduction of

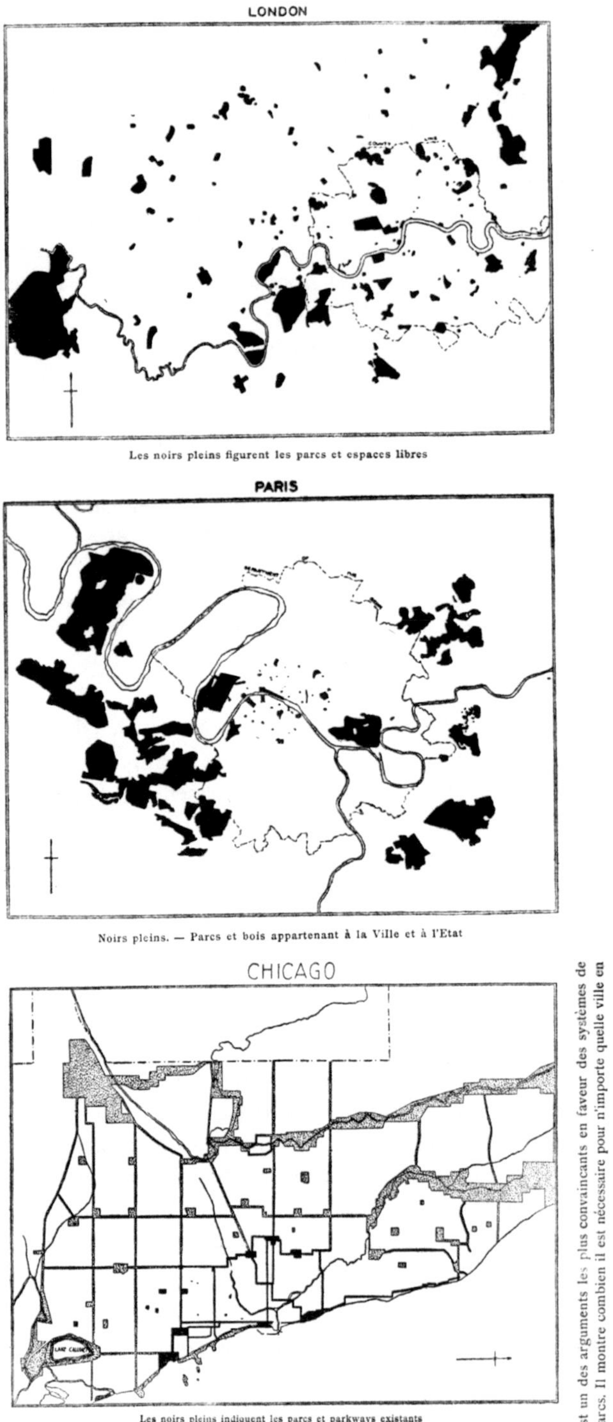

LONDON
Les noirs pleins figurent les parcs et espaces libres
PARIS
Noirs pleins. — Parcs et bois appartenant à la Ville et à l'Etat
CHICAGO
Les noirs pleins indiquent les parcs et parkways existants
Les parties grises sont les réserves, parcs et parkways en projet ou en voie d'acquisition
est un des arguments les plus convaincants en faveur des systèmes de
parcs. Il montre combien il est nécessaire pour n'importe quelle ville en

Figure 5.1 Study of the structuring role of open landscapes in Chicago, London, New York, and Paris, 1908. © Jean-Claude-Nicolas Forestier, Grandes villes et systèmes de parcs (Paris: Hachette, 1908).

nature into the city. Forestier told the Resident General that he proposed to apply the concept of the 'parks system' that he had theorized in 1906 in his book *Grandes villes et systèmes de parcs*.[14]

In this book, Forestier claimed that in the middle of the nineteenth century a new sensibility had emerged in large metropoles such as Paris and London which was geared towards green stages (*stades verts*) and green walkways (*promenades verts*). He referred to the example of Adelaide in Australia and its parklands as a precedent in which various green landscapes, ranging from gardens to sporting and cultural terrains, had been part of the planning of the new city from the inception, and even a structuring principle. Forestier claims: 'In this way the city is always assured of a large belt of air and of green. [...] The big city is composed of a series of neighbourhoods that are each surrounded by promenades and green, of playing and sporting grounds in the open air.'[15] In the example of the English garden city, Forestier points to 'the general preoccupation to mix roads with public gardens as much as possible, and to position the relaxing notes of plants somewhat everywhere'.[16] Forestier situates these tendencies in relation to the ongoing urbanization and the interest in maintaining a certain 'picturesque' and living quality (Figure 5.2):

> The majority of the large cities in America and Europe [...] have understood how a developing city across centuries can find hygiene and beauty in numerous and well-distributed 'reserves'. They have understood that an 'urban plan' is insufficient if it is not complemented by a general plan of free interior and exterior spaces for the present and the future: by a system of parks.[17]

Forestier imagined that new cities would be structured by such a park system, which would entail a broad spectrum of green spaces ranging from green 'avenues' and 'children's playgrounds' to 'neighbourhood parks', 'recreation grounds', 'suburban parks' and 'large natural reserves'.[18] Together these spaces would create a hierarchical and continuous network of open spaces, ranging from public inner-city parks to peri-urban green belts. Forestier maintained that these 'parks and free spaces are indispensable for the life and growth of the large city'.[19] The green landscape was the harbinger of the good life in the city, and therefore he claimed that it was necessary to 'acquire the terrains for parks and the connections between them well in advance, and to foresee the necessary protection measures'.[20] Forestier believed that in order to articulate the green landscape, and thus the quality of the city, for future generations, 'the parks and new park projects need to be the subject of a larger general programme.'[21] Although Forestier's theory was not being applied in France, the colonial territory of the Moroccan Protectorate seemed the ideal testing ground.

The landscape of 'discreet separation'

The landscape was not only considered a site of progress and emancipation, however. In his 1913 report on the territory of the Moroccan Protectorate, Forestier had already made some recommendations concerning a *zone non-aedificandi* (non-constructed zone), which took the form of a fifty-metre-wide green landscape and was positioned between the urban neighbourhoods of colonizer and colonized.[22] According to Forestier, this green ribbon would play an important role in the coexistence of different groups and urban fabrics in the city. Prost and Forestier did not believe in a melting pot model for their new cities, but rather maintained that the two parts of the city should be kept in 'separated proximity'. Prost, however, drew the reader's attention to the fact that this was not a question of radical separation: 'This is not a remoteness, if I may say so, a sort of attitude of contempt towards the native city (an attitude which I believe to be the English method), but on the contrary, a discreet separation of two cities, which are also closely

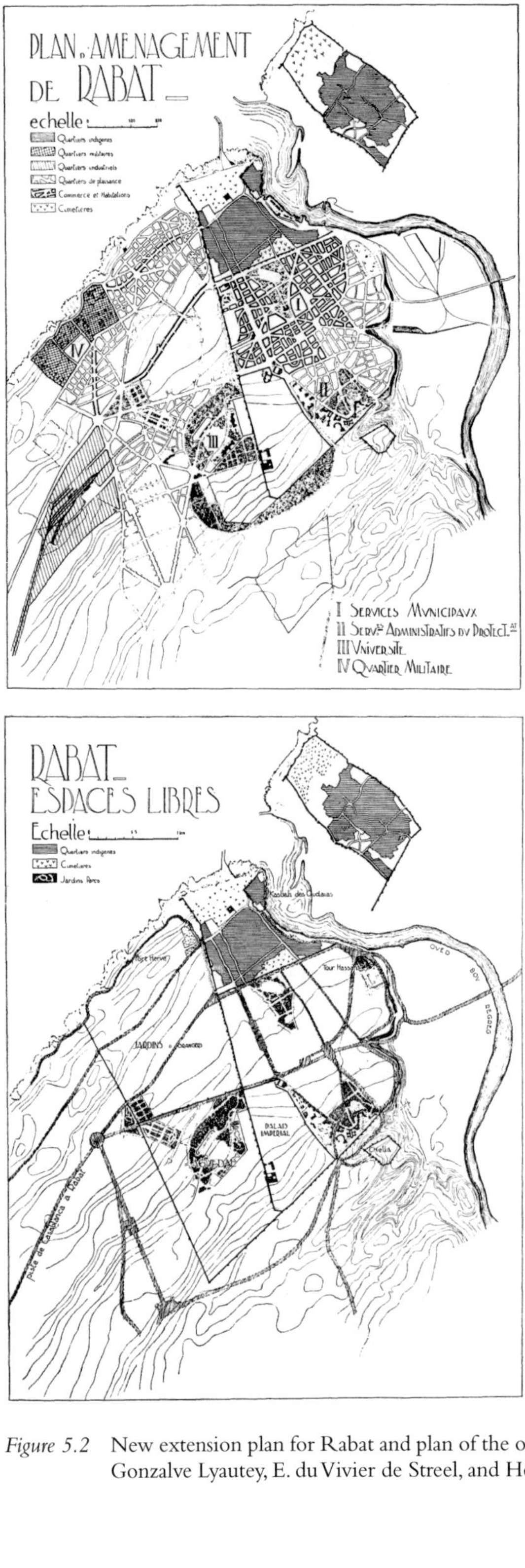

Figure 5.2 New extension plan for Rabat and plan of the open spaces, 1932. © Royer, Jean, Louis Hubert Gonzalve Lyautey, E. du Vivier de Streel, and Henri Prost, 1932.

united.'[23] The landscape was thought to have the capacity to fulfil this role of 'discreet separation' (see Figure 5.3).

This landscape of 'discreet separation' was argued for by Prost and Forestier from a variety of perspectives. First, they maintained that the indigenous city was not culturally adapted to Europeans. Prost wrote:

> We came to Morocco to install a collaboration that should not disturb the living conditions of its inhabitants. [...] Muslim women have only the terraces of their houses to expose themselves to light and to live. The result is a formal obligation: any window overlooking the terraces and courtyards is forbidden. It is easy to understand how such a rule is inapplicable to European customs.[24]

From this perspective, the green landscape would become the buffer between the two different cultures inhabiting the city. It was literally conceived as a transition zone between a European way of urban life and a Muslim way of practising urban life, with very strong boundaries between the private and public realms, especially for women.

Second, the landscape was also believed to play a hygienic role. Prost and Forestier maintained that the water and sewage systems in the indigenous city were not usable by Europeans, because 'the health situation of these is often faulty in terms of hygiene. We need to combat the

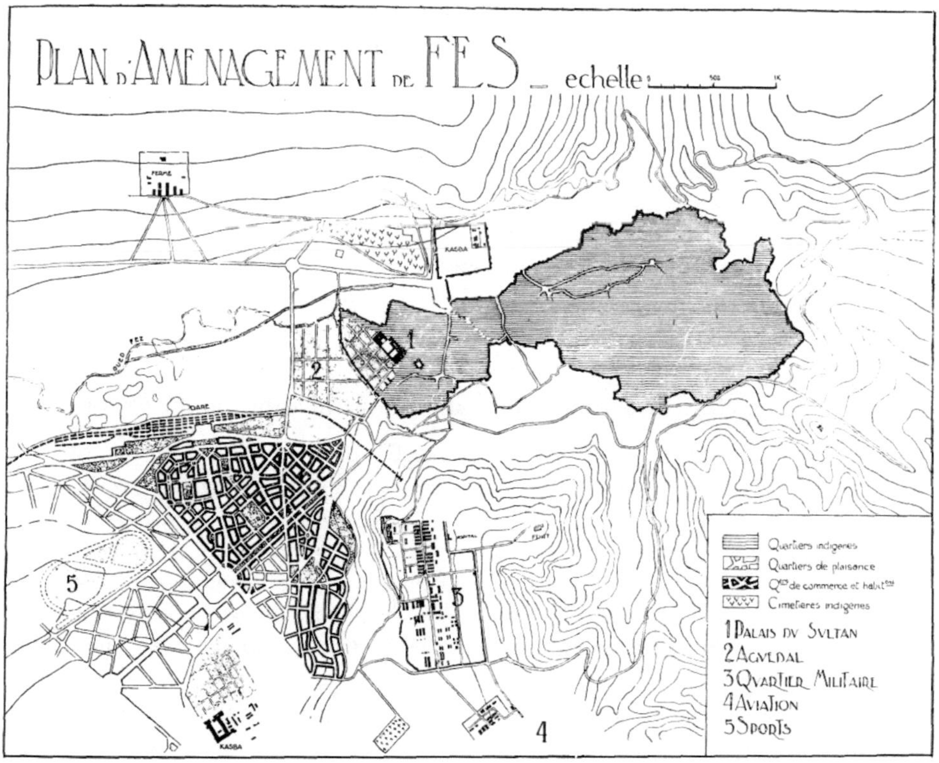

Figure 5.3 The landscapes as dividing element between the indigenous settlement and the new European city, in the extension plan for Fez, 1932. © Royer, Jean, Louis Hubert Gonzalve Lyautey, E. du Vivier de Streel, and Henri Prost, 1932.

practices of a population which does not always accept the advice of modern science.'[25] They were supported by the arguments of doctors such as Marcel Léger, who stressed

> the need to remove the indigenous element from the European element, or at least to diminish as much as possible the contact between them, especially during the night, [by practising] the segregation of the natives in villages, or at least neighbourhoods, coupled with the Europeanized cities. The natives are, in fact, for certain local diseases, reservoirs of viruses which it is well to remove. [...] The dwellings which are found in the colonies are still far from answering to the most basic hygiene rules. They were generally established haphazardly, outside of any hygienic supervision, the administration being disinterested. Thus 'natives" lodgings, generally poorly kept, attract rats, whose fleas are the main agents in the transmission of the plague.[26]

The green landscape in the middle of the city was defined as a key strategy to improve these conditions.

Finally, there was another argument for the separation of the two cities by a green middle landscape. Prost and Forestier also looked upon the indigenous city as local heritage that was still unspoiled by twentieth-century modern civilization: 'We want to preserve the appearance of the native cities, the historical and religious monuments, the picturesque old walls [...] The autonomy of the indigenous towns will enable them to preserve their characteristic features and their wonderful panoramic aspects, which offer such superb views from our modern cities.'[27] Prost and Forestier regarded the separation of the indigenous town and the European city as an urban planning principle that was capable of ensuring hygiene, order, security and social peace.

From this perspective, the landscape no longer appears as a site of personal emancipation and modernization, but rather as a razor-sharp instrument of governance. The urban landscapes of Morocco emerge as strategic elements within a larger colonial project of cultural domination, as well as sites of experimentation for new rational planning techniques that might contribute to the construction of a 'science of cities'. They offered professional opportunities and functioned as real-life laboratories where landscape architects and architects could test new procedures, new materials and new design approaches. This understanding of the landscape as an important spatial and material element of the colonial project reveals the dominant, oppressive and restrictive character of urban landscapes.

Challenges for research on colonial urban landscapes, and beyond

In hindsight, the role of landscape architecture in the planning of the *villes nouvelles* in Morocco poses a set of theoretical and methodological challenges that reach far beyond the temporal and spatial borders of the French Protectorate and concern the ways that we conceive and investigate urban landscapes in general.

First, the story of the design of colonial urban landscapes in the Protectorate in Morocco invites us to retheorize the very character of landscape architecture design. It illustrates that the design of landscape architecture can simultaneously be a generous accommodator of possibilities and also a sharp and restrictive instrument of control. Theorist Alexander Tzonis has qualified the latter aspect as the 'power through oppression' that designers exercise over societies and nature in order to quell the anxiety created by 'the unknown, the unexplained, the unstructured'.[28] Architect Robert Goodman points out that this oppressive character of landscape architects, architects or urban planners is anything but visible, because they 'aren't the visible symbols of oppression, like the military or the police. We're more sophisticated, more educated, and more socially conscious. We're the soft cops.'[29] Acknowledging the simultaneity of this 'softly oppressive' and 'explicitly accommodating' character of landscape architecture design remains one of the biggest theoretical challenges that contemporary research faces.

Second, the study of the *villes nouvelles* in Morocco reveals that it is fruitful to broaden the scope of our landscape studies beyond the Euro-American, towards different cultural and political geographies, in order to understand some of the hidden political dimensions of urban landscapes. Colonial landscapes are peerless in their ability to demonstrate that landscape design is not an innocent activity, but a charged practice that strongly depends upon dominant political regimes. In the past decades scholars have scrutinized colonial urban landscapes of the nineteenth and twentieth centuries as efficient political tools to assert European power, and as sites of social and cultural experiments that were sometimes brought home by the colonizer after being tested on colonial ground.[30] Recent research has also focused on the longevity of the political dimension, arguing that urban landscapes should not only be conceived as the 'application of principles of governance and power, but as "common place", a material and spatial reality that has persisted in anti-colonial and postcolonial eras—in original, modified or altered forms'.[31]

Extremely explicit and recognizable in colonial urban landscapes, these political regimes play a decisive role in all urban landscapes. In other words, colonial landscapes invite us to rethink the various political dimensions of landscape architecture—not as exceptional qualities, but as fundamental aspects of every landscape design. Janet Abu-Lughod's notion of 'urban apartheid' and Paul Rabinow's concept of the urban landscape as 'the localization and operation, the micro-politics, of a specific form of power'[32] were developed in close relation to the colonial urban landscape, but they also have the capacity to uncover important dimensions of other landscapes—think for instance of the numerous welfare landscapes—composed of new housing estates and collective buildings—in Europe.[33]

A third challenge that the colonial urban landscapes of the Moroccan *villes nouvelles* pose is how we relate to the materiality of the landscape. Colonial conditions make us aware that powerful political and cultural constructs can be projected onto urban landscapes, but very often the material construct of the landscape survives these political and cultural discourses. Paradoxically, our studies of urban landscapes have often solely focused on the political, cultural and designerly discourses of urban landscapes. They often go no further than 'macro-theorizing', for instance by connecting urban landscape forms directly to colonial power schemes. One of the consequences of this 'macro-theorization' is a reduced attention to the role of the physical object—to the landscape as an artefact. Richard Sennett has emphasized that this perspective on 'the shaping of physical things as mirrors of social norms, economic interests, religious convictions' often implies that 'the thing itself is discounted.'[34] Indeed, many of our studies of urban landscapes do not engage with what Pierre Bourdieu has called the 'oeuvres': the concrete spatial and material presence of neighbourhoods and cities.[35] The 'matter' of landscape architecture is too often silenced.

The reasons for this macro-theorization and silencing of the landscape artefact can in my opinion be found in the predominance of discursive hypotheses (or discursive meanings) in explanations of urban landscapes. Discursive hypotheses are meanings attributed to the built environment and expressed in verbal form (eventually translated in figures or illustrated by graphics) by landscape architects, politicians and developers. Discursive hypotheses are the expressions of certain thought frames that have explicitly or implicitly been generative of the specific articulation of certain urban landscapes. Discursive hypotheses can be analysed through a discursive methodology in which textual, oral and graphic sources are analysed. As Tim Ingold has observed, the problem with such a discursive methodology is that 'culture is conceived to hover over the material world but not to permeate it.'[36]

Alongside discursive hypotheses, one can distinguish embedded hypotheses (or embedded meanings) that emerge from the specific temporal and spatial ordering of the material 'artefact' and its location in a particular space and time. As such, embedded hypotheses are not necessarily part of larger cultural schemes or thought frames, and are not necessarily in line with the discursive hypotheses. From this perspective, the very materiality of the urban landscape appears as an active agent in the mediation and transformation of meaning. However, the landscape as agent operates in an implicit 'embedded mode'; only fragments of its action are made explicit in 'discursive mode'. Through careful analysis, observation and inquiry, it is possible to make part of these embedded hypotheses explicit. However, with the exception of the most obvious ones, embedded hypotheses are hardly ever taken into consideration in studies of urban landscapes using postcolonial theory.

Needless to say, in urban landscapes, discursive and embedded dimensions—that is, ideas and matter—are intimately linked, as they ceaselessly constitute one another. The point here is not to denigrate the need to study discourses on urban landscapes, but to critique what appears to be a distorted balance between discursive and substantive aspects in our studies of urban landscapes. This distorted balance is troubling, not only because knowledge needs to be accumulated and theorized regarding all aspects of urban landscapes, but also because political, ideological and designerly discourses may change or be forgotten, while their material legacy remains for generations. Speaking from a postcolonial vantage point, this is one of the main challenges that colonial urban landscapes pose to the study of landscape architecture: how to account in a nuanced way for the relation between discursive and embedded dimensions, for the field of tension between the landscape as conceptual and material fact.

Notes

1 For this reform movement, see for instance Rajesh Heynickx and Tom Avermaete, *Making a New World: Architecture & Communities in Interwar Europe* (Leuven: Leuven University Press, 2012).

2 Janet L. Abu-Lughod, *Rabat: Urban Apartheid in Morocco* (Princeton: Princeton University Press, 1980).

3 An introduction to this urban venture in the city of Rabat can be found in Mounia Bennani and Augustin Berque, *Le système des parcs et jardins publics du début du Protectorat français au Maroc: Rabat, le prototype de la ville-paysage idéale (1912–1930)* (Paris: École des hautes études en sciences sociales, 2006).

4 Jean-Claude-Nicolas Forestier, *Grandes villes et systèmes de parcs* (Paris: Hachette, 1908). For a general introduction to the work of Forestier, see Bénédicte Leclerc, *Jean Claude Nicolas Forestier, 1861–1930: Du jardin au paysage urbain: Actes du colloque international sur J.C.N. Forestier, Paris, 1990* (Paris: Picard, 1994). For his mission to Morocco, see Jean-Claude-Nicolas Forestier, Bénédicte Leclerc and Salvador Tarragó, 'La mission de Jean Claude Nicolas Forestier au Maroc', in *Grandes villes et systèmes de parcs, suivi de deux mémoires sur les villes impériales du Maroc et sur Buenos Aires* (Paris: Norma, 1997), 178–196.

5 Letter of Minister of Foreign Affairs to the prefect of the Seine of 6 January 1913, cited in Leclerc, Bénédicte. *Jean Claude Nicolas Forestier: 1861–1930: Du Jardin Au Paysage Urbain* (Paris: Picard, 1994), 189.

6 For the role of the Musée Social in French society and culture, see Janet R. Horne, *A Social Laboratory for Modern France: The Musée Social and the Rise of the Welfare State* (Durham: Duke University Press, 2002) and Paul Rabinow, *French Modern: Norms and Forms of the Social Environment* (Chicago: University of Chicago Press, 1995).

7 Rabinow, *French Modern*, 186.

8 Ibid.

9 Forestier, Leclerc and Tarragó, *Grandes villes*, 51, my translation.

10 Ibid., 57.

11 Royer, Jean, Henri Prost et al., eds., *L' Urbanisme aux colonies et dans les pays tropicaux: Communications et rapports du congrès international de l'urbanisme aux colonies et dans les pays de latitude intertropicale.* (La Charité-sur-Loire: Delayance, 1932), 32.

12 Guillaume de Tarde, 'L'Urbanisme en Afrique du Nord: Rapport general', in Royer, Prost et al., *L'Urbanisme*, 27.

13 Forestier, Leclerc and Tarragó, *Grandes villes*, 27–43.
14 Forestier, *Grandes villes et systèmes*.
15 Ibid., 8.
16 Ibid., 10.
17 Ibid., 13–14.
18 Ibid., 16.
19 Ibid., 53.
20 Ibid., 54.
21 Ibid., 53.
22 In the planning of the extension of the city of Rabat, this non-constructed zone was finally set at 250 metres wide.
23 Prost as paraphrased by de Tarde, 'L'Urbanisme en Afrique du Nord: Rapport general', in Royer, Prost et al., *L'Urbanisme*, 26.
24 Henri Prost, 'Le développement de l'urbanisme dans le protectorat du Maroc, de 1914 à 1923,' in Prost et al., *L'Urbanisme*, 60.
25 Ibid.
26 Marcel Leger, 'L'habitation coloniale du point de vue medical', in Prost et al., *L'Urbanisme*, 46.
27 Henri Prost, 'Le développement de l'urbanisme dans le protectorat du Maroc, de 1914 à 1923,' in Royer, Prost et al., *L'Urbanisme*, 60.
28 Alexander Tzonis, *Towards a Non-Oppressive Environment: An Essay* (Boston: George Braziller, 1972), 11.
29 Robert Goodman, *After the Planners* (New York: Simon and Schuster, 1985), 47.
30 Abu-Lughod, *Rabat*; Rabinow, *French Modern*; Joe Nasr and Mercedes Volait, *Urbanism: Imported or Exported?* (Chichester: Wiley-Academy, 2003); Carlos N. Silva, *Urban Planning in Sub-Saharan Africa: Colonial and Post-Colonial Planning Cultures* (New York: Routledge, 2015).
31 Tom Avermaete, Serhat Karakayali and Marion von Osten, *Colonial Modern: Aesthetics of the Past, Rebellions for the Future* (London: Black Dog, 2014), 47.
32 Setha M. Low and Denise Lawrence-Zúñiga, *The Anthropology of Space and Place: Locating Culture* (Oxford: Blackwell, 2012), 354.
33 This is for instance the argument of Fassil Demissie, *Colonial Architecture and Urbanism in Africa: Intertwined and Contested Histories* (Farnham: Ashgate, 2012).
34 Richard Sennett, *The Uses of Disorder: Personal Identity and City Life* (New Haven: Yale University Press, 2008).
35 Pierre Bourdieu et al., *Language and Symbolic Power* (Cambridge: Polity Press, 1994).
36 Tim Ingold, *The Perception of the Environment: Essays on Livelihood, Dwelling and Skill* (London: Routledge, 2011), 340.

Bibliography

Abu-Lughod, Janet L. *Rabat, urban apartheid in Morocco*. Princeton, N.J.: Princeton University Press, 1980.

Avermaete, Tom, Serhat Karakayali, and Marion von Osten. *Colonial Modern: Aesthetics of the Past, Rebellions for the Future*. London: Black Dog, 2010.

Bénédicte Leclerc. *Jean Claude Nicolas Forestier, 1861–1930: Du Jardin au Paysage Urbain: Actes du Colloque International sur J.C.N. Forestier, Paris, 1990*. Paris: Picard, 1994.

Bennani, Mounia, and Augustin Berque. *Le Système des Parcs et Jardins Publics du Début du Protectorat Français au Maroc: Rabat, Le Prototype de La Ville-PaysageIidéale (1912–1930)*. Paris: École des Hautes Études en Sciences Sociales, 2006.

Bourdieu, Pierre, Matthew Adamson, Gino Raymond, and John B. Thompson. *Language and Symbolic Power*. Cambridge: Polity Press, 1999.

Demissie, Fassil. *Colonial Architecture and Urbanism in Africa Intertwined and Contested Histories*. Farnham: Ashgate, 2012.

Forestier, Jean-Claude-Nicolas. *Grandes Villes et Systèmes de Parcs*. Paris: Hachette, 1908.

Heynickx, Rajesh, and Tom Avermaete. *Making a New World: Architecture & Communities in Interwar Europe*. Leuven: Leuven University Press, 2012.

Horne, Janet R. *A Social Laboratory for Modern France: The Musée Social & The Rise of The Welfare State*. Durham: Duke University Press, 2002.

Ingold, Tim. *The Perception of the Environment Essays on Livelihood, Dwelling and Skill*. London: Routledge, 2011.

Low, Setha M., and Denise Lawrence-Zúñiga. *The Anthropology of Space and Place: Locating Culture*. Oxford: Blackwell, 2012.

Royer, Jean, Prost, Henri, et al., eds., *L' Urbanisme aux Colonies et Dans Les Pays Tropicaux: Communications et Rapports du Congrès International de L'Urbanisme aux Colonies et Dans Les Pays de Latitude Intertropicale*. La Charité-sur-Loire: Delayance, 1931.

Rabinow, Paul. *French Modern: Norms and Forms of the Social Environment*. Chicago: University of Chicago Press, 1995.

Sennett, Richard. *The Uses of Disorder: Personal Identity and City Life*. New Haven [Conn.]: Yale University Press, 2008.

Silva, Carlos Nunes. *Urban Planning in Sub-Saharan Africa: Colonial and Post-Colonial Planning Cultures*. New York: Routledge/Taylor & Francis Group, 2015.

PART II

The art of archiving landscapes

Tools for capturing moving relationships

6
SCENES FROM AN ANTHROPOCENIC ARCHIVE

Christina Capetillo

The contemporary landscape is a composite landscape in which culture has developed into a force that counterbalances nature: in the words of Michel Serres, 'at last, we exist in nature's scale.'[1] Nature has become framed by culture, as a conversion has taken place—from civilizing islands in an all-powerful, untouched nature, to fragments of nature surrounded by a cultural condition.

The 'Anthropocenic Archive' is an open-ended collection of sites across this cultural condition. It is a photographic narrative ranging in scale from monumental post-tsunami earthworks in Japan to minimal fauna crossings in Denmark, portraying anonymous places such as soil depots, fire ponds and highway basins, alongside landscapes of vast dimensions such as sea walls, mining pits and infrastructural relics.

The archive maps phenomena and conditions which we have created in the Anthropocene, and makes visible how a new landscape is being formed by infrastructure and land works, as we embank our coastlines and erect sea walls to protect ourselves against hyperobjects such as climate changes.

Donna Haraway, among others, considers the Anthropocene to be a boundary event, not an epoch: 'What comes after will not be like what came before.'[2] In my photographic work I investigate and record this event, in the hope that narratives like these will be part of a telluric current that makes sure the Anthropocene becomes the threshold where we decide which way to go.

Notes

1 Michel Serres, *The Natural Contract* (Ann Arbor: University of Michigan Press, 1995), 19.
2 Donna Haraway, 'Anthropocene, Capitalocene, Plantationocene, Chthulucene: Making Kin', *Environmental Humanities* 6 (2015): 160.

Bibliography

Haraway, Donna. 'Anthropocene, Capitalocene, Plantationocene, Chthulucene: Making Kin'. *Environmental Humanities* 6 (2015): 159–65. http://www.environmentalhumanities.org/arch/vol6/6.7.pdf.
Serres, Michel. *The Natural Contract*. Anne Arbor: University of Michigan Press, 1995.

Figure 6.1 Terrain 110317. Rødby, Lolland, Denmark. © Capetillo, C.

Figure 6.2 Basin 060415. Skelvej, Greve, Denmark. © Capetillo, C.

Figure 6.3 Plantation II 280316. Arahama, Sendai, Tohoku, Japan. © Capetillo, C.

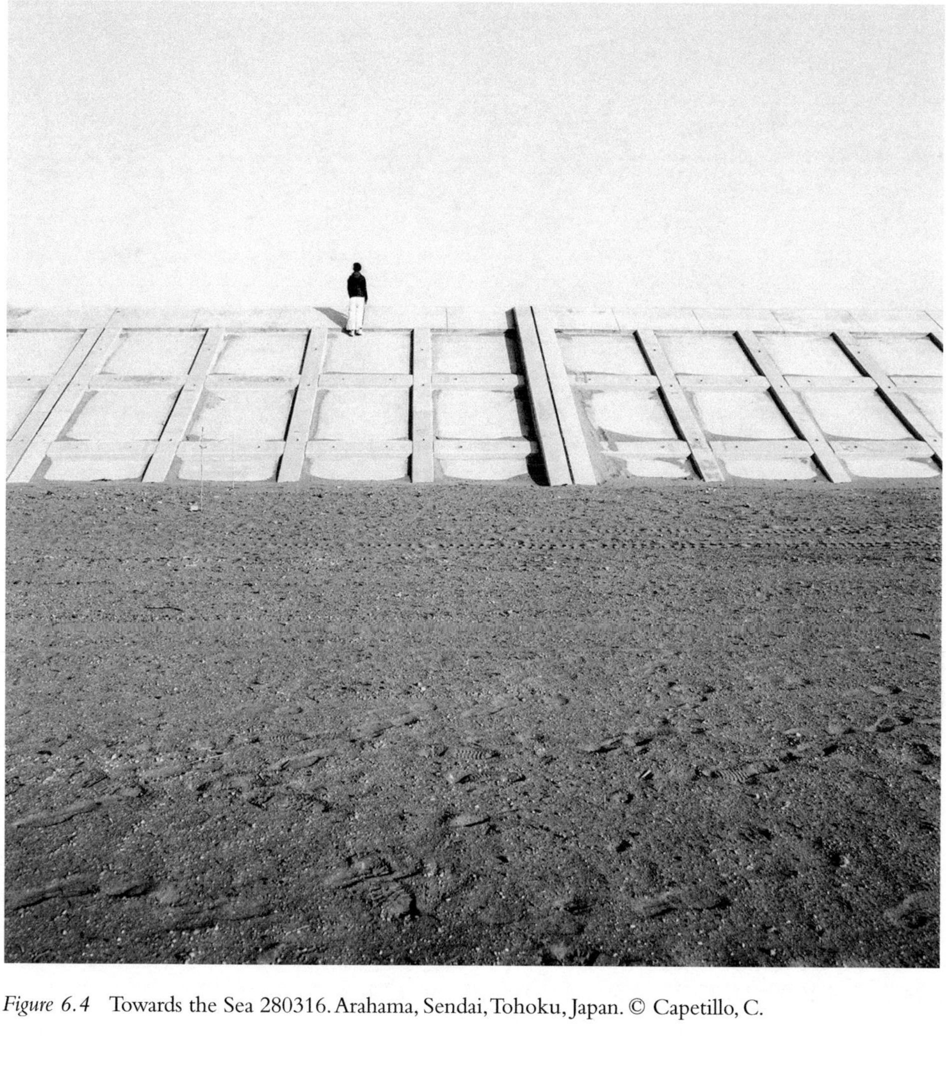

Figure 6.4 Towards the Sea 280316. Arahama, Sendai, Tohoku, Japan. © Capetillo, C.

Figure 6.5 Fire Pond II 290317. Kallemærsk Hede, Blåvand, Denmark. © Capetillo, C.

Figure 6.6 Pit I 100716. Welzov, Lausitz, Germany. © Capetillo, C.

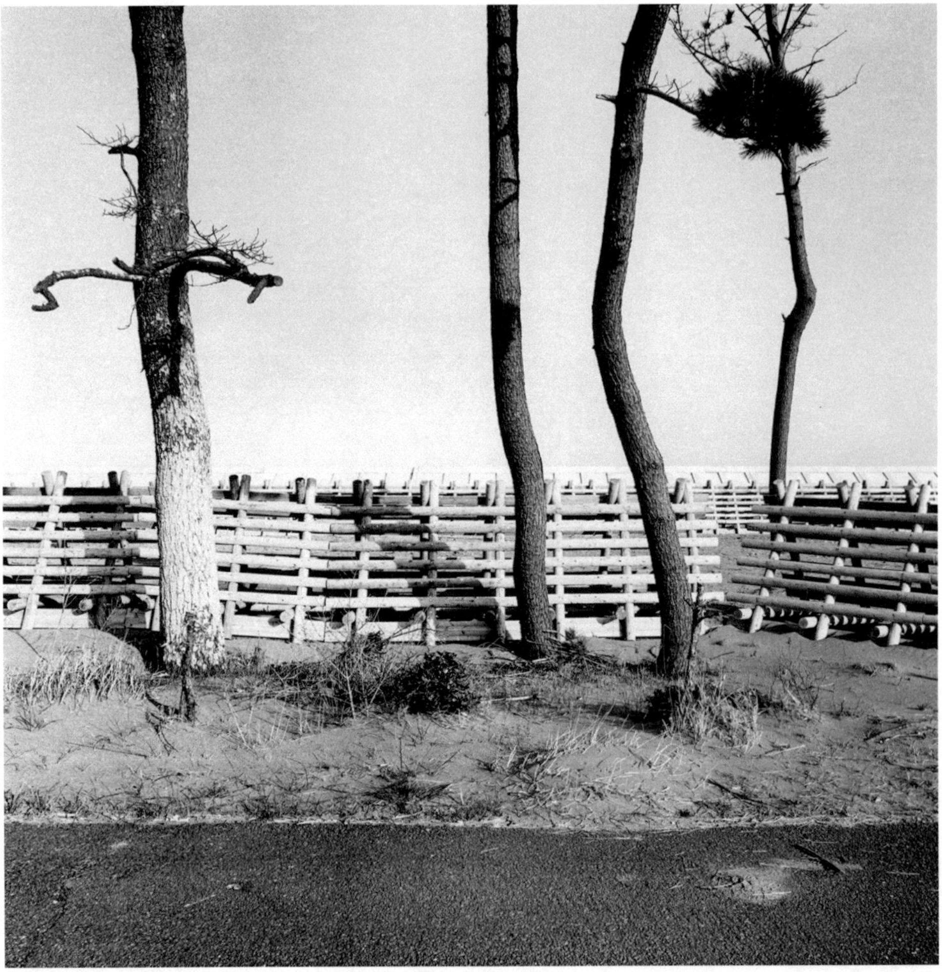

Figure 6.7 Plantation I 280316. Arahama, Sendai, Tohoku, Japan. © Capetillo, C.

Figure 6.8 Reserve 270717. Nors, Thy, Denmark. © Capetillo, C.

Figure 6.9 Burning II 300317. Tane Moor, Fanø, Denmark. © Capetillo, C.

Figure 6.10 Pit 140699. Ejerslev, Mors, Denmark. © Capetillo, C.

7

TRANSAREAL EXCURSIONS INTO LANDSCAPES OF FRAGILITY AND ENDURANCE

A contemporary interpretation of Alexander von Humboldt's mobile science

Gini Lee and Lisa Diedrich

One constant that is ever present in the practice of landscape design and representation is the mutability of the environment and the landscapes that characterize human-nature relationships. Aiming to share tangible and intangible knowledge to support design-based projects for changing landscapes, we present this transareal excursion research through exposing narratives drawn from fragile site situations on the relational margins between land and water, historical exploration and method precedents, and approaches to place documentation and representation. Our investigations into Alexander von Humboldt's *tableau physique* representations of the landscapes through which he travelled, and which he subsequently recorded and annotated in immense detail, are composed alongside adopting Ross Gibson's changescape theoretical propositions. We initially retrace Humboldt's Canary Islands excursions alongside subsequent new works in eastern Australia brought to visibility through the medium of the cartographic diary and utilizing material amassed during fieldwork excursions known as travelling transects. Focusing on the water-land intersections of the rocky coasts and tidal pools across hemispheres, these transareal projects seek out the edge conditions most subject to the vagaries of change brought about through climatic and environmental effects, social use and ritual, and economic and political pressures. The rocksect project is an ongoing transareal transect exploration aimed at uncovering the dynamics confronting the design and management of these places, which are simultaneously heritage sites and also fundamental expressions of peoples' and ecosystems' affinities with the sea, viewed from the perceived safety of the land-water interface. The stability of these places was ever subject to weathering and material erosion, but increasingly they are emblematic of where critical pressures brought by global and local economic and political regimes affect the endurance of their occupation. The stories that underpin their agency as places of critical importance to human well-being emerge at the interface between the (dis)integration of culture and nature in the time of the Anthropocene.

Landscape architecture is the discipline tasked with the responsibility to design the spaces and places that will accommodate the challenges of a sustainable future, employing forms of life that sustain environmental health, economic prosperity and social equity. In the early years of the twenty-first century, landscape scholar Elizabeth Meyer called for reinserting the aesthetic dimension into the sustainability triad of environment, culture and economy.[1] She saw landscape designers as well positioned not only to provide functionally correct spaces fit for the future but also—in the very present, through landscape research and design—to create wider public awareness of, acceptance of, and engagement in tackling the wicked ecological, social and political problems of this Anthropocene epoch. Meyer references recent thought in social science to posit that changes in behaviours and policies require more than data and rational thought, suggesting that aesthetic experiences of landscape could help inform larger publics. In order to perceive common societal goals, and to establish insight into the interdependence of socioecological values, a new activism is needed, based in affective encounters with landscapes to establish the missing link between overarching networked surroundings and everyday personal interest, imaginary and engagement.[2]

This resonates with both the European Landscape Convention[3] and the Faro Convention,[4] two treaties that stress the importance of people identifying with their lifeworlds, where landscapes are the fundamental groundscapes for action. In all their complexity, landscapes provide the grounding opportunity for people to bring together all sorts of interrelated, even if potentially disassociated, aspects of their everyday environments and living systems in a place. Indeed, the suffix *-scape* promotes the apprehension of an extensive view or a representation, pictorial or otherwise, of land-based conditions. In recent theoretical examinations, -scape has been utilized to highlight more abstract aspects beyond the purely physical and visible, to embrace the intangible forces that impact upon our aforementioned lifeworlds.

As early as the 1990s, the anthropologist Arjun Appadurai referred to the all-embracing and dynamic qualities of 'disjuncture and difference' in landscape and proposed five -scapes to develop a meta-theory of the complex, overlapping order of the 'global cultural economy', which is composed of different interrelated yet disjunctive global cultural flows.[5] The first three of these 'perspectival landscape sets' are explicated as 'ethnoscapes'—the landscape of mobile and displaced people in a shifting world; 'technoscapes'—where high-speed movement across once closed boundaries is driven by complex relationships between economies, politics and labour availability; and 'financescapes'—which are similar to technoscapes in terms of rapid global capital dispersal. These are landscapes of unreliability and unpredictability, impacting on physical territories, their material fabric and their peoples. The final two -scapes are even more intangible, operating as modes of representation of global flows in local contexts: 'mediascapes' are the modes of information production and dissemination, and 'ideoscapes' convey the ideologies of political, economic and territorial states and movements, where ideas are presented through narrative forms of imagery, text and performance to reimagine, reconstruct and potentially overwrite significant historical, ecological and physical landscapes.[6]

Such conditions of fluidity and change are characterized by creative and cultural researcher Ross Gibson in the appearance of another -scape, advanced in the concept of changescapes. He suggests that 'a changescape is a special kind of artwork—dynamic, tendency-governed, ever-reactive, never finished [...] Changescapes are everywhere in contemporary culture.'[7] We adopt Gibson's changescape methodology, which combines narrative in travelling through a landscape situation with a return to the studio to annotate the experience in writing, film or performative expression. Changescape concepts resonate with our landscape research, driving this aesthetic-political standpoint into an aesthetic-affective research practice, reaffirmed through Gibson's words:

> Some aesthetic forms 'dramatise' change. I call them 'changescapes'. They help us know mutability by immersing us in it, by letting us be with it. Change is their theme and it is often their matter too, for they are usually of fragile and ephemeral stuff that reacts to

altering conditions in the larger world. Transformations happen at their boundaries, at the limits between the inside and the outside of their systems, and then the symptoms of change become manifest in them, palpably available for our contemplation.[8]

Based in Europe and Australia, we pursue affective encounters with these changescapes as a shared aesthetic practice in landscape research across continents and areas of concern, a transareal enterprise inspired by the contemporary reinterpretation of Humboldtian science; and we forward this practice to others in proposing a fieldwork methodology termed the travelling transect. Our objects of study are ordinary landscapes we have inherited from the past, sometimes the very remote past, at the margin of land and water, exposed to extreme forces of nature and featuring human intervention, both fragile and having endured until the present moment because people cared for them. This is the reason we expect them to offer us insight into their qualities as changescapes—landscapes that accommodate change in the form of *force majeure* and human responses to it, and from which we can devise design guidelines for design interventions understood as transformation of that which already exists.[9] This also positions us within a new understanding of heritage in the cultural landscape as a living set of complex systems,[10] updating the preservation and protection regimes of the 'grand conservation project' of the nineteenth century[11] into a more mobile twenty-first-century 'translation project'.

Alexander von Humboldt's mobile science: A theoretical foundation for transareal journeys

Our enquiry commenced some years ago through a series of travelling fieldwork experiences with students and researchers across Europe and Australia, and through a series of related publications and exhibitions.[12] The urge to refine this practice methodologically and theoretically was inspired by recent reinterpretations of the early nineteenth-century traveller, writer, explorer and polymath Alexander von Humboldt. Contemporary scholars in many disciplinary fields are currently rediscovering Humboldt's understanding of science as a mobile, transareal enterprise that moves across disciplinary and geographical boundaries and territories.[13] His work operated within an environment characterized by intense global movement through seafaring and increased trade with the colonies. We now find similar conditions of movement driven by the globalized economy, the enormous changes inflicted by climate change, and the resulting demographic shifts and human imaginaries.

In response to a changing worldview in his time, Humboldt advanced two 'epistemological revolutions'. The first consisted in the rejection of pure reflection at a distance (epitomized by the encyclopaedic knowledge of the French philosophers of the eighteenth century) and posited empirical exploration on-site as the new authority for reliable knowledge generation. Humboldt's two great travels, to the Americas via the Canary Islands as a test site (1799–1804) and to Central Asia (1829), adeptly depict his mode of practice through reliance on fieldwork and immediate observation by (his) subject-observer. Upon returning home he eventually related his findings through critical thought, leading to writing and publishing visual impressions, sections and maps in an ever-evolving process of knowledge generation. This leads precisely to his second epistemological revolution: Humboldt posited knowledge as an open work, driving his research across boundaries between areas of study, exploring their interrelatedness and relational dynamics, and regarding science as a transareal pursuit. He was a pioneer of this approach in his day, in opposition to the existing or emerging intellectual boundaries between disciplines and territories, many of which evolved into the specialized disciplines and defined area studies we still know today.[14]

As landscape researchers, we appropriate Humboldt's empiricism and his transareal approach. We employ relational thinking through an expanded redefinition of local site observation and investigations across disciplinary, geographical and cultural boundaries and territories—this is relational knowledge gained from one place to inform understanding of another while travelling across. In a contemporary reinterpretation of Humboldt's epistemology, this knowledge is understood to arise 'on the move', through bodily immersion in the fields of study, including accidental deviation from planned itineraries, identifying immersion, motion and deviation as constitutive for knowledge production—creating a form of mobile, relational and open-ended knowledge particularly apt to tackle twenty-first-century challenges, precisely because of their interrelatedness and changeability (Figure 7.1).

Ottmar Ette and Eberhard Knobloch at Berlin-Brandenburg's Academy of Sciences and Humanities direct the long-term research project 'Travelling Humboldt: Science on the Move' (2015–33) and contend that Humboldt's time has come again. The merit of Humboldt's work as a travelling scientific figure is being rediscovered through contemporary digitization, publication and research on the documents and texts he wrote during his field work and upon returning home, with the purpose of further investigation and documentation of aspects of his still unknown scientific understanding. These scholars' thinking delivers the foundation which we propose to translate into landscape architecture through our travelling transect research. We regard Humboldt's itineraries as historic travelling transects, from Europe to the Americas and Asia, across sea and land, across settlements and jungle, from deepest depths to highest peaks.[15] In Humboldt's times, the synthesis of a travelling transect resulted in compellingly illustrated landscape artworks described as *tableaux physiques*. They appear to enhance the

Figure 7.1 A. von Humboldt's Tableau Physique of Teide Volcano, Tenerife, 1831. © Smith, J. David Rumsey Historical Map Collection, Paris.

narrative and spatial aspects of landscapes and provide inspiration for the pictorial representation of complexity—which inspires us to work in our research with contemporary versions of *tableaux physiques*.

Travelling transect: A methodology for engagement with changescapes

In the natural and social sciences, the transect is known as a fieldwork method for collecting empirical data. It makes use of the transverse section across a territory, along which points are located for observation and/or measurement. In the natural sciences this form includes scrutiny of the distribution of the object, element or creature of study, whereas the social sciences aim to reveal community practices along a transect line inscribed across a site of study. For our landscape research purpose, we draw from Humboldtian science and suggest a dynamization of the transect method so that it allows mobile, relational and open-ended knowledge generation, by adding 'travelling' to 'transect'. The travelling transect fieldwork method presents an alternative approach to current time-short yet distant, big-data-based 'site analysis', which in the design disciplines is increasingly commonly seen as preceding the very act of 'design'. We seek to appropriate and adjust Humboldt's ways of rigorously capturing material and immaterial site qualities through fieldwork while 'designing' their interpretation in a journey encompassing field preparations, being in the field and refining field findings, in various acts of 'thinking together' pre-, during, and post-journey discoveries. We seek to encourage landscape researchers to perform similar fieldwork as an aesthetic-affective landscape scrutiny on the move.

The travelling transect consists of the transect travel (which is composed of the itinerary as planned and the deviation as travelled), the cartographic diary (as an ongoing individual record and interpretation by each researcher, and an ongoing conversation and composition between researchers) and the *tableau physique* (as a co-created time-specific artwork fed from the cartographic diary's resources and aiming to reach wider audiences post-travel) (Figure 7.2). The travelling transect is organized in four sequences.

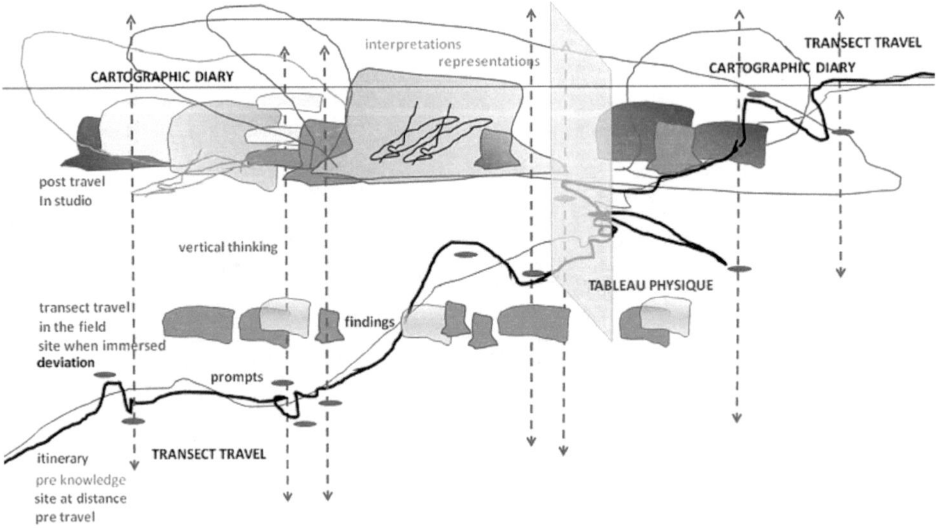

Figure 7.2 The Travelling Transect's methodological principles pre-travel, in the field, and post-travel, 2016. *Relationship between transect travel, cartographic diary and tableau physique.* © Lee, Gini and Lisa Diedrich.

First, pre-travel and at distance, we trace an itinerary to transect a landscape identified as a changescape according to knowledge gained from distant sources or previous studies, promising to lead us to sites where we expect discoveries of qualities and conflicts on the level of territorial scales and at the micro-scale. We do not expect that our transect line will remain the fine line of our pre-travel map; we expect it to spatially thicken in order to embrace an expanded geography of local narratives and forms.

Second, we embark on our journey, and take with us a set of tools from the landscape-architectural repertoire to guide our capture of site particularities. While travelling along the line prescribed by the itinerary, we immerse ourselves in our changescapes, and we amass digital images, hand-drawn sketches, real-time videos, samples captured in plastic bags, conversations with and offerings from others, water samples, historical documents and temporal mappings. We get our feet wet, are windblown and sunburnt. We take time to wander with intent regarding the distant view and the stuff under foot. And we are always receptive of deviation. Where deviation is prompted, we engage in a process of 'vertical thinking', an activity termed after the artist William Kentridge and designating the intellectual process of relating site discoveries with issues 'below' or 'above' the site, further up or down the road, or from other geographical or intellectual areas.[16] Deviation opens onto topics of all sorts, taps into discourses, launches transareal thinking and triggers interpretation of site findings—this is where we start to 'think together', in Humboldt's words,[17] and where new knowledge starts to arise. As in Humboldt's 'tropic(al) constructions',[18] the shift between the planned itinerary and the factual on-site experience enables discovery.[19]

Third, post-travel and in studio, we develop our collection of raw materials—photos and small films, sketches and annotations, models, material samples and interview notes—which we rigorously sort, combine, interpret and synthesize. Returning to our respective home continents, we lack easy face-to-face collaboration, and we need to correspond across the ether to process our physical and mental experiences. Our fieldwork conversations evolve far from the field; the travel itinerary thickens further into a thread of thinking which intertwines individual recollection and exchange at distance, and involves formalization of thought into artefacts that range from photocollages to epistemic drawings, annotations and writings, both analogue and (mostly) digitized (to be sent from computer to computer), all feeding into our cartographic diary, an ever-evolving, never-ending form, or journey-form as art critic Nicolas Bourriaud called this new genre of time-specific, dynamic artworks of the twenty-first century.[20] The cartographic diary is both an account of a fieldwork itinerary and at the same time an open work that provokes further journeys made in the mind after the initial landing on-site. In terms of eidetic operations, this cartography fulfils the condition of landscape conceptions beyond the purely pictorial. A spatial review in real time, it opens up to the next iteration, annotation and/or journey. It expresses an open-ended method of communicating a site after the first event to make subsequent events in abstract time.[21]

Fourth and finally, again post-travel, we seek the opportunity to publish and exhibit the travelled changescapes far from where we travelled them, to raise awareness of their qualities and conflicts through reproducing the affective encounter we experienced on-site, for audiences off-site. Through the process of writing up and fleshing out our findings from the field, we sharpen our interpretations and refine our designerly representations. In so doing, we seek to represent and communicate on-site experiences in distant off-site spaces—and we have come to call the result a *tableau physique*, after Humboldt's historical model.[22] Media are juxtaposed to make a tableau, and each tableau is co-made: one researcher's drawings, models and photos, the other researcher's films, collections and story-gathering. The tableau is a graphic narrative form that conveys the spatial, material and atmospheric qualities of places to communicate critical systems and values to locals, designers and visitors—it conveys meaning to assist in framing where and how to (or not to) act and intervene. The tableau is a momentarily synthetic composition of the ongoing cartographic diary's resources, produced at particular opportunities for larger audiences whom we want to sensitize to similar discoveries or engage in the design of changescapes.

Coastal margins: Changescapes par excellence

This writing now traces a series of transect itineraries known as rocksects across changescapes at the critical land-water interface of coastal environments. Mapping travel across multiple tidal bathing pools formed in landscapes of material continuity across coastal rock-water interfaces, we seek to convey theoretical approaches to landscape and heritage studies. Landscapes of fragility are present at the margins between physical systems and sociocultural systems—the pre-existing landscapes that offer an occupiable and enduring interface between land and ocean systems—and they offer natural opportunities both for interaction with nature and for human cultural ritual and everyday activities.

The tidal pools and their rocky coasts have been occupied, enjoyed and exploited by humans everywhere since time immemorial. In the course of our transareal excursions, we first came upon the rocky pools of Tenerife, the largest island in the Canary Islands archipelago, and found them to be fertile sites for investigations into human-nature activities at the margins of land-water interfaces. The places are sites of diverse intervention, as a direct response to the opportunities found in natural coastal structures that provide safe havens for organisms that inhabit these interstitial places. Scientists and naturalists have long studied these pools as indicators of resilience in the face of dynamic systems:

> Tidepools and rocky shores are among the most physically stressful environments on earth. When the tide is high, waves can sweep over plants and animals at velocities as high as 60 miles per hour, while at low tide, the same organisms dry up and bake in the sun. Yet despite this seeming inhospitality, tidepools and rocky shores are exceptionally complex and biologically diverse.[23]

This multiplicity of diverse animate and inanimate features also frames human experience and activity on the coast. The act of seeking to bathe in nature results in rock pools that have been formed over time, and then moulded and appropriated as places of occupation when weather conditions permit. At each bathing place a narrative thread is built up in time to emerge as cultural ritual alongside the spatial, temporal, atmospheric and emotional qualities of immense value to coastal inhabitation. For designers interested in the entropy of changescapes, and in the design vocabularies needed to either preserve current conditions or formulate programmes that recognize and work with change, alternative design terms can arise: deepening, smoothing, staining, moulding, submerging, parasitic attachment and assemblage, to name a few, may provide clues as to useful operations for changescapes in an environment where sea level rise and increasingly intense weather systems place these pools and their human activities at great risk.

Upon hearing in the popular press that many tidal pools are subject not only to weather and oceanic perils, but also to economic, safety and political risks arising from development and maintenance regimes, we expanded our transect explorations to investigate a sequence of tidal pools located in the cities of Sydney and Newcastle along the eastern Australian coast. While these are some of the most aesthetically regarded places in these cities, recorded though atmospheric photography, writings and stories over time, local councils have deemed them too problematic to maintain, despite their association with pre-colonization and settler regimes of occupation and deep cultural connection. As these pools appeared to be at extreme risk, we were motivated to document the major pools through a landscape-architecture-focused coastal transect that traversed many spatial and material landscapes, framed through an itinerary informed by the ability to bathe. We found places that were easy to access and map, those that were made inaccessible due to weather and/or the ocean swell, places under construction and half-formed, and places where wandering and recording the land, water, bio-organism and human interfaces brought a sense of calm alongside intense noticing of the small narratives unfolding beneath one's feet. From these recordings we returned to the studio and made an exhibition through a deep mapping of the spatial, material and temporal narratives that defined each place. We noticed that within the similar landscape elements of rock, sand, sea and marine organisms, each place is a place of distinction, made so by the variety of human responses through constructed landscapes formed through moulding and making places suitable for bathing with others.

We continued our mapping, collecting and temporal recording of the rock pools in Tenerife, Canary Islands, building our archive of tidal pool landscapes, atmospheres and qualities. While weathering and climatic forces impact upon the ongoing changescapes of the Tenerife tidal pools, the intensity of tourism activity and overpopulation of the land-sea interface is a direct result of the economic and political forces that require intensive infrastructural regimes to manage the environmental and social forces that make the coastal places such magnets for recreation and occupation. The naturally occurring tidal pools are subject to intense use at peak times, and therefore require greater modification of the rocky shorelines to facilitate tourism economies and developer-led construction inland. Landscape researchers Juan Manuel Palerm and Ángela Ruiz have extensively mapped the Canary Islands' coastlines to reveal the designed qualities of the places in context with their surroundings while also offering a catalogue of parts for each beach, including the tidal pools and their environs.[24] Palerm makes the point that there is currently an imbalance between the 'heavily constructed coast' and its tourist influx and the 'local social fabric' which is being undermined or even destroyed, and calls for 'greater equilibrium with the territory' and for 'contextualised solutions and not to import aesthetics models, rather to make places and to listen to the genius loci before acting'.[25]

Mindful of Palerm's concerns, and armed with his guidebooks, we transected the coast, seeking tidal pools at places large and small, intensely occupied and abandoned, to uncover the ephemeral, material and spatial qualities of these bathing places. We were seeking to assess responses to the impacts of weather, ocean and people, alongside the inherent qualities and cultures of each place. In each pool, a mix of resilience against forces that wear down the material form of the landscape pushes back against an overlay of commercial and touristic ephemera that contributes to interrupting the ritual of the bathing experience. Weathering here is both a natural and a cultural effect, able to be mapped as layers of occupation and give-and-take over time.

Challenging anthropocentrism in the design of sustainable futures for coastal landscapes and their tidal pools means challenging conventional development and management projects geared to the exploitation of natural ecologies, such as standard waterfront complexes or overly engineered and regulated management on the coastal margins. Heritage concerns are predicated on design motivations that are frequently insensitive to the ecological complexity of these human-water landscapes. Our aim is to bring the essence of the places travelled into presence and endurance in the face of unrelenting change dynamics.

Rocksect: A transareal exploration of Canarian and Australian tidal pools

We reiterate that tidal pools on the Canary Islands and bathing pools in coastal Australia (and indeed the world over) are fragile and eroding examples of human intervention at the critical edge between nature and culture; they are deemed too unsafe and expensive to retain, or are overrun because they are too popular, and mostly both conditions apply. At these intersections it is clear that human intervention challenges the natural order of material decay through weathering and wearing. Yet the alchemy of rocky reaction through weathering, staining and eroding means that the pools are continually remade over time—through natural processes, as much as affected by human interaction. The issue today is that these processes are accelerating.

Our travelling transect works not only seek to record these places at this point of time, but also attempt to uncover regimes of care that convey an understanding of the enduring qualities of these places even if they disappear under the oceans. We trace a line along coastlines, embracing an itinerary of pool-spotting wherever pools arise or where there seems to be a likely arrangement of cliff, pool and ways of getting down to the water. We notice qualities of typical human reactions to the proximity of rocky structures and ocean water according to form, scale, ease of bathing in and out of the water, material comfort and temporal accessibility. Back in the studio we arrange a cartographic diary out of our collection of images, samples and videos across many forms according to certain thematics. We propose applying a heritage lens to the landscape forms of the collection of tidal pools under examination as changescapes to expose an alternative language of intervention. Currently we are working on a *tableau physique* that narrates the carved, moulded and fashioned rocky coast of six tidal pools in New South Wales and Tenerife through three degrees of design intervention in changescapes: modified nature ('as found'), mediated nature ('as tuned'), and assemblages of forms from the natural to the constructed ('as made') (Figure 7.3).

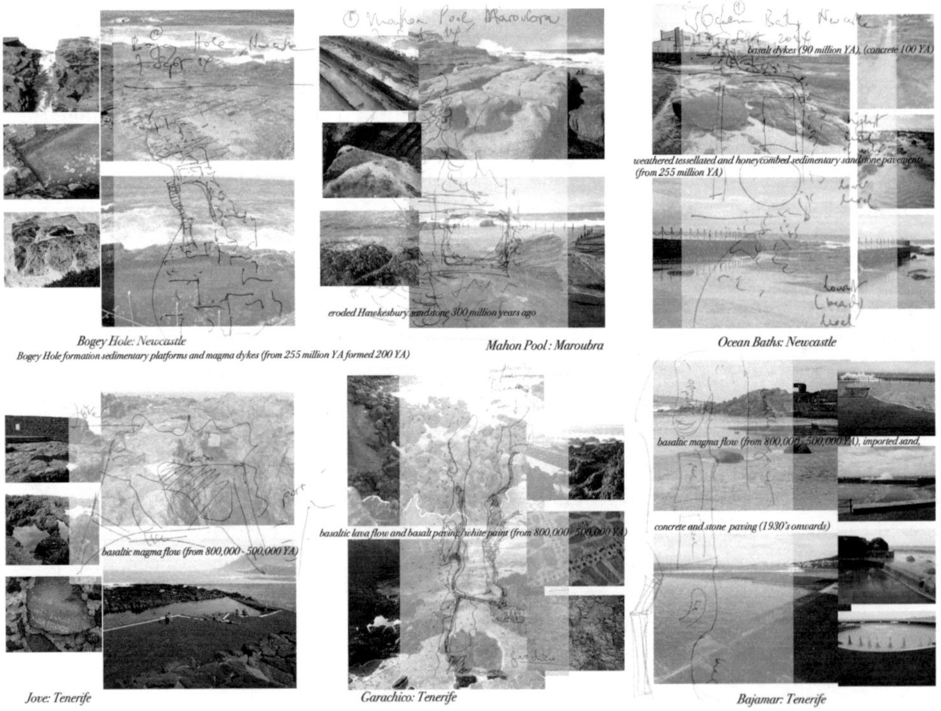

Figure 7.3 Tableau Physique in the making, 2017. Narratives drawn from fieldwork (site notes, geologic annotations, material imagery) uncover design intervention in six tidal pools in New South Wales and Tenerife. © Lee, Gini and Lisa Diedrich.

Tableau physique in the making: Design intervention in six tidal pools in New South Wales and Tenerife

As found: carved, moulded from the rocky coast

(1) Bogey Hole, Newcastle, New South Wales, and (2) Jover pool, Tenerife

It is written in Newcastle tourist literature that the rocky shoreline was reworked by hand with convict labour to make today's Bogey Hole, in order to service the bathing needs of the local commandant; yet others were there before him, as the term 'bogey hole' derives from Aboriginal words meaning bathing place. This bogey was smoothed and formed from the magma dyke at the bottom of a steep cliff, looking out to sea without interruption by beach or rocky ledges. Metal stakes and chains at the margins afford protection from the roiling sea, but today it is deemed too dangerous to access beyond marvelling at the restless interaction of water, land and a few sturdy plants. The Canarians regard the compact Jover pool as a natural place emerging from the intersection between magma breakwater and the more recently formed sunbathing plaza adjacent to the coastal buildings. It feels local and loved as a meeting place of clean waters able to be traversed from its rocky perimeter or through the clear azure water landscape. At Jover there is space for being inside the pool and outside as well, where rock walkers can edge towards the sea to fish and regard the ocean.

As tuned: mediated extension as coastal parasite

(3) Mahon Pool, Maroubra, Sydney, New South Wales, and (4) Garachico pools, Tenerife

Coming to Mahon ocean beach means navigating through wondrous shaped, carved and moulded sandstone forms that resonate with colours from the palest white-yellow to dark red-brown. Watery forms flow in and over rocky platforms that appear and disappear into the sea in a concert of water play, from waves to tiny dripping fountains emerging from the rocks. Constructed forms subtly enclose a bathing place among the rocks, as do the constructed platforms that fill the gaps between volcanic magma flows on re-formed coastline at Garachico, a relic from the last great eruption (1706). Stone paving with designed intent reveals pathways folded between rocky outcrops, yet there is apparent danger at this pool, resonating in whitened edges that confirm natural and constructed interfaces so that the unwary will not trip or fall into edged crevices. On the day we came, the weather and water stopped any access to the sea paths, closed off by plastic ropes of the type you might find at crime scenes.

As made: Assemblage of forms along a rocky plane

(5) Ocean Baths, Newcastle, New South Wales, and (6) Bajamar pools, Tenerife

At the Ocean Baths, metal stakes driven into the sandstone in the nineteenth century mark the bathers' and waders' territory, and appear to hold back the dangers of the nature of the swell, although challenged by increasingly turbulent seas. Much later, a new pool built adjacent to the sandy beach allowed laps to be swum, and platforms and steps enabled the display of swimming prowess. The rocky coast is displaced to outside the modern concrete geometric form, painted ocean blue. Yet, when the sea is high and the swell stormy, waves still intrude as they break over the edges, and the salty water slowly drains back to where it came from. When the sea takes over at Bajamar, the pool waters merge back into nature in a rush of spray, wild water and encroaching tide. Bajamar also transitions from a 'natural' yet unusual sandy beach, with calm waters sheltered by rocky forms fashioned into seating and bathing steps, into constructed pools and terraces for swimming and sunbathing. Grey stone materiality and constructed geometries tame the ocean waves and spray with the formidable breakwater, an abrupt closure to the land-sea interface.

Enduring ways of seeing and knowing: Mobile science meets landscape change

The travelling transect supports the conception of fluid, fragile, yet durable landscapes as changescapes, which we see as a prerequisite to reconcile heritage with design. This novel approach refutes design as creation on a *tabula rasa*, rather translating the pre-existent into future forms through deepening, smoothing, staining, moulding, submerging, attaching or assembling what is in place. Our interpretation of Alexander von Humboldt's mobile science has enabled us to theoretically ground the fieldwork-based methodology of the travelling transect as a more dynamic and deviant version of conventional scientific transects, yet it aims at gathering empirical data. Our transect travels along the fragile and enduring coasts of the Canaries and the Australian east have identified water landscapes designed 'as found', 'as tuned' and 'as made'. Their qualities and conflicts are recorded in a cartographic diary, then developed into a *tableau physique* to engage wider audiences affectively and aesthetically. Parallels of use and concern for water exploitation are told in alternative hemispheres on islands and at the edges of continents over various timescales, forms and processes—yet subject to the same weathering and occupation where humans occupy land-sea margins, with the same need for affective engagement catering for their future endurance.

Notes

1 E. Meyer, 'Sustaining Beauty: The Performance of Appearance', *Journal of Landscape Architecture* 3, no. 1 (2008).
2 E. Meyer, 'Landscape Entanglements: Aesthetic Practices in a Networked World' (conference keynote, 'Beyond Ism: The Landscape of Landscape Urbanism', Swedish University of Agricultural Sciences, Alnarp, 2016).
3 Council of Europe, 'The Faro Convention, Council of Europe Framework Convention on the Value of Cultural Heritage for Society, Treaty no. 199', 1 June 2011, http://www.coe.int/en/web/conventions/full-list/-/conventions/treaty/199.
4 Council of Europe, 'The European Landscape Convention, Treaty no. 176', 20 October 2000, https://www.coe.int/en/web/conventions/full-list/-/conventions/treaty/176.
5 A. Appadurai, *Disjuncture and Difference in the Global Cultural Economy* (New York: Sage, 1990); A. Appadurai, *Modernity at Large: Cultural Dimensions of Globalisation* (Minneapolis: University of Minnesota Press, 1996).
6 Appadurai, *Disjuncture and Difference*, 297–9.
7 R. Gibson, *Changescapes: Complexity, Mutability, Aesthetics* (Perth: UWA Publishing, 2015), vii.
8 (R. Gibson, 'Changescapes', IDEA Journal 2005 (Brisbane: QUT, 2005), 195–206, http://idea-edu.com/wp-content/uploads/2013/01/2005_IDEA_Journal.pdf, accessed1/12/2017. Presented at INSIDEOUT IDEA Symposium, Melbourne in April 2005, co-convenors Suzie Attiwill, RMIT University and Gini Lee, University of South Australia.
9 E. Braae, *Beauty Redeemed: Recycling Post-Industrial Landscape* (Risskov: Ikaros, 2015).
10 G. Fairclough, R. Harrison, J. Schofield and J.H. Jameson Jnr, eds., *The Heritage Reader* (London: Routledge, 2008).
11 J. Kolen, 'The Rejuvenation of Heritage', *Scape: The International Magazine for Landscape Architecture and Urbanism*, no. 2 (2006).
12 L. Diedrich, G. Lee and E. Braae, 'The Transect as a Method for Mapping and Narrating Water Landscapes: Humboldt's Open Works and Transareal Travelling', *NANO: New American Notes Online*,

no. 6: Cartography and Narratives, (2015); E. Braae, L. Diedrich and G. Lee, 'The Travelling Transect: Capturing Island Dynamics, Relationships and Atmospheres in the Water Landscapes of the Canaries', in E. Brandt, P. Ehn, T.D. Johansson, M. Hellstrom Reimer, T. Markussen and A. Vallgarda, eds., *Nordes 2013, Proceedings of Nordic Design Research Conference 2013*, (Copenhagen: Royal Danish Academy of Fine Arts, Schools Architecture, Design and Conservation, 2013), 191–200; L. Diedrich, G. Lee and J. Raxworthy, 'Transects: Developing an Experience-Based Methodology for Design Education and Design Research', in *Exposure/00*, eds. M. Jonas and R. Monacella (Melbourne: Melbourne Books, 2012).

13 O. Ette, 'Alexander von Humboldt and Hemispheric Constructions', in *Alexander von Humboldt and the Americas*, eds. V. Kutzinski et al. (Berlin: Walter Frey, 2012), 209–36; O. Ette, *Alexander von Humboldt und die Globalisierung: Das Mobile des Wissens* (Frankfurt aM: Insel, 2009); V. Kutzinski et al., *Alexander von Humboldt*.

14 L. Diedrich, E. Braae and G. Lee, 'Traversing Humboldt's Transareal Landscapes: Explorations in Fieldwork Travels and the Tableau Physique as Design Method', in *III. Potsdamer Alexander von Humboldt-Symposion 'Landschaften und Kartographien der Humboldtian Science'* (Potsdam: Universität Potsdam, 2018), 147–66; Diedrich, Lee and Braae, 'Canarysect'; Diedrich, Lee and Braae, 'Transect as a Method'.

15 A. von Humboldt, *Ansicht der Kordilleren und Monumente der eingeborenen Völker Amerikas* (Frankfurt aM: Eichborn, 2004).

16 W. Kentridge, *Six Drawing Lessons* (Boston: Harvard University Press, 2014).

17 A. von Humboldt, *Kosmos: Entwurf einer physischen Weltbeschreibung* (Frankfurt aM: Eichborn, 2004).

18 Ette, 'Humboldt and Hemispheric Constructions'.

19 L. Diedrich and G. Lee, 'Disruption Onsite: Weather, the Tyranny of Distance and Archipelagic Approaches to Fieldwork', in *Fieldwork in Landscape Architecture: Methods Actions Tools*, eds. P. Horrigan and T. Oles (London: Routledge, forthcoming); Diedrich, Braae and Lee, 'Traversing Humboldt's Transareal Landscapes'.

20 N. Bourriaud, *The Radicant* (New York: Lukas and Sternberg, 2009).

21 Diedrich and Lee, 'Disruption Onsite'.

22 Diedrich, Braae and Lee, 'Traversing Humboldt's Transareal Landscapes'; M. Farso and A. Henriksson, 'Defined by Deviations: The Travelling Transect as a Bodily Research Approach to Appropriate and Disseminate Places', *Spool* 3, no. 1 (2016).

23 M.W. Denny and S. Gaines, eds., *Encyclopedia of Tidepools and Rocky Shores* (Berkeley: University of California Press, 2007).

24 Á. Ruiz Martínez, *Paisaje Litoral Atlántico Insular: Arquitecturas Frente al Horizonte, Atlas de las Charcas Mareales* (Las Palmas: Universidad de Las Palmas de Gran Canaria, 2016); J.M. Palerm, *Paisaje Litoral de Canarias / Coastal Landscape of the Canary Islands* (Tenerife: Observatorio del Paisaje Bienal de Canarias, 2011).

25 J.M. Palerm, *Paisaje Litoral de Canarias/Coastal Landscape of the Canary Islands* (Tenerife: Observatorio del Paisaje Bienal de Canarias, 2011).

Bibliography

Appadurai, A. *Disjuncture and Difference in the Global Cultural Economy*. New York: Sage, 1990.

Appadurai, A. *Modernity at Large: Cultural Dimensions of Globalisation*. Minneapolis: University of Minnesota Press, 1996.

Bourriaud, N. *The Radicant*. New York: Lukas and Sternberg, 2009.

Braae, E. *Beauty Redeemed: Recycling Post-Industrial Landscape*. Risskov/Basel: Ikaros Press/Birkhäuser Verlag, 2015.

Braae, E., L. Diedrich and G. Lee. 'The Travelling Transect: Capturing Island Dynamics, Relationships and Atmospheres in the Water Landscapes of the Canaries'. In *Nordes 2013, Proceedings of Nordic Design Research Conference 2013*, edited by E. Brandt, P. Ehn, T.D. Johansson, M. Hellstrom Reimer, T. Markussen and A. Vallgarda, 191–200. Copenhagen: Royal Danish Academy of Fine Arts, Schools Architecture, Design and Conservation, 2013.

Council of Europe. 'The European Landscape Convention, Treaty no. 176'. 20 October 2000. https://www.coe.int/en/web/conventions/full-list/-/conventions/treaty/176.

Council of Europe. 'The Faro Convention, Council of Europe Framework Convention on the Value of Cultural Heritage for Society, Treaty no. 199'. 1 June 2011. http://www.coe.int/en/web/conventions/full-list/-/conventions/treaty/199.

Denny, M.W., and S.D. Gaines, eds. *Encyclopedia of Tidepools and Rocky Shores*. Berkeley: University of California Press, 2007.

Diedrich, L., and G. Lee. 'Disruption Onsite: Weather, the Tyranny of Distance and Archipelagic Approaches to Fieldwork'. In *Fieldwork in Landscape Architecture: Methods Actions Tools*, edited by P. Horrigan and T. Oles. London: Routledge, forthcoming.

Diedrich, L., and G. Lee. 'Rocksect: Profiling Site Processes Through Translation of Deep Fieldwork in the Fluid Margins of Land/Water'. Exhibition at Nordic Design Research Conference 2015. www.nordes.org.

Diedrich, L., E. Braae and G. Lee. 'Traversing Humboldt's Transareal Landscapes: Explorations in Fieldwork Travels and the Tableau Physique as Design Method'. In *III. Potsdamer Alexander von Humboldt-Symposion 'Landschaften und Kartographien der Humboldtian Science'*, 147–66. Potsdam: Universität Potsdam, 2018.

Diedrich, L., G. Lee and E. Braae. 'The Transect as a Method for Mapping and Narrating Water Landscapes: Humboldt's Open Works and Transareal Travelling'. *NANO: New American Notes Online*, no. 6: Cartography and Narratives (2015) (no pagination).

Diedrich, L., G. Lee and J. Raxworthy. 'Transects: Developing an Experience-Based Methodology for Design Education and Design Research'. In *Exposure/00*, edited by M. Jonas and R. Monacella, 150–64. Melbourne: Melbourne Books, 2012.

Ette, O. *Alexander von Humboldt und die Globalisierung: Das Mobile des Wissens*. Frankfurt aM: Insel, 2009.

Ette, O. 'Alexander von Humboldt and Hemispheric Constructions'. In *Alexander von Humboldt and the Americas*, edited by V. Kutzinski O. Ette and L. Dassow Walls, 209–36. Berlin: Walter Frey, 2012.

Fairclough, G., R. Harrison, J. Schofield and J.H. Jameson Jnr, eds. *The Heritage Reader* (London: Routledge, 2008).

Farso, M., and A. Henriksson. 'Defined by Deviations: The Travelling Transect as a Bodily Research Approach to Appropriate and Disseminate Places'. *Spool* 3, no. 1 (2016): 5–22.

Gibson, R., 'Changescapes', *IDEA Journal* 2005 (Brisbane: QUT, 2005): 195–206. http://idea-edu.com/wp-content/uploads/2013/01/2005_IDEA_Journal.pdf.

Gibson, R. *Changescapes: Complexity, Mutability, Aesthetics*. Perth: UWA Publishing, 2015.

Humboldt, A. von. *Ansicht der Kordilleren und Monumente der eingeborenen Völker Amerikas*. Frankfurt aM: Eichborn, 2004.

Humboldt, A. von. *Kosmos: Entwurf einer physischen Weltbeschreibung*. Frankfurt aM: Eichborn, 2004.

Kentridge, W. *Six Drawing Lessons*. Boston: Harvard University Press, 2014.

Kolen, J. 'The Rejuvenation of Heritage'. *Scape: The International Magazine for Landscape Architecture and Urbanism*, no. 2 (2006): 50–3.

Kutzinski, V., O. Ette and L. Dassow Walls, eds. *Alexander von Humboldt and the Americas*. Berlin: Walter Frey, 2012.

Meyer, E. 'Sustaining Beauty: The Performance of Appearance'. *Journal of Landscape Architecture* 3, no. 1 (2008): 6–23.

Meyer, E. 'Landscape Entanglements: Aesthetic Practices in a Networked World'. Keynote speech delivered at conference 'Beyond Ism: The Landscape of Landscape Urbanism', Swedish University of Agricultural Sciences, Alnarp, October 19–21, 2016.

Palerm, J.M. *Paisaje Litoral de Canarias/Coastal Landscape of the Canary Islands*. Tenerife: Observatorio del Paisaje Bienal de Canarias, 2011.

Ruiz Martínez, A. *Paisaje Litoral Atlántico Insular: Arquitecturas Frente al Horizonte, Atlas de las Charcas Mareales*. Las Palmas: Universidad de Las Palmas de Gran Canaria, 2016.

8
SMART NATURE?
Views from the cyborg tree

Natalie Marie Gulsrud[1]

Mounting pressure from urbanization and overall challenges of climate resilience have elevated cities to the top of the global political agenda.[2] The UN sustainable development goals in tandem with the Habitat III new urban agenda outline cities not only as sites of environmental challenges but also as critical sites of environmental solutions. Along these lines, transnational political bodies have called for landscape planners and managers to develop nature-based solutions for urban climate resilience with a focus on an integrated approach to balancing ecological, social, and digital methods. Urban nature, such as urban trees, parks, and blue and green open space, has come to the forefront of these discussions as many have underlined the multifunctional ecological and social benefits provided by resilient urban landscapes that can accommodate cycles of disruption and reorganisation, such as increased air and water quality, improved storm water management, social cohesion, and increased human health and well being.[3] Policymakers are demanding not only 'smart' cities managed through information and communication technology, but also 'smart' urban nature with enhanced ecological and social benefits delivered through digital solutions.

The urban landscape design community is ripe with discussion regarding utopian visions for digital approaches to urban nature.[4] For example, landscape architects Bradley Cantrell and Justine Holzman outline a design methodology for 'synthetic ecologies', complex and autonomous ecological landscapes mediated in real time by unlimited data simulations, presenting the possibility of removing humans entirely from landscape decision-making processes.[5] Such visions promise not only enhanced performance and efficiency but also potential agency for more-than-human actors, and are not far from reality. Landscape planners and managers currently possess an advanced suite of computational programmes that can semi-autonomously govern and rewild landscapes, such as urban forest inventories where tagged trees transmit information to smartphone platforms,[6] biodiversity assessments through gaming,[7] urban foraging with community-developed semi-autonomous drones,[8] and autonomous underwater vehicles that monitor the growth and decline of extensive coral reefs.[9] Taken together, these examples represent rapid technological development that will generate disruptive changes in the field of landscape planning and management. Moving forward, it is critical to consider not only the opportunities but also the challenges that accompany such profound change, and specifically concerns regarding the transparency, fairness, and technical proprietorship of autonomous urban natures.[10]

This chapter takes up this discussion by asking: what is 'smart' urban nature? How does a digital approach to landscape planning and management mediate the configuration and development of urban nature and to whose benefit? This is done through a review of key issues and trends in digital approaches to urban nature planning and management to uncover the systems, networks, and ways of doing that accompany computational technologies.[11] Principal points are elucidated through three digital case studies outlining major themes in the nature-based solutions agenda: sustainable food production, citizen engagement, and green energy and the circular economy. The chapter is concluded with a discussion regarding the social and material implications of smart urban natures, outlining potential implications for the field of landscape planning and management moving forward.

Everyware: From smart cities to smart natures

Computational technology such as wireless broadband, analytical software, real-time sensing and feedback, and the Internet of Things is ubiquitous in our cities today.[12] 'Smart' cities deploy such computational technology as a network of information and control systems to respond to large-scale problems of climate change, urbanization, citizen engagement, and resource efficiency.[13] This proliferation of networked computational technology has two effects, according to human geographers Rob Kitchin and Martin Dodge.[14] First, there is the aspect that urban environments are composed of 'everyware':[15] computational fabric such as fixed wireless networks, digitally controlled transport infrastructure, sensor and camera networks, etc. This fabric is used to monitor and manage real-time urban flows, coupled with the mobile computing (smartphones) of everyday users of the city that produce user data regarding location and activity. The second aspect relates to the socio-material implications of how the software, or code, underpinning computational technology impacts on the 'spatialities and governance of everyday life'.[16] Kitchin and Dodge describe this aspect as the computational performances between people, and between people and code. Uniting these two aspects is the argument that digital technology, through systems, networks, and ways of doing, broadly impacts on the configuration, development, and management of cities and urban regions and thereby mediates almost every aspect of everyday life, including the social-ecological and biophysical contexts of urban landscapes.

Landscape planning and management has long grappled with the potentials of computational technologies to manage large-scale and complex urban nature challenges with efficiency, process, and design.[17] As innovations in computational technology such as artificial intelligence and deep-machine learning accelerate, some landscape designers have called for an integration of 'responsive technologies' to account for non-linear design processes with greater consideration of the more-than-human.[18] Along these lines, some have called for a move away from the top-down mono-cultural programming of master plans toward a collaborative community- and stewardship-driven design methodology that acknowledges social-ecological change and uncertainty.[19] These calls are taken up in a project from Het Nieuwe Instituut at the World Expo in Milan in 2015, *Garden of Machines*, where curator Klaas Kuitenbrouwer poses machines as organic elements of ecosystems capable of learning to communicate, adapt behaviour, and become energy self-sufficient. In this vision, digital technologies are not only integrated into the computational fabric of the classic urban landscape, the garden, but also mediate the computational performances in the garden, breaking down hierarchical relationships between people, machines, and nature through code. Such computational advances call for an assessment by landscape planners and managers regarding the potentially radical consequences of autonomous landscapes, moving beyond the technology itself to new intellectual, political, and material pathways.

Conceptualizing smart urban nature

I now turn to three digital case studies outlining major themes in the nature-based solutions agenda—sustainable food production, citizen engagement, and green energy and the circular economy—to elucidate key issues and trends in digital approaches to urban nature planning and management. Key questions to consider are: what kind of work is being performed by synthetic and autonomous ecosystems? What new relational and contingent spaces are opened or closed through these systems? What are the potential implications moving forward?

Automated urban food production

In densely populated urban settings across the globe, entrepreneurs have launched various renditions of automated indoor vertical farming facilities to produce salad greens solely with the aid of robots, drastically reducing labour costs while increasing the quality and nutrient value of the lettuce. For example, the Japanese company Spread has recently opened a facility outside of Kyoto, while similar facilities have sprung up in the United Kingdom, United States, Canada, Singapore, Israel, and European countries.[20] Automated vertical farming uses advanced sensors, LED lighting, and hydroponic systems combined with machine learning to grow crops without sunlight or soil, and thereby maximizes production while minimizing labour costs. Individuals can purchase high-tech urban farming boxes for use in offices or homes.

Automated vertical farming is a response to market demand for fresh local produce and growing concerns regarding urban food security and overarching challenges of urban resilience. Precision farming can greatly reduce water and fertilizer usage and has a higher yield than traditional farming. In an urban context, automated vertical farming has the potential to shield consumers from major disruptions in the food system due to increasingly unpredictable weather and/or labour shortages. However, this promise is limited in that indoor farming is largely constrained to greens and mushrooms and is not suitable for other crops grown outdoors.[21] Robots present an efficiency advantage on automated indoor farms, but as humans are pushed out of food production, local environmental knowledge might be lost. Additionally, automated indoor farming has the potential to disrupt current land use patterns and open up new and novel uses of land.

Citizen scientists as human sensors

Citizen science empowers individuals to contribute to scientific data sets that impact on critical arenas of sustainable urban landscape planning and management, such as water and air quality, noise levels, species identification, and nature perception. Self-organized grassroots citizen scientists are recognized as networks of sensory nodes linked through geotagged data points crowdsourced from individual smartphones and social media.[22] In Oakland, California, low-income youth of colour have created a citizen sensor network to map, track, and monitor neighbourhood access to environmental services such as clean water, safe air quality, healthcare, wellness, transport, and food access.[23] Using point-algorithm technology, they have combined paper maps with smartphone geotagging inventories to analyse community environmental and health services. Their data reveal that East Oakland has a disproportionate allocation of 'environmental bads' such as poor air and water quality and a lack of well-stocked grocery shops with healthy food. As a result of this human sensor network, youth involvement in local urban nature planning and implementation processes has increased.

Citizen sensor networks have the opportunity to greatly democratize urban nature planning processes by legitimizing diverse local knowledge.[24] Citizen science apps aimed at youth can support environmental education and initiate curiosity and community engagement. Digital community planning processes have the potential to include more varied perspectives in urban nature monitoring and planning, which could improve the allocation of environmental goods across socio-economic boundaries. Challenges associated with citizen sensor networks touch on racial and socio-economic digital divides which favour citizens with agency and access to smartphones. Citizen sensory networks can over-represent white, socio-economically advantaged voices, presenting a racial and economic 'digital divide' in crowdsourced scientific data sets for urban nature management.[25] This barrier can undermine the co-creation and equity of urban nature planning and policy, and raises questions about who stands to gain from the computational performances of citizen science.

The circular economies of e-plants and cyborg trees

The concept of green energy now has a literal application, as scientists in Sweden have successfully implanted basic computational technology into living plants, transforming autonomous more-than-human beings into electronic circuits or e-plants.[26] E-plants have the potential to sense and display environmental changes as well as to generate fuel cells by converting photosynthetic sugars into electricity.[27] The technology behind these smart plants builds from fundamental aspects of plant physiology by using the plant's vascular transport channels (xylem) to suck up a synthetic polymer that turns into a self-assembling wire conducting electronic signals.[28] In Sweden, scientists transformed a rose into an e-plant, connecting polymer wires with naturally occurring electrolytes in the plant's tissue, turning the rose into an 'electrochemical transistor' as well as a 'digital logic gate', basic components of a computer system.[29] Scientists are working on implanting this technology into trees.

Singapore has produced a type of hybrid cyborg tree with the advancement of the 'supertree', where biophysical trees have been replaced with treelike cement structures covered in a 'living skin' of over 200 native plant species controlled by photovoltaic technology. This technology can harvest solar power, feeding into a daily night-time illumination show.[30] While the supertrees are not equipped with digital sensors, they are one of the most 'Instagrammable' geocoded sites in Singapore.[31] Their physicality lives on in the Internet of Things as an online memory and virtual representation of a tree.

A circular economy approach promises a closed-loop energy production system in which resource input and waste is minimized. Plants and trees equipped with electronic interfacing could provide opportunities for autonomous energy systems with increased reliance on local resources, leading to greater efficiency in the use of energy and materials.[32] These plants and trees could assume a monetized value if they were able to contribute back to the energy grid, elevating their biophysical status in the eternal fight for urban space. Additional opportunities could arise if plants and trees wired with computational logic hardware were able to transparently execute higher-level functions previously only associated with human decision-making. Digital archives of trees and perceptions of trees could provide a lasting ecological memory of biocultural preferences and knowledge.[33] Major risks exist in terms of vesting biophysical material with a higher utilitarian or economic value over other species. In Singapore, the cyborg supertrees have replaced indigenous forests, illustrating the risk that ecosystems with lower economic value could be replaced with landscapes that can literally pay the rent.[34] Additionally, there is uncertainty regarding the technological proprietorship of e-plants and cyborg trees. The proprietors of such technology stand to gain enormous financial sums, an interest that most likely will clash with those of citizen and public-sector consumers.[35] If we read the impacts of Monsanto on the agricultural sector as a cautionary tale, we know that patenting biophysical

Figure 8.1 Supertrees in Singapore's Gardens by the Bay, 2016. Photo from Natalie Gulsrud © Gulsrud, Natalie.

materials can inadvertently erode and weaken the socio-economic agency of small farming families. Who owns smart nature? And who benefits from the economic and social gains from the technological outputs? (Figure 8.1).

Bounded cyborg natures

What is 'smart' nature? Seen from the cyborg tree, the entanglement of computational fabric and computational performance that emerges in the digital natures reviewed above reveals a tight and potentially discomfiting kinship between machines, humans, and more-than-humans. Food production is rethought as a relationship between sensors, seeds, and LED lighting, with human ecological memory of local food production traditions captured in farming algorithms. New pathways for knowledge co-creation are realized in human sensor networks capturing and legitimating diverse and frequently unheard vulnerable voices; however, digital pathways might still be reinforcing well-known socio-economic and racial divisions. The biophysical is revealed to be computational, rational, and transactional, granting Donna Haraway's once-fictional cyborg nature economic potency and political agency. In these cases, computational technology blurs the boundaries in landscape planning and management between the sentient and the non-sentient, and opens up a Latourian reassembling of a post-humanist public sphere where all actors (rivers, trees, pipes, wires, the climate, a mayor, a town) have the potential to be given equal agency.[36] Smart nature is a techno-ecological assemblage of code and space radically altering the conditions through which social and economic relations take place: smart natures are reconfiguring the way in which value is generated by inviting businesses/communities/individuals/the more-than-human to reorganize their operations.[37] Yet on the boundaries of this techno-ecological assemblage many unanswered questions loom regarding how the social and material work that the software is doing in smart nature determines, disciplines, and potentially discriminates.[38]

A reading of Timothy Morton's *The Ecological Thought* helps us open up a discussion of these questions by analysing the reconfiguration of traditional binaries (human/machine, nature/culture, human/nature), including the very definition of nature itself. Morton argues that now is the time for humans to rethink relationships not only with a 'romanticized' and Western notion of nature as an 'Other' but also with each other, with the more-than-human, with the climate, and with society as a whole. In this regard he aims to cultivate a 'radical openness' or 'ecological' approach to thought, providing a new way forward in the Anthropocene amidst the enormity of the sixth mass extinction, climate change, and increasingly violent weather events. He writes:

> Ecology shows us that all things are connected. *The ecological thought* is the thinking of interconnectedness [...] The ecological thought doesn't just occur in the mind. It's a practice and a process of becoming fully aware of how human beings are connected with other beings—animal, vegetable, or mineral. Ultimately this includes thinking about democracy. What would a truly democratic encounter between truly equal beings look like, what would it be—can we even imagine it?[39]

Morton equips us with a new metaphor, 'the mesh', to help us understand how our ecological thinking shapes relations. The idea of the mesh helps us understand the 'interconnectedness' of existence that both protests at the rigid ideological binaries of human/animal and nature/culture and demonstrates how what we as humans consider to be the natural and unnatural (trees, ponds, birds, windmills, pipes, cement), or what he calls the 'strange strangers', is actually all very human.[40] The mesh of ecological thought therefore allows an 'expanded unboundedness of relations', not unlike a Latourian interpretation, which helps us identify with the coexistence of the 'strange strangers' as well as the 'strange' depths of our subjectivities.[41]

If we read smart natures through the mesh of ecological thought, we become hyper-aware of the potentially unlimited expansion of our relations with everything human and more-than-human, including the computational. Smart natures are one concrete step toward a manifestation of the Mortonian mesh, melding relational experience into code and space, thus generating mysterious and unknowable 'unboundedness'. Morton argues that the unbounded relational reassembly of smart natures presages a positive democratic and ethical future of queer and hierarchy-free societal transactions. The mesh helps us understand the material and social potential of computational technology in smart natures, as objects and infrastructures can be transformed, and new forms of creativity and empowerment can be engendered.

We get a glimpse of this hopeful future through the case examples presented in this chapter, yet we also are confronted with the limitations of Morton's mesh. The technologies that make possible the 'diffused microcircuits of power' that result from citizen activism and localized resistance in the human sensor networks in Oakland also are implicated in digital segregation that impact traditionally oppressed communities.[42] E-plants and other cyborg natures pose a challenge to human hegemony but the technology is at risk of proprietary exclusion by large corporate powers. Automated vertical farming could open up entirely new modes of economic production and political organization; yet studies show that robotic labour threatens the livelihoods of millions of citizens.[43] Smart natures in their post-human amalgamation support Morton's claim that absolutely everything is related to absolutely everything else; however we are left wondering how, if we radically reassess our coexistence with everything, we can use the flattened vocabulary of the mash to navigate the unequal allocation of power, capital, and resources in our urban landscapes, smart or not.

Moving forward there is a need to integrate the liberating flatness of Morton's mesh into the code and space of the computational reconfigurations that we have traced. Kitchin and Dodge remind us that the computational technologies driving smart natures are bound up in socio-spatial practices and institutional arrangements and thus contribute to and potentially limit complex discursive and material practices, relating to both living and non-living.[44] As such, the computational blurring of understandings of the natural, of species, of living and non-living, of nature and machine has loaded implications for the short- and long-term governance of urban nature. Key issues that emerge concern the role of humans in designing and managing the computational systems driving autonomous landscapes. Such intelligent landscapes build on algorithms, step-by-step sequences of tasks which execute automated decisions. While the computational performances of smart natures are frequently innovative and well executed, they can also lead to damaging social and environmental consequences.[45] The computational technology behind intelligent systems such as automated vertical farming, citizen science apps, and e-plants are dependent on the assumptions inserted into the algorithms and the data that provide the foundation for their training. If such data are flawed or biased, which has been shown in both the finance and crime prevention sectors, the results of the algorithm can be damaging, with widespread consequences. For example, the traditions, knowledges, and cultures represented in heavily automated urban landscapes will be dependent upon who controls the technology of automation in urban ecosystem management. Landscape planners and managers also need to be aware that automation which actively engages local communities may not necessarily serve social justice or public health and well-being outcomes at the city scale, because only local interests, as programmed into the algorithms, will be considered. Important cultural and social justice issues also need to be addressed to build trust in automated systems, such as whose knowledge and customary traditions will be drawn on to manage urban landscape and how smart natures will be designed to manage issues of social exclusion and allow a range of voices to be considered in ecosystem management.

The computational advances described and reviewed in this chapter are a response to economic incentives and disincentives bound up in neoliberal logics that demand high utility, high efficiency, and productivity hand in hand with the prosperous development of people and nature. The most powerful and promising algorithms used to design and manage autonomous landscapes are called neural networks; yet these algorithms are the least transparent in terms of their operational logic. As some scholars have suggested, landscape planners and managers might confront a strange paradox in the management of smart natures: the higher the tendency for computational control through artificial intelligence and robotics, 'the stronger the need for human supervision'.[46] As a result, landscape planners and managers might need to negotiate complex trade-offs between computational effectiveness and algorithmic transparency. As we rely more on techno-ecological assemblages to steer the planning and management of our urban landscapes, we will need more of a socio-spatially-rooted interpretation of Morton's ecological thought as a form of ethical and democratic oversight.

Notes

1 I would like to acknowledge the contributions of the following colleagues to the development of the ideas presented in this book chapter: Christopher Raymond, Rebecca L. Rutt, Anton Stahl Olafsson, Tobias Plieninger, Mattias Sandberg, Thomas H. Beery, K. Ingemar Jönsson.

2 Xuemei Bai et al., 'Six Research Priorities for Cities and Climate Change', *Nature* 555, no. 7694 (2018): 23–5, doi:10.1038/d41586-018-02409-z.

3 Carsten Nesshöver et al., 'The Science, Policy and Practice of Nature-Based Solutions: An Interdisciplinary Perspective', *Science of the Total Environment* 579 (2016): 1215–27, doi:10.1016/j.scitotenv.2016.11.106.

4 'Garden of Machines/Tuin van Machines', *Garden of Machines*, accessed 3 April 2018, https://tuinvanmachines.hetnieuweinstituut.nl/en; Bradley Cantrell, Laura J. Martin, and Erle C. Ellis, 'Designing Autonomy: Opportunities for New Wildness in the Anthropocene', *Trends in Ecology and Evolution* 32, no. 3 (2017): 156–66, doi:10.1016/j.tree.2016.12.004.
5 Bradley Cantrell and Justine Holzman, 'Synthetic Ecologies: Protocols, Simulation, and Manipulation for Indeterminate Landscapes', *Acadia* (2014), accessed 3 April 2018, http://papers.cumincad.org/data/works/att/acadia14_709.content.pdf.
6 Andrea Luvisi and Giacomo Lorenzini, 'RFID-Plants in the Smart City: Applications and Outlook for Urban Green Management', *Urban Forestry and Urban Greening* 13, no. 4 (2014): 630–7, doi:10.1016/j.ufug.2014.07.003.
7 Chris Sandbrook, William M. Adams, and Bruno Monteferri, 'Digital Games and Biodiversity Conservation', *Conservation Letters* 8, no. 2 (2015): 118–24, doi:10.1111/conl.12113.
8 Carl DiSalvo, Tom Jenkins, and Thomas Lodato, 'Designing Speculative Civics', *Proceedings of the 2016 CHI Conference on Human Factors in Computing Systems: CHI '16* (2016): 4979–90, doi:10.1145/2858036.2858505.
9 Cantrell, Martin, and Ellis, 'Designing Autonomy'.
10 Gulsrud, Natalie Marie, Christopher M Raymond, Rebecca L Rutt, Anton Stahl, Tobias Plieninger, Mattias Sandberg, Thomas H Beery, and K Ingemar Jönsson. "'Rage against the Machine'? The Opportunities and Risks Concerning the Automation of Urban Green Infrastructure." *Landscape and Urban Planning* 180, no. August (2018): 85–92. doi:10.1016/j.landurbplan.2018.08.012.
11 Rob Kitchin and Martin Dodge, *Code/Space: Software and Everyday Life* (Cambridge, MA: MIT Press, 2011), doi:10.7551/mitpress/9780262042482.001.0001.
12 Rob Kitchin, 'The Real-Time City? Big Data and Smart Urbanism', *GeoJournal* 79, no. 1 (2014): 1–14, doi:10.1007/s10708-013-9516-8.
13 Nick Taylor Buck and Aidan While, 'Competitive Urbanism and the Limits to Smart City Innovation: The UK Future Cities Initiative', *Urban Studies* 54, no. 2 (2017): 501–19, doi:10.1177/0042098015597162.
14 Kitchin and Dodge, *Code/Space*.
15 Adam Greenfield, *Everyware: The Dawning Age of Ubiquitous Computing* (Berkeley, CA: New Riders, 2006).
16 Kitchin and Dodge, *Code/Space*, 18.
17 Jillian Walliss and Heike Rahmann, *Landscape Architecture and Digital Technologies: Re-Conceptualising Design and Making* (New York: Routledge, 2016).
18 Bradley Cantrell and Justine Holzman, *Responsive Landscapes: Strategies for Responsive Technologies in Landscape* (New York: Routledge, 2016).
19 Kate Orff, *Toward an Urban Ecology* (New York: Monacelli Press, 2016).
20 'Techno Farm', *Spread: A New Way to Grow Vegetables*, accessed 4 April 2018, http://www.spread.co.jp/en/technofarm/.
21 Devi Buehler and Ranka Junge, 'Global Trends and Current Status of Commercial Urban Rooftop Farming', *Sustainability* 8, no. 12 (2016): 1108, doi:10.3390/su8111108.
22 Linda Carton and Peter Ache, 'Citizen-Sensor-Networks to Confront Government Decision-Makers: Two Lessons from the Netherlands', *Journal of Environmental Management* 196 (2017): 234–51, doi:10.1016/j.jenvman.2017.02.044.
23 Antwi Akom et al., 'Youth Participatory Action Research (YPAR) 2.0: How Technological Innovation and Digital Organizing Sparked a Food Revolution in East Oakland', *International Journal of Qualitative Studies in Education* 29, no. 10 (2016): 1287–307, doi:10.1080/09518398.2016.1201609.
24 Carton and Ache, 'Citizen-Sensor-Networks'.
25 Alec Foster and Ian M. Dunham, 'Volunteered Geographic Information, Urban Forests, and Environmental Justice', *Computers, Environment and Urban Systems* 53 (2015): 65–75, doi:10.1016/j.compenvurbsys.2014.08.001.
26 Maddie Stone, 'Scientists Unveil the World's First Cyborg Plant', *Gizmodo*, 21 November 2015, https://gizmodo.com/scientists-have-created-a-cyborg-rose-1743933339.
27 Ibid.
28 Eleni Stavrinidou et al., 'Electronic Plants', *Science Advances* 1, no. 10 (2015): e1501136, doi:10.1126/sciadv.1501136.
29 Stone, 'Scientists Unveil'.
30 Natalie Marie Gulsrud and Can-Seng Ooi, 'Manufacturing Green Consensus: Urban Greenspace Governance in Singapore', in *Urban Forests, Trees, and Green Space: A Political Ecology Perspective*,

eds. Anders Sandberg, Adrina Bardekjian, and Sadia Butt (New York: Routledge, 2014), 77–92, doi:10.4324/9781315882901.

31 'Singapore's Most Instagram-Worthy Attractions', *Visit Singapore*, 28 June 2017, http://www.visitsingapore.com/editorials/singapores-most-instagram-worthy-attractions/.

32 Eleni Stavrinidou et al., 'In Vivo Polymerization and Manufacturing of Wires and Supercapacitors in Plants', *Proceedings of the National Academy of Sciences* 114, no. 11 (2017): 2807–12, doi:10.1073/pnas.1616456114.

33 *Treebank: A Digital Forest Inspired by Hollow*, accessed 12 March 2018, http://treebank.online/.

34 Gulsrud and Ooi, 'Manufacturing Green Consensus'.

35 Taylor Buck and While, 'Competitive Urbanism'.

36 Donna J. Haraway, *Simians, Cyborgs, and Women: The Reinvention of Nature, Contemporary Sociology* 21, no.3 (1992), doi:10.2307/2076334; Bruno Latour, *Politics of Nature: How to Bring the Sciences into Democracy* (Cambridge, MA: Harvard University Press, 2004).

37 Kitchin and Dodge, *Code/Space*.

38 Ibid.

39 Timothy Morton, *The Ecological Thought* (Cambridge, MA: Harvard University Press, 2010), 7.

40 Ibid., 15.

41 Peter Gratton, 'Review of *The Ecological Thought*', review of *The Ecological Thought*, by Tim Horton, *Speculations* 1, no. 1 (2010): 196.

42 Kitchin and Dodge, *Code/Space*.

43 Carl Benedikt Frey and Michael A. Osborne, 'The Future of Employment: How Susceptible Are Jobs to Computerisation?', *Technological Forecasting and Social Change* 114 (2017): 254–80, doi:10.1016/j.techfore.2016.08.019.

44 Ibid.

45 Galaz and Mouazen, 'New Wilderness'.

46 Ibid., 629.

Bibliography

Akom, Antwi, Aekta Shah, Aaron Nakai, and Tessa Cruz. 'Youth Participatory Action Research (YPAR) 2.0: How Technological Innovation and Digital Organizing Sparked a Food Revolution in East Oakland'. *International Journal of Qualitative Studies in Education* 29, no. 10 (2016): 1287–307. doi:10.1080/09518398.2016.1201609.

Bai, Xuemei, Richard J. Dawson, Diana Ürge-Vorsatz, Gian C. Delgado, Aliyu Salisu Barau, Shobhakar Dhakal, David Dodman, et al. 'Six Research Priorities for Cities and Climate Change'. *Nature* 555, no. 7694 (2018): 23–5. doi:10.1038/d41586-018-02409-z.

Buehler, Devi, and Ranka Junge. 'Global Trends and Current Status of Commercial Urban Rooftop Farming'. *Sustainability* 8, no. 12 (2016): 1108. doi:10.3390/su8111108.

Cantrell, Bradley, and Justine Holzman. 'Synthetic Ecologies: Protocols, Simulation, and Manipulation for Indeterminate Landscapes'. *Acadia* (2014). http://papers.cumincad.org/data/works/att/acadia14_709.content.pdf.

Cantrell, Bradley, and Justine Holzman. *Responsive Landscapes: Strategies for Responsive Technologies in Landscape*. New York: Routledge, 2016.

Cantrell, Bradley, Laura J. Martin, and Erle C. Ellis. 'Designing Autonomy: Opportunities for New Wildness in the Anthropocene'. *Trends in Ecology and Evolution* 32, no. 3 (2017): 156–66. doi:10.1016/j.tree.2016.12.004.

Carton, Linda, and Peter Ache. 'Citizen-Sensor-Networks to Confront Government Decision-Makers: Two Lessons from the Netherlands'. *Journal of Environmental Management* 196 (2017): 234–51. doi:10.1016/j.jenvman.2017.02.044.

DiSalvo, Carl, Tom Jenkins, and Thomas Lodato. 'Designing Speculative Civics'. *Proceedings of the 2016 CHI Conference on Human Factors in Computing Systems: CHI '16* (2016): 4979–90. doi:10.1145/2858036.2858505.

Foster, Alec, and Ian M. Dunham. 'Volunteered Geographic Information, Urban Forests, and Environmental Justice'. *Computers, Environment and Urban Systems* 53 (2015): 65–75. doi:10.1016/j.compenvurbsys.2014.08.001.

Frey, Carl Benedikt, and Michael A. Osborne. 'The Future of Employment: How Susceptible Are Jobs to Computerisation?' *Technological Forecasting and Social Change* 114 (2017): 254–80. doi:10.1016/j.techfore.2016.08.019.

Galaz, Victor, and Abdul M. Mouazen. '"New Wilderness" Requires Algorithmic Transparency: A Response to Cantrell et al'. *Trends in Ecology & Evolution* 32, no. 9 (September 1, 2017): 628–9. doi:10.1016/j.tree.2017.06.013.

Gratton, Peter. 'Tim Morton, *The Ecological Thought*: Book Review'. *Speculations* 1, no. 1 (2010): 192–9.

Greenfield, Adam. *Everyware: The Dawning Age of Ubiquitous Computing*. Berkeley, CA: New Riders, 2006.

Gulsrud, Natalie Marie, and Can-Seng Ooi. 'Manufacturing Green Consensus: Urban Greenspace Governance in Singapore'. In *Urban Forests, Trees, and Green Space: A Political Ecology Perspective*, edited by Anders Sandberg, Adrina Bardekjian, and Sadia Butt, 77–92. New York: Routledge, 2014. doi:10.4324/9781315882901.

Haraway, Donna J. *Simians, Cyborgs, and Women: The Reinvention of Nature*. New York: Routledge, 2013.

Kitchin, Rob. 'The Real-Time City? Big Data and Smart Urbanism'. *GeoJournal* 79, no. 1 (2014): 1–14. doi:10.1007/s10708-013-9516-8.

Kitchin, Rob, and Martin Dodge. *Code/Space: Software and Everyday Life*. Cambridge, MA: MIT Press, 2011. doi:10.7551/mitpress/9780262042482.001.0001.

Latour, Bruno. *Politics of Nature: How to Bring the Sciences into Democracy*. Cambridge, MA: Harvard University Press, 2004.

Luvisi, Andrea, and Giacomo Lorenzini. 'RFID-Plants in the Smart City: Applications and Outlook for Urban Green Management'. *Urban Forestry and Urban Greening* 13, no. 4 (2014): 630–7. doi:10.1016/j.ufug.2014.07.003.

Morton, Timothy. *The Ecological Thought*. Cambridge, MA: Harvard University Press, 2010.

Nesshöver, Carsten, Timo Assmuth, Katherine N. Irvine, Graciela M. Rusch, Kerry A. Waylen, Ben Delbaere, Dagmar Haase, et al. 'The Science, Policy and Practice of Nature-Based Solutions: An Interdisciplinary Perspective'. *Science of the Total Environment* 579 (2016): 1215–27. doi:10.1016/j.scitotenv.2016.11.106.

Orff, Kate. *Toward an Urban Ecology*. New York: Monacelli Press, 2016.

Sandbrook, Chris, William M. Adams, and Bruno Monteferri. 'Digital Games and Biodiversity Conservation'. *Conservation Letters* 8, no. 2 (2015): 118–24. doi:10.1111/conl.12113.

Stavrinidou, Eleni, Roger Gabrielsson, Eliot Gomez, Xavier Crispin, Ove Nilsson, Daniel T. Simon, and Magnus Berggren. 'Electronic Plants'. *Science Advances* 1, no. 10 (2015): e1501136–e1501136. doi:10.1126/sciadv.1501136.

Stavrinidou, Eleni, Roger Gabrielsson, K. Peter R. Nilsson, Sandeep Kumar Singh, Juan Felipe Franco-Gonzalez, Anton V Volkov, Magnus P Jonsson, et al. 'In Vivo Polymerization and Manufacturing of Wires and Supercapacitors in Plants'. *Proceedings of the National Academy of Sciences* 114, no. 11 (2017): 2807–12. doi:10.1073/pnas.1616456114.

Stone, Maddie. 'Scientists Unveil the World's First Cyborg Plant'. *Gizmodo*, 21 November 2015. https://gizmodo.com/scientists-have-created-a-cyborg-rose-1743933339.

Taylor Buck, Nick, and Aidan While. 'Competitive Urbanism and the Limits to Smart City Innovation: The UK Future Cities Initiative'. *Urban Studies* 54, no. 2 (2017): 501–19. doi:10.1177/0042098015597162.

Walliss, Jillian, and Heike Rahmann. *Landscape Architecture and Digital Technologies: Re-Conceptualising Design and Making*. New York: Routledge, 2016.

9

"CLOUDISM"

Towards a new culture of making landscapes

Christophe Girot

A quiet "cloudist" revolution has taken place in the culture of making landscape architecture. This has come with a profound conceptual shift in aesthetic representation, brought about by new digital tools and methods pertaining to landscape analysis and design. Landscape projects are now conceived digitally as physical entities in the full topological sense of the word. Various aspects of a terrain can be worked on as a body and put into relationship with one another. The environment is changing rapidly and requires a different approach to problem-solving and the material culture and fabrication of landscapes. We are currently being asked to change the shape of natural things through a mix of science and artistry. The fact that nature is now understood as a rapidly evolving global phenomenon has marked our awareness of the world with a sense of urgency, one that calls for an immediate response with tools of a different kind. Point cloud tools can help to rapidly replicate any physical situation and are capable not only of embodying a landscape virtually, but also of achieving greater congruence transversally across disciplinary barriers.[1] This calls for a critical revision of current modes of conceiving landscape in design, analysis and practice.[2] Schools of engineering and architecture as well as studios worldwide have been testing the possibilities of computer-aided design (CAD) and analysis over the past decades. But methods relying on high-resolution digital surveys made of point clouds have been the most helpful in advancing procedural design methods and modelling linked to geographical information.[3] "Cloudism" is the term invented to describe this new art of thinking and making landscape architecture that uses point cloud modelling as a base.

Currently, only a small portion of the data in a point cloud scan (less than one per cent) is used to produce conventional surveys or documents such as contour plans, elevations and sections. The remaining ninety-nine per cent of the scanned data is seldom used and usually discarded. Even advanced engineering agencies currently use laser scans at best to reproduce series of photographic documents that represent parts of existing engineering works, which may be compared over time to obtain very accurate clash detection data.[4] At no time are point clouds ever used to generate new forms of design or projects at a territorial scale. The geographically positioned models can be used for various applications in landscape design, analysis and simulation. Each model can comprise up to one billion points of information. Sometimes these can be quite unwieldy, but programmes now exist that enable the operator to reduce their density according to the task at hand.

Point cloud models are now finding a broader range of applications in the practice of architecture, urban design, landscape architecture and engineering. "Cloudism" has changed the way we approach landscape architecture projects and studies in a broad range of cases, from the scale of an entire territory to that of a small garden. Research and teaching has shifted from conventional contour-modelling and GIS overlay mapping, towards more dynamic and versatile forms of landscape exploration within the cloud. This method promotes thinking and interaction with the full physical body of a landscape setting, in ways that simply were not possible before. The new modelling tools are capable not only of defining but also of modifying real-world surfaces embedded in a geographical reality. "Cloudism" enables design simulations in space and time to be staged in ambient conditions, regardless of difficulty. A conceptual revolution in landscape architecture and engineering is leading towards an entirely new mode of design, one that facilitates the apprehension, assessment and conception of landscapes through all scales.[5]

Building a cloud

Point cloud modelling was originally meant as a technique to heighten topological precision in land surface and atmospheric measurements. It is now becoming ubiquitous in most fields of research at the Institute of Landscape Architecture at ETH Zurich.[6] The "cloudist" technique may still be in its infancy, but it is spreading rapidly to most domains of planning and design. Current CAD software provides a point cloud function in addition to standard vectorial and mesh applications. This enables a designed object such as a building or a terrain modulation to be embedded precisely into geographical reality while maintaining topological integrity. Designers and engineers can thus develop variants of a given project and precisely calculate different forms of optimization.

Point cloud technology has its roots in the early development of laser scanners and their application to satellite reconnaissance and terrestrial mapping. Lidar (the term is a blend of 'light' and 'radar') is used for both airborne surveys and mobile terrestrial scans. It was developed in the 1960s at the US Center for Atmospheric Research for applications in meteorology, and by NASA as an altimeter to map the surface of the moon on the Apollo 15 mission of the early 1970s. The Experimental Cartographic Unit at the University of Cambridge's Cavendish Laboratory pioneered a prototype laser scanner called the Sweepnik for particle research in the 1960s.[7] As mainframe computers and lasers evolved, so did landscape feature detection and measurement techniques in the environment.[8] Both the precision and the range of laser scanning devices have improved, as have their potential applications. Terrestrial laser scanners (TLSs) function with a narrow infrared or ultraviolet ray and are able to map distant features at a very high resolution, with hundreds of points per square metre. Several publications heralded the birth of this new technique decades ago. The term 'point cloud', for instance, was first coined in the mid-1980s by a team of researchers at Chapel Hill.[9] But it was really at the turn of the millennium that cloud computing began to develop and spread more widely in the architectural and engineering fields, with the rapid digitization of groundbreaking surveying and modelling techniques.

Currently, point cloud data sets can be sampled with either lidar or TLS technologies. Every device produces a spherical pixel cloud of information around a recording point. The device maps all visible features within a given range of the source. It is ideal for landscapes where data is not easily available, but it is also suited to complex urban projects where more specific site data may be required. Point cloud models allow various data sets with varying point densities to be combined. The different data sets all align within a three-dimensional coordinate system, which precisely positions them relative to each other in geographical space. The laser scanner operates by measuring the distance to the first object on its path and then returning the information to a sensor at the speed of light, giving a precise reading of the height, depth and position of the pixel received. With this incredible degree of precision, an entire landscape can be apprehended and recorded.

Each model comprises millions of pixel points that reproduce the exact surface conditions of the terrain under study. Scans taken at different locations on a site can be assembled to produce a complete point cloud model.[10] Depending on the scale and intricacy of the landscape under study, a couple of scans may suffice, but sometimes up to 100 scans are needed to precisely depict a given landscape context. A terrestrial laser scanner typically pivots around a 360-degree vertical axis. It may take up to thirty minutes to complete a single scan rotation, depending on the resolution and the complexity of the surroundings. A mobile lidar typically operates at a lower density of points than a fixed terrestrial laser scanner, because it operates while in motion on a vehicle or aircraft. But a lidar survey by air or road is able to cover vast areas of land in a relatively short amount of time, making it highly efficient for large-scale urban and territorial projects. For instance, Zurich has deployed a lidar-generated point cloud model of the entire city, as many other cities in the world are also doing at present.

For many beginners, the technique still produces an inordinate amount of data that makes design computation and analysis with current software rather unwieldy. Nonetheless, the precision of point cloud modelling is gradually becoming more accessible and easier to handle. Within the next decade, new "cloudist" techniques will become commonplace in landscape architectural research, teaching and practice, as is already the case in the realms of architecture, archaeology and engineering. Thanks to significant advances in computer processing and data storage capacities, what would have been an exorbitant experiment a few years ago is now becoming readily available to all. One might even say that "cloudism" has generated an aesthetic of its own.

Towards a new culture of making landscapes

The digital revolution has had a direct impact on the way we analyse, plan and conceive landscapes today. Just as previous eras were influenced by the advent first of perspective and then of photography, our era will be shaped by the digital. From analytics to robotics and procedural design, most practices will use a laser scanning device on a daily basis to take advantage of the extraordinary physical precision and plasticity on offer.[11] The technology is inviting practitioners to think and conceive of landscape architecture differently, broadening its conceptual scope, scale and sphere.

Landscape architecture studios taught at ETH over the past decade have tested (albeit not without difficulty) new methods that employ point cloud data as a basis for analysis and design.[12] The effort of changing the established mode of design overlays and breaking free from old mapping habits has been hard, but after many successive studios, the methodological results are all pointing in the right direction.[13] Landscape can now be observed, modelled and simulated at various scales of design, in ways that were previously inconceivable. The method enables projects to be modified in geographical reality, both physically and spatially, following a design. A project can be embedded, tested and amended within the model, as the designer can refine the scheme from its preliminary stages to the final output. In the process of design, the model adapts to projected objectives by moving back and forth seamlessly between discrete designed units and the general context of the site. Although such manipulations are also possible using conventional CAD software, the quantum leap in precision, definition and spatial performance is phenomenal. A point cloud model remains dimensionally stable at all scales, regardless of how it is used and modified. The designer imprints onto a site a projected physical reality that can be verified and tested in geographical reality. This is where the "cloudist" method diverges considerably from more conventional mapping and design approaches. The model can also serve as the basis of a much broader interdisciplinary inquiry, which in turn informs the designer about various physical and natural phenomena affecting a given location. Using design variants in the model, one is able to predict a range of physical events such as floods, surges, heat island effects, wind and sound. Interdisciplinary teams can discuss, quantify, measure and test pertinent design decisions, which can then be fed back into the project.[14]

Methodologically speaking, any part of a point cloud model can be subdivided, extracted and developed locally to test specific design implementations in the physical realm. The extracted piece can then be fitted back into the overall site model and integrated into a larger system for simulation and evaluation purposes. "Cloudism" accompanies the landscape architect throughout the design process and at all stages of project development: it is not so much the final render that matters, but rather the extraordinary versatility that the model offers to depict the projected situation in great detail. A larger point cloud model can also be used to embed a project for further assessment; this works particularly well in cities like Zurich that provide a point cloud model of the entire conurbation. A new design proposal can thus be embedded in the larger city reality, to be viewed from without and within. Vantage points chosen at different locations immediately show how a project bears impact on the site far beyond its actual location. This opens up a field of relativity and reciprocity between a given site and any new project; with this intricate level of precision, one can, for instance, access each building and window in an urban model to question participants in the assessment and visual impact of a project. This understanding of a site as a whole context will considerably affect decision-making processes in the future, and point cloud models will take on much more significance not only in early design decisions, but also in the pre-emptive evaluation of projects.[15] The intricate complexity of the modern city can now be revealed spatially and operated upon within the digital realm. Through "cloudism", interrelations and interdependencies between buildings and sites can be further studied, revealed and understood, at various scales that can be compared and compounded seamlessly.

The degree of accuracy and potential interplay of a design embedded in a point cloud model is impressive: it not only changes our gaze, but also transforms the material, historical and spatial meanings of the landscape we are working in.[16] Each time a feature is introduced, whether a contemporary design or some archival reference, the sheer materiality, constitution and accuracy of things is questioned. New situations appear where the relative interdependences of artefacts can be examined and interrogated.

Progress in artificial intelligence and robotics is pointing the way towards revolutionary techniques in design automation and execution at a territorial scale.[17] The phenomenon has long been operating in agriculture, with large-scale industrial farming now using self-guided, GPS-equipped machines to till and harvest the land. The Swiss National Centre of Competence in Research's current research programme on digital fabrication at ETH's Department of Architecture is developing a self-driving intelligent excavator called the *Menzimuck*, which will perform independent tasks in remote, risk-prone alpine areas such as landslips and river banks. Developed by a multidisciplinary team of engineers, architects and landscape architects, the machine operates with precise GPS sensors that log into a site in real time through a point cloud model: robotic "cloudism", if you will. The point cloud model helps to guide the *Menzimuck* excavator through the steps of its 'intelligent' driverless operations. The time is approaching when landscape projects will be driven by machines directly on-site, with only a minimal form of design preamble required.[18] Depending on the task, the excavator will be further informed about the physical difficulties of a given terrain by drones that hover above and are equipped with small, lightweight Velodyne lidar devices similar to the artificial eyes used in self-driving cars. The excavator is meant to develop an 'intelligent' response to the terrain in each situation. In the coming decade, self-driving earth movers will become ubiquitous and will proactively change the way we conceive and design our landscapes. Possible applications will range from flood management and landslip control in remote areas, to large-scale coastal protection in and around large urban areas. The tasks at stake in this rapidly changing world are simply immense, and we will be glad to have these robotic 'friends' around to help us work out the best possible topologies. The "cloudist" prospect means that landscape assessment and production will take on dramatically different dimensions and purposes in years to come.[19]

A pioneering experimental studio was held at ETH in autumn 2017 on the topic of robotic landscapes. With support from Fabio Gramazio and Matthias Kohler (professors of Procedural Design Fabrication and Research at the Department of Architecture) and Marco Hutter (professor of Robotic Engineering at the Department of Mechanical Engineering), architecture students were asked to work on a large-scale point cloud model using artificial intelligence as a design approach. The project, located on a five-kilometre stretch of land between the River Ticino and the motorway above Bellinzona, studied the possibility of designing effective sound barrier mounds using robotic terrain-modelling techniques. The results were challenging, and quite different: a lot of the ego of the designer was substituted by the logic of the robot, which was not always necessarily better. Preliminary design work was scaled down to a sandbox in which a three-axis robot arm operated. The sandbox results were scanned and then scaled up to a procedural code for the entire project. There was, so to speak, no prior sketching or drawing of the final terrain-modelling throughout the entire design process. The design forms remained unknown until they were delivered on the terrain, but they followed a precise topological path

Figure 9.1 'Robotic Landscapes' Design Studio, Prof. Girot with Prof. Gramazio and Prof. Kohler, ETH Zürich Fall 2017. © ETH LVML Chair of Landscape Architecture Christophe Girot with Ilmar Hurkxkens, Fojan Fahmi and students Maximilien Durel, Shohei Kunisawa, Nicolas Wild.

generated by ground data retrieved from the point cloud base. A few height, width and length instructions sufficed to set the rest of the design in motion; the results were principally machine generated. Without a point cloud model of the entire valley, the robotic design would have lacked the precision basis that enabled machine operations to modify the terrain accurately while balancing remnant material accordingly (Figure 9.1).

The following examples discuss point cloud models and their adaptation to landscape analysis and design. Some projects have been developed in design studios at ETH's Landscape Visualizing and Modelling Laboratory (LVML) over the past decade. The selection shows a range of experiments that have used point clouds in site reconnaissance, modelling and design. Point clouds have been around for over three decades in engineering and procedural design circles, but it is only recently that the "cloudist" technique has been transferred to the field of landscape architecture. The examples are far from exhaustive, but they show what has been tested in terms of possibilities. Some field experiments are truly exemplary, in that they challenge established design methods with a broad array of questions. Resiliency has become a top priority in predictive landscape planning, and the "cloudist" design process seems best adapted to a world of changing circumstances in need of immediate spatial and topographic accuracy while maintaining conceptual flexibility.

One important aspect of this approach is the active return of the designer to the site, and the permanent immersion that follows with the model in studio thereafter. During recording on the terrain, the landscape can be discovered differently, particularly during the scanning and droning process, which can take a day or two. This is something that conventional planning and design practices, which often rely on ready-made maps and plans, tend to overlook and disconnect from. This on-site immersion during scanning is paramount for garnering unique field observations that become an integral part of the feeling for a place. The follow-up in studio through different stages of design enables a better comprehension of the environment as well as the physical and spatial conditions involved. The point cloud model enables the discretionary development of design solutions that can be repeatedly tested and adapted to a broader context. In the "cloudist" approach, there exists no separation between a model, a section and a plan: they all stem from the same cloud of design information. Separate renderings or visualizations become quite unnecessary, since the views generated are directly derived from the model, with their own singular aesthetic.

Field experiments: Scanning the territory

With rapid progress in airborne lidar technology, entire cities and regions are now being modelled in open-source point clouds. Only a few years ago this was still far from the case. Some of the early experiments in scanning an extensive urban context in Switzerland using a long-range Riegl TLS were conducted at LVML. This started in 2012 as part of a Commission for Technology and Information project. I co-directed the project, called '4D Sites', with Pascal Werner in close collaboration with Marc Pollefeyes, professor at ETH's Department of Computer Imaging, together with Octagon, the company that runs Leica Systems, as industrial partner. The site in question was the town of St Moritz in Engadin, Switzerland. It was chosen because it was very well documented, with all kinds of historical pictures.

Only a partial point cloud model of the city and its alpine surroundings was developed for the purpose of the research. This was done by overlapping a series of terrestrial scans in and around the town. The most interesting aspect of this project was the elaboration of a series of precise transects that showed the exact relationship of the buildings to the topography by cutting sections through the model at various locations. There followed an attempt to embed historical photographs and documents into the point cloud model, in order to reveal the chronological development of St Moritz. However, the results of this attempt were mitigated and only partially successful. The problem was that the software for large-scale point cloud modelling at that time was still quite rudimentary, cumbersome and poorly adapted to the overlaying of other kinds of data set. The historical material embedded in the model looked rather askew in the end, but the results, although not really conclusive, helped to further the enquiry methodologically (Figure 9.2).

The Swiss Cooperation Project started in 2010 under the leadership of Christian Sumi and Marianne Burckhalter from the Academy of Architecture in Mendrizio. The research project, entitled 'The St Gotthard: Landscape, Myths and Technology', brought together competences from various disciplines, ranging from history and technology to architecture, engineering, landscape architecture, literature and art.[20] My team at ETH was asked to join in order to scan a vast stretch of mountainous folds, roughly eighty kilometres long and fifteen kilometres wide, comprising the entire St Gotthard pass and road. The first forays required intensive field operations at high altitude, involving a mix of mobile lidar data collection, photogrammetric recordings by fixed-wing drones, and long-range TLS data-sampling. The data was then set on a GIS mesh base of the Alps provided by Swiss Ordnance Survey, which worked as both canvas and background. A mix of point cloud data was compiled into a single landscape model. It took a long time to register and combine all the points, not only because of the complexity of the terrain, but also because some of the data sets were registered in different geographical coordinate systems. My team learned as they went along how to deal with such an unwieldly model. In fact it took several years to complete, but the end result

Figure 9.2 Point Cloud Section through the town of St Moritz, Canton of Grizons Switzerland in 4D Sites Research, 2012. © ETH LVML Chair of Landscape Architecture, Christophe Girot with Pascal Werner.

was astounding. Data sets of the railway and motorway tunnels have subsequently been incorporated into the model. Some test runs of the model were shown at the Architecture Biennale in Venice in 2014, and a more elaborate version of it was delivered in a TEDx talk in 2016 that was later shown at the World Economic Forum in Davos in 2017. The model revealed the incredible infrastructural complexity of this mythical landscape, and juxtaposed aspects of the rugged terrain in a most brilliant and unexpected way. The facility with which one was able to fly in and out across scales of alpine territories revealed the mountain in a new and fascinating "cloudist" light. The diaphanous, semi-transparent granite surface of the mountain enabled one to see through to the tunnel infrastructures deep below the surface at a glance. Although the Gotthard project still only serves a general informational purpose, it has permitted my team to assemble underground elements such as the new Gotthard Base Tunnel (currently the longest tunnel in the world), as well as the older nineteenth-century spiral railway tunnels and 1970s motorway tunnel, with a high degree of precision. This modelling feat shows how "cloudist" technology can serve the extraordinary complexity of the Swiss alpine context, by showing and demystifying an extremely complex terrain, simply, beautifully and poetically.

In a more tropical register, two projects developed over the past five years, one in Jakarta and the other in Singapore, have used large-scale point cloud modelling technology similar to the Gotthard project. It was important to transfer the point cloud technology out of the pristine ice fields and peaks of Switzerland to the sweltering flood-prone slums of smoggy Jakarta. The River Ciliwung project, located in the megacity of Jakarta, was certainly my most challenging research assignment in terms of data collection, processing and problem-solving. We arrived in Jakarta accustomed to data-rich Switzerland and were confronted with data-poor Indonesia. We were forced to use only local sampling means and drones to generate a rudimentary point cloud model. The goal of the project was to restore the River Ciliwung for flood protection by developing proposals into a workable model. The project's location in the Kampung Melayu and Bukit Duri slums of Jakarta was completely unprecedented. The slum dwellings covered almost the entire surface area of the settlement, meaning that it was particularly difficult to process data on the ground surface and all along the river bank. Once the data had been collected, great skill was required to triangulate the model accurately, so that it could effectively transcribe the actual ground topography of this flood-prone area. The model was used by an interdisciplinary team of doctoral researchers to simulate the major flood events that recur each year in this part of Jakarta. A group of ETH studio students were brought to Jakarta to learn how to tackle the site and design on the basis of the model. The design results were unanimously acclaimed, and several design proposals were subsequently tested by doctoral researchers specialized in hydrology, landscape architecture and planning, to see how new riverbed typologies might possibly work. This was in 2013, and to my knowledge it was the first time that "cloudist" landscape designs had been embedded in a model for the purpose of flood simulations and experimentation.

The Jakarta experiment definitely opened the way for further applications in large-scale landscape studios. The research project currently underway at the Future Cities Laboratory in Singapore attempts to tackle the elusive urban heat island effect that plagues many tropical downtown areas. The goal of the studio is to develop and enhance a design for a landscape corridor twelve kilometres long capable of conveying cooler air from the hills of Bukit Timah down towards the Central Business District using passive convection. This will entail the creation of a set of continual landscape structures and terrain-modelling for shading and wind conveyance. The project is ambitious and far more risk-taking than the Jakarta project, because it speculates on the role of vegetation as a potential factor in the temperature regulation of tropical urban centres.[21] The simulation will entail modelling cool airflow coming down from the hills by testing various landscape configurations at different times of day, to find out which would best alleviate the heat island problem.

What is interesting with the "cloudist" method is that it is not so much the final rendered project that matters to researchers, but rather the repeated testing and optimizing of a concept throughout a project. With this new approach, landscape design education is entering a phase of development where the designs will effectively be tested, evaluated and understood prior to any form of implementation. Natural phenomena such as climatic events, temperature surges, wind, floods and vegetative growth can be simulated in point cloud models. Landscape projects conducted at the scale of the territory will be modelled, experimented on and tested, before they get implemented in reality. The "cloudist" design studios that I have been directing over the past five years develop large-scale landscape design in models for predictive purposes. From flash floods in the Sonora Desert of Arizona, to the artificial island of La Certosa in the *acqua alta* of the Venice Laguna, from lake and river floods in the Magadino Plain of Ticino to the landscape regeneration of Klybeck Island in Basel Harbour, all studios show an uncanny ability to operate designs in the physical reality of a site.

What has become particularly interesting in teaching is the use of modelling to test the pertinence of the design throughout the semester. Even during early reviews, each design set in the point cloud model can not only be viewed through a series of screenshots, but can also be 'visited' and 'walked through' at any scale and in any desired direction. One can, for instance, walk with the student during a desk critique to the corner of a river bend, to see how the project feels spatially and aesthetically as a whole. This form of studio teaching and critiquing through point cloud models is proving to be very effective on large-scale landscape designs. It shows how intensive topological training can indeed become an integral part of the education of the landscape architect in this era of resilient design. All these "cloudist" studios developed at ETH have significantly helped to establish and shape an entirely new design methodology in landscape architecture.

It would be a mistake, however, to think that the "cloudist" method applies only to large-scale environmental and urban projects, and that the scale of an intimate garden somehow eludes the grasp of this technique. A joint research and teaching programme on Japanese gardens between ETH and the Kyoto Institute of Technology (KIT) over the past four years has yielded results of extraordinary precision and delight, combining point clouds with geolocated sound samples.[22] I still marvel at the extraordinary versatility of the work done in these short one-week workshops, where KIT students encapsulate all the wonder, culture and mystery of a traditional Japanese garden in the most exquisite detail. The sound samples, which are located in the model precisely where they were taken, enhance a sense of space and time of the garden.[23] This acoustic dimension adds a sensory layer to a point cloud model; it also reanimates decades of dormant research in landscape acoustics (Figure 9.3).

There is definitely a new aesthetic arising from the "cloudist" experiments I have listed. Their digital form and appearance will depend on the tasks and processes at hand. A broad range of projects, from the most engineered and technical to the most artistic and poetic, is opening an array of new possibilities in our discipline. "Cloudism" is the new horizon that will bring much-needed changes to the design, analysis and production of landscape architecture. I am convinced that it is here to stay. Far from being a flash in the pan, it is something fundamentally new that is going to contribute significantly to the field of landscape architecture. We just need to learn how to be more creative and grow with it in the decade to come.

Figure 9.3 Point Cloud Section of the Ninigi House in Kyoto Japan with KTI, 2014. © ETH LVML Chair of Landscape Architecture, Christophe Girot with Matthias Vollmer.

Notes

1 Jillian Wallis and Heike Rahmann, *Landscape Architecture and Digital Technologies: Re-Conceptualizing Design and Making* (New York: Routledge, 2016).
2 Christophe Girot and Dora Imhof, *Thinking the Contemporary Landscape* (New York: Princeton Architectural Press, 2017).
3 Ranjith Unnikrishnan, 'Statistical Approaches to Multi-Scale Point Cloud Processing' (PhD diss., Carnegie Mellon University, 2008).
4 The Swiss Federal Railway Company, for instance, makes point cloud scans of every tunnel it manages, but does not transform this data into three-dimensional models. Instead, the scans of the tunnels are looked at as flattened two-dimensional segments of the tunnel vault, laid out so as to compare previous campaigns sector by sector to detect any clashes.
5 Christophe Girot et al., 'Scales of Topology in Landscape Architecture', *New Geographies*, no. 4 (2011).
6 The point methodology introduced ten years ago has become the basis of series of studio projects and research projects I am conducting at ETH.
7 The Sweepnik used a laser with a set of movable mirrors and became the first known laser scanning device on the research market.
8 Riegl patented its first laser scanner in the early 1970s, followed in the early 1990s by Cyra Technologies, which was bought by Leica in 2001.
9 Marc Levoy and Turner Whitted, 'The Use of Points as a Display Primitive' (Technical Report 85-022, Computer Science Department, University of North Carolina at Chapel Hill, 1985).
10 TLS devices have a range radius between 100 metres and a few kilometres. Depending on the model used, they are capable of scanning vast surfaces of terrain with a single swipe. Beyond its specific range, the scanner remains completely blind to the environment. The LVML at the ETH currently operates with a RIEGL TLS that has a 1.5-kilometre range; other Riegl devices can have a range up to five kilometres, but require very stable conditions to operate.
11 Ilmar Hurkxkens and Georg Munkel, 'Speculative Precision: Combining Haptic Terrain Modeling with Real Time Digital Analysis for Landscape Design', *Peer Reviewed Proceedings of Digital Landscape Architecture 2014 at the ETH Zurich*, eds. U. Wissen Hayek et al. (Heidelberg: Wichmann, 2015): 399–405.
12 Christophe Girot et al., *Pamphlet 15: Topology* (Zurich: GTA Publishers, 2012).

13 Christophe Girot, 'The Elegance of Topology', *Topology: Topical Thoughts on the Contemporary Landscape, Landscript 3*, eds. Christophe Girot et al. (Berlin: Jovis, 2013): 79–115.
14 Adrienne Gret-Regamey et al., 'River Rehabilitation as an Opportunity for Ecological Landscape Design', *Sustainable Cities and Society*, no. 20 (2016).
15 Maarten J. van Strien et al., 'An Improved Neutral Landscape Model for Recreating Real Landscapes and Generating Landscape Series for Spatial Ecological Simulations', *Ecology and Evolution* 6, no. 11 (2016).
16 Christophe Girot, 'Immanent Landscape', *Harvard Design Magazine*, no. 36 (2013).
17 Fabio Gramazio and Matthias Kohler , eds., *Made by Robots: Challenging Architecture at the Large Scale* (London: Wiley, 2014).
18 Ben Popper, 'Robo-Bulldozers Guided by Drones are Helping Japan's Labor Shortage', *The Verge*, 13 October 2015, https://www.theverge.com/2015/10/13/9521453/skycatch-komatsu-drones-construction-autonomous-vehicles.
19 Wallis and Rahmann, *Landscape Architecture*.
20 Marianne Burkhalter and Christian Sumi, *Der Gotthard/Il Gottardo: Landscape, Myths and Technology* (Zurich: Scheidegger and Spies, 2016).
21 Xiao Ping Song et al., 'The Economic Benefits and Costs of Trees in Urban Forest Stewardship: A Systematic Review', *Urban Forestry and Urban Greening*, Elsevier, 29, (2018): 162–170.
22 Christophe Girot, *Pamphlet 21: Sampling Kyoto Gardens* (Zurich: GTA Publishers, 2017).
23 Nadine Michèle Schütz, 'The Acoustic Dimension of Landscape Architecture' (PhD diss., ETH, 2018).

Bibliography

Burkhalter, Marianne and Christian Sumi. *Der Gotthard/Il Gottardo: Landscape, Myths and Technology*. (Zurich: Scheidegger and Spies, 2016).

Carpo, Mario. *The Second Digital Turn: Design Beyond Intelligence*. Cambridge. MA: MIT Press, 2017.

De Jong, T.M. and D.J.M. Van der Voort, eds. *Ways to Study and Research*. Delft: Delft University Press, 2002.

Desimini, Jill and Charles Waldheim. *Cartographic Grounds: Projecting the Landscape Imagination*. New York, NY: Princeton Architectural Press, 2016.

Girot, Christophe. 'Immanent Landscape'. *Harvard Design Magazine*, no. 36 (2013). http://www.harvard-designmagazine.org/issues/36/immanent-landscape.

Girot, Christophe. 'The Elegance of Topology'. In *Topology: Topical Thoughts on the Contemporary Landscape*. Landscript 3, edited by Christophe Girot, Anette Freytag, Albert Kirchengast and Dunja Richter, 79–115. Berlin: Jovis, 2013.

Girot, Christophe. *The Course of Landscape Architecture: A Natural History of our Designs on The Natural World, From Prehistory to The Present*. London: Thames & Hudson, 2016.

Girot, Christophe. *Pamphlet 21: Sampling Kyoto Gardens*. Zurich: GTA Publishers, 2017.

Girot, Christophe and Dora Imhof. *Thinking the Contemporary Landscape*. New York: Princeton Architectural Press, 2017).

Girot, Christophe et al. 'Scales of Topology in Landscape Architecture'. *New Geographies*, no. 4 (2011): 156–163.

Girot, Christophe et al. *Pamphlet 15: Topology*. Zurich: GTA Publishers, 2012.

Gramazio, Fabio and Matthias Kohler, eds. *Made by Robots: Challenging Architecture at the Large Scale* (London: Wiley, 2014).

Gramazio, Fabio, Matthias Kohler and Jan Willmann. 'Procedural Landscapes'. In *The Robotic Touch: How Robots Change Architecture: Gramazio & Kohler, Research ETH Zurich 2005–2013*. Zurich: Park Books, 2014.

Gret-Regamey, Adrienne et al. 'River Rehabilitation as an Opportunity for Ecolgical Landscape Design'. *Sustainable Cities and Society*, no. 20 (2016): 142–146.

Haupt, S., A. Pasini and C. Marzban. *Artificial Intelligence Methods in the Environmental Sciences*. Amsterdam: Springer Netherlands, 2009.

Hurkxkens, Ilmar and Georg Munkel. 'Speculative Precision: Combining Haptic Terrain Modeling with Real Time Digital Analysis for Landscape Design'. *Peer Reviewed Proceedings of Digital Landscape Architecture 2014 at the ETH Zurich*, edited by U. Wissen Hayek et al. Heidelberg: Wichmann, 2015: 399–405.

Hutton, Jane. *Material Culture, Landscript* (2017).

Lee, Michael J. and Kenneth I. Helphand. *Technology and The Garden*. Dumbarton Oaks: Harvard University Press, 2014.

Levoy, Marc and Turner Whitted. 'The Use of Points as a Display Primitive'. *Technical Report 85–022.* Computer Science Department, University of North Carolina at Chapel Hill, 1985.

Lin, Shengwei Ervine. 'Point Clouds as a Representative and Performative Format for Landscape Architecture–A Case Study of the Ciliwung River in Jakarta, Indonesia.' PhD diss., ETH Zurich, 2016.

Petschek, Peter. *Grading: Landscape SMART 3D Machine Control Systems Stormwater Management.* Basel: Birkhäuser, 2014.

Popper, Ben. 'Robo-Bulldozers Guided by Drones are Helping Japan's Labor Shortage'. *The Verge*, October 13, 2015. https://www.theverge.com/2015/10/13/9521453/skycatch-komatsu-drones-construction-autonomous-vehicles.

Unnikrishnan, Ranjith, 'Statistical Approaches to Multi-Scale Point Cloud Processing'. PhD diss., Carnegie Mellon University, 2008.

Schütz, Nadine Michèle. 'The Acoustic Dimension of Landscape Architecture'. PhD diss., ETH Zurich, 2018.

Song, Xiao Ping et al. 'The Economic Benefits and Costs of Trees in Urban Forest Stewardship: A Systematic Review', *Urban Forestry and Urban Greening*, Elsevier, 29, (2018): 162–170.

Strien, Maarten J. van et al. 'An Improved Neutral Landscape Model for Recreating Real Landscapes and Generating Landscape Series for Spatial Ecological Simulations'. *Ecology and Evolution* 6, no. 11 (2016): 3803–3821.

Verburg, P.H., J.A. Dearing, J.G. Dyke, S.v.d. Leeuw, S. Seitzinger, W. Steffen, and J. Syvitski. 'Methods and Approaches to Modelling the Anthropocene.' *Global Environmental Change* 39 (2016): 328–340.

Wallis, Jillian, and Heidi Rahmann. *Landscape Architecture and Digital Technologies: Re-Conceptualizing Design and Making.* New York: Abington (Oxon) Routledge, 2016.

10
THE MARNAS DIGITAL ARCHIVE
Exploring practice, theory, and place in space and time

Anne Whiston Spirn

Garden design is a temporal art. Composed of materials that grow and decay, subject to changing needs and fashion, gardens cannot be frozen in time. How, then, can they be preserved? How can they be designed to accommodate change? These conundrums were a preoccupation of Scandinavian landscape architect and designer/theorist Sven-Ingvar Andersson (1927–2007), which he explored in writing and built work throughout his career, particularly at Marnas, his own summer place. Andersson described Marnas as an open experiment steeped in tradition, a celebration of transience, changing as the plants grew, as he aged. He documented his experiments there from the 1950s until his death in 2007 through thousands of photographs. This unusually rich documentation is the wellspring of the Marnas project, a digital archive that includes hundreds of photographs and a series of multimedia videos embedded in an interactive website, which was designed as an immersive, three- and four-dimensional experience of the place and the interplay that exists there among history, theory, practice, form, and function. It is now possible to journey through the garden in space and time: to walk down leafy tunnels, through diverse spaces; to travel across time in successive views of the same space from morning through evening, from winter through spring, summer, and fall, across decades. The goal was to create a visit to Marnas as it existed during Andersson's lifetime, over the course of more than fifty years, conveying the experience of theory and practice embodied in the garden. This project exemplifies what the digital humanities, a field at the intersection of digital technology and humanities disciplines, has to offer landscape architecture and landscape history.

Sven-Ingvar Andersson and Marnas: Theory and practice in place

Sven-Ingvar Andersson (SIA), one of Scandinavia's most important landscape architects, was both an author and a practitioner. From 1963 to 1994, he was Professor of landscape and garden art in the School of Architecture at the Royal Danish Academy of Fine Arts in Copenhagen, Denmark. Among his best-known professional works are the restoration of Sophienholm (1967), a historic landscape garden in Denmark; the design for Karlsplatz (1972) in Vienna; the landscape of Tête Défense (1984–6) in Paris; and the Museumplein (1992–9) in Amsterdam. His writings include books on garden history and articles on preservation, design theory, landscape architecture, gardens, and plants. Andersson's writings are in dialogue with his built works.

Nowhere was that dialogue more intimate than at Marnas, his own garden and summer house in southern Sweden.

In 1953, SIA's parents gave him a wedding gift: an old three-room farmhouse and small plot of land adjacent to their own farm, near the Swedish university town of Lund. From then on, it became SIA and his family's summer place. He called it Marnas, after the farm woman who had lived there until her death in 1951. In 1954, he graduated with degrees in horticulture and landscape architecture from the agricultural college at Alnarp, and in art history from Lund University, where he focused on garden history and also studied botany and genetics. He worked in other parts of Sweden until 1964, when he moved to Copenhagen, after his appointment as Professor at the Royal Danish Academy of Fine Arts. On weekends and vacations, he replaced the overgrown meadow at Marnas with lawn, incorporated parts of the barn's ruins (a remnant wall, the cobble floor) into a new patio, and restored the exterior of the house. In the late 1960s, he built a simple one-storey building with a guest room, outhouse, and shed. The new structure, perpendicular to the old house, created a long wall between the old garden and a new one to the north, as can be seen in Figure 10.1.

In 1965, SIA expanded the original property across the driveway onto an adjacent field, where he laid out a grid of seven garden 'rooms', each enclosed by its own hawthorn hedge, with the gaps between forming passages. Each room had a different function: a kitchen garden, a flower garden, a garden for his daughter Beata, a compost area, a room for barbecues, and one for sunbathing. In the largest garden room, he planted fourteen small, egg-shaped hawthorns. These 'eggs' would eventually grow into twelve-foot-high topiary 'hens'. In 1964, he had written about predecessors to these hens, a brood of several privet shrubs in the garden of his house at Helsingborg, which had 'suddenly gotten the desire to become long-necked birds'. 'We helped

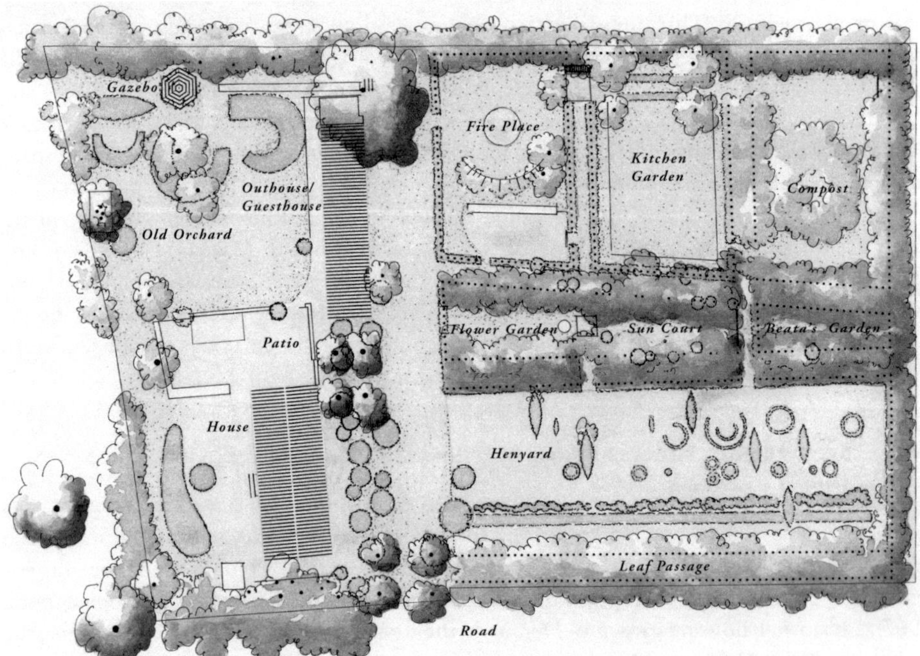

Figure 10.1 Plan of Marnas, Sven-Ingar Andersson, 2008. © Andersson, Beata Engels.

them by pruning and clipping,' he wrote, 'but they aren't yet ready for flight.'[1] Several years later, in 1967, SIA published 'A Letter from My Henyard', a witty manifesto in which he criticized designers and planners for conceiving 'finished' works with fixed details. Instead, he advocated the design of a framework within which details would be able to evolve in response to changing circumstances, needs, and desires.[2]

The framework of garden rooms at Marnas accommodated much change over the years: the room for sunbathing became a shady haven; the flower garden, a room full of boxwood; Beata's garden, a dark den full of ivy. At the end of 'A Letter from My Henyard', SIA imagines himself as an old man at Marnas. Too weak to clamber up ladders and wield clippers, he sits in a garden that has transformed into a grove of tall hawthorns. Indeed, by the time of his death in 2007, the central portion of the gridded garden had evolved from clipped hedges into a freely growing grove, as he had envisioned forty years earlier, and the hens, which had sprouted plumy tails and combs, had become 'cocks'.[3] Over the course of four decades, he documented these transformations in photographs, and reflected on what he learned from this process in publications about landscape and urban design and Marnas.

Marnas: A garden laboratory

SIA was both a modernist and an advocate for preservation. In 1958 he published a plea for the preservation of old gardens. Many others would follow, including 'Principper for bevaring' ('Principles of Preservation', 1993), where he compared three approaches: reconstruction, renovation, and free renewal.[4] SIA preserved the old house at Marnas much as it was a century ago, with no indoor plumbing or central heating. He reconstructed the exterior half-timbered walls, but introduced a large new window in the kitchen. In the garden, he combined historic garden motifs and contemporary forms, experimenting with practices required by renovation and free renewal. In renovation, one proceeds 'with an eye toward conserving the garden's form and spirit by [...] preserving the spatial structure'. 'What you are conserving with free renewal is not the form. Neither is it the spirit, as when you renovate. What you are conserving is the dignity, the artistic quality.'[5] He defines renovation and free renewal as 'artistic projects', in contrast to reconstruction, where 'the aim is historical accuracy.'[6] The former requires not only a deep knowledge of and aesthetic affinity with the forms and materials of historic gardens, but also the experience of working with them, especially the cultivation of plants.

For forty years, Marnas was a laboratory not only for landscape renovation and free renewal (particularly in the grid of garden rooms where SIA adapted traditional motifs from historic gardens of France and Italy), but also for new garden forms more closely allied with modern art. In the old orchard near the house, he experimented with modernism's free forms, garden versions of painting or sculpture by artists such as Hans Arp (1886–1966), in clipped privet rather than paint or stone. The overarching experiment, however, was an exploration of the design of change over time and the roles of circumstance, serendipity, and improvisation. At the end of a 1966 article where he argued for the preservation of the old and the pursuit of the new, SIA juxtaposed two commemorative postage stamps ('Danish Preservation' and 'Niels Bohr's Atomic Theory').[7] These two images could stand for Marnas.

In 1990, when I first met Sven-Ingvar Andersson, visited Marnas, and read 'A Letter from My Henyard' (1967), I recognized the kinship between his research and my own. In 1987, I had launched the West Philadelphia Landscape Project (WPLP), where I was exploring ideas about design and change, framework and improvisation, across scales of community gardens and urban neighbourhoods in inner-city Philadelphia.[8] At first glance, Marnas and WPLP seem so disparate: Marnas is a small garden of 'high' design; WPLP includes community gardens and plans for a large watershed. But the central questions of Marnas and WPLP are the same: how to create dynamic designs and plans that will guide change over time, accommodate unanticipated developments, and invite improvisation by those who inhabit the place? Both are longitudinal research projects (four decades in the case of Marnas, three in WPLP), and both are participatory action research projects where the designer/planner tests ideas through practice, reflects on the process and result, then formulates further action, in an iterative, open-ended process. The participants in the case of WPLP are people; in Marnas they include plants (SIA regarded plants as actors and responded to their spontaneous appearance, growth, and decay). Both Marnas and WPLP proceed from an understanding of history and treat the past as ever-present. From 1990 to SIA's death in 2007, I visited Marnas fifteen times, often staying for several days, studying it through my camera's lens, discussing ideas with SIA and hearing from him the history of Marnas, his ideas and practices, and how they were embodied in the garden.

After SIA's death in 2007, his daughter Beata and her son inherited Marnas, and change has been inevitable. Marnas still resembles the place it once was during its designer's lifetime, but, with new guiding minds and hands, it is different from what it once was or what it would be with an elderly SIA still in charge. And, of course, the inclination of his heirs to honour him through preservation is a new element. Marnas was a dynamic laboratory, shaped by SIA's ideas and experiments, his experience, knowledge, sensibility, and aesthetic judgement. Someone else might replicate his lines of investigation, but it would gradually become a new place. So Marnas is no longer entirely SIA's garden, and today a first-time visitor has no way to know what has changed. But this need not be. SIA's photographs permit the reconstruction of what the place looked like through time.

Marnas: The photographic record

A few plans of Marnas exist, but they are schematic, not to scale, and not entirely accurate; each was made to accompany a publication, the first in 1967, another in 1976, and the last in 2008.[9] At Marnas, SIA did not proceed from a detailed plan. He envisioned the goal, designed a structure to hold the vision, and then improvised within that framework. 'I have a definite idea of how my henyard will end,' he wrote in 1967, 'but that which lies between *now* and *then* is an open plan. [...] A lot can happen before the henyard becomes a hawthorn grove.'[10]

We know what happened, thanks to the photographs. SIA made photographic sequences of passage through the garden and diverse views from the same locations in different seasons and across the years. He documented new additions (a gazebo built from discarded seventeenth-century windows, a sculpture made from found objects, a pyramid of red sandstone, two black ceramic cubes) and his decisions (to permit a few leaders to grow up from the hawthorn hedges and become trees; to cut off the side branches of willows, leaving only the uppermost to grow freely). During the early 1980s, when he was writing a chapter on Karen Blixen's garden, SIA experimented with elaborate bouquets of flowers at Marnas, which he photographed.[11] 'To make a bouquet is to assemble a perennial bed in miniature', he would write a quarter-century later.[12]

I had seen SIA's photographs from the 1960s, 70s, and 80s in the collection of the Royal Danish Academy of Fine Arts. More than 300 large-format transparencies (2.25 inches square), encased in glass with labels and dates, starting in 1965, portray the newly planted hedges and garden rooms, as well as the older orchard garden. To view these images in chronological order is to be propelled forward through time: to see the hawthorn 'eggs' hatch into birds and then gradually grow long necks; to watch the hedge walls grow higher, the functions of the garden 'rooms' change. I was determined to obtain high-resolution digital scans in order to study how the garden had evolved prior to my first visit in 1990. MIT Libraries agreed to scan these transparencies, along with a selection of photographs from 1980 to 2007 in the possession of SIA's daughter, and to make them available in their online collection of architectural images.[13] Once the photographs arrived at MIT, I began the task of curation. The large collection from the 1980s to the 2000s consisted mostly of thirty-five-millimetre transparencies, prints, and negatives, which SIA had inserted into plastic holders and organized into binders by time period. Initially, the goal was to select a few hundred photographs that would be useful to me and other scholars, but, after sifting through thousands of images, I realized that the material was far richer than I had imagined. There was the potential to go beyond a static collection and try something new.

'Marnas': A journey through space, time, and ideas

Might it be possible, I wondered, by linking hundreds of images, to chart paths through the garden, a virtual recreation of Marnas? Working with the garden plan of 2008 and from memory, I selected images to create those paths and filled the gaps with my own photographs from 1990 to 2012. To convey the intangible—the garden's magical qualities and the ideas behind the forms—I composed a series of short multimedia videos from many still photographs, which explored the interplay among theory, practice, form, and function, providing tours of ideas in place. The videos could 'show' ideas because SIA had used photography as a medium of research, as had I. Our photographs freeze a moment, but they also capture thoughts. We had photographed picturesque views and recorded change, but we also probed more deeply. The photographs focus on significant details, features that convey SIA's practices, ideas, and aspirations (Figure 10.2).[14]

The Marnas website[15] invites the visitor to embark on a series of journeys. It not only transports the visitor through space and time, but also aspires to evoke the magical quality of Marnas and bring alive the ideas that shaped it. Like Marnas, it has both a clear formal structure and one that is freely growing, with no one beginning, and no single end. The homepage unfolds via a vertical scroll through a series of gateways. One gateway invites you to take a guided tour of the garden, others to explore on your own, meet Sven-Ingvar Andersson, or travel in time. The goal was to stimulate exploration and interaction, the viewer an active participant, not a passive observer (Figure 10.2).

'Take a tour': The multimedia videos

Take a tour through the garden with Sven-Ingvar Andersson as guide. Hear his stories. Watch the garden change through seasons and years. Follow the ideas that drive the form. You can choose among seventeen short episodes or view the full-length tour (twenty minutes), which is a composite of the separate tours. First, you meet Seven-Ingvar Andersson, hear his words ('My garden is a green den with a little house, the remains of a small farmstead, where, in summer, I live a simple life') against a background of birdsong. You see in succession: SIA standing, enveloped by green foliage; the old half-timbered house and cobbled courtyard; SIA in the kitchen garden, planting yellow yarrow (sound of digging); SIA by an espaliered pear tree, threading a tiny, newly formed pear through the mouth of a bottle and securing it with twine; later, the mature pear in a bottle of brandy; a small round table in the garden with the remains of lunch, SIA emerging from the house in the background. Then SIA continues with his introduction: 'The first time I saw the place, which Marna, an old farmer's wife, left after her death in 1951, it was a wildflower meadow, the kind that grows up on abandoned land. Gradually, the meadow became a lawn enclosed by the garden's old trees.' And you see: the old house as it was in the 1950s, the courtyard overgrown; Marna in a kerchief, standing alongside the road, against distant fields; the shadow of an apple tree cast on clipped grass; the trunk of an old pear tree and fallen pears on grass and cobbles.

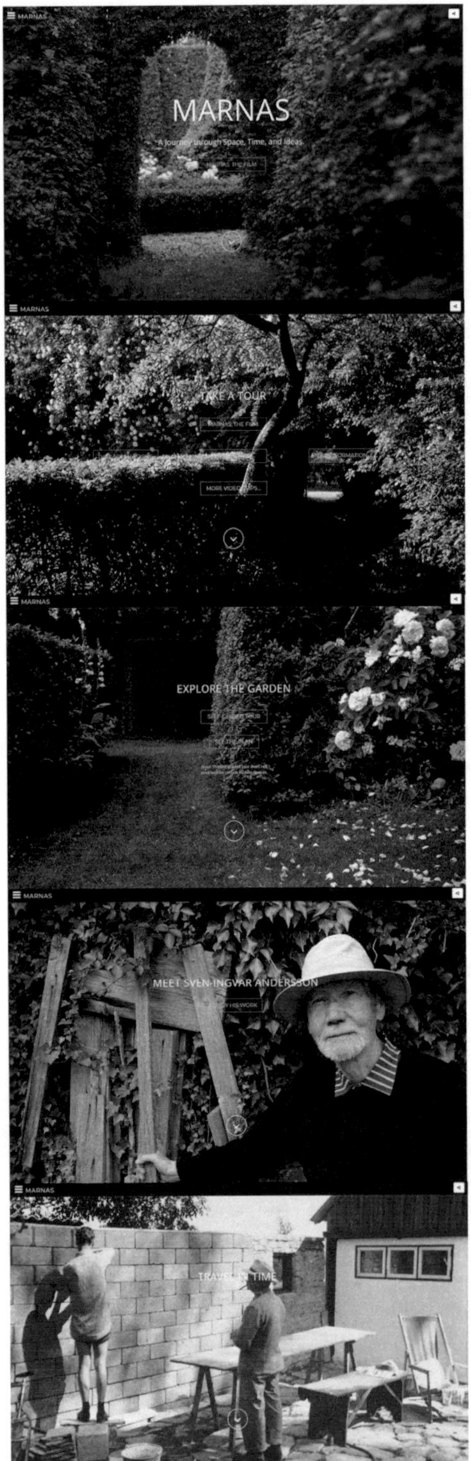

Figure 10.2 Marnas Homepage with Gateways, 2017. © Spirn, Anne Whiston.

Thus the long tour begins. In the first fifty-one seconds, you have seen twelve images, some for only two to three seconds, the longest for six. And thus it ends, eighteen minutes later: 'The inevitability of transformation sets garden art apart from all other art forms. My garden develops, culminates, tires, and deteriorates in step with me. The garden also shows life's natural vitality—and the sure way to the peace of death—which makes way for new life.' Four series of photographs appear in succession: purple allium and red tulips enclosed by curving borders of boxwood fade to the same view in autumn; a circle of purple allium fades, replaced by a circle of white flowers; a view through hedges to a small topiary hen fades to the same view many years later, the hen now twelve feet tall; a succession of five views moves forward in time and backward in space, down a long leafy passage. Then the final words: 'It will take fifty years more for my garden of hawthorn to reach the planned goal. By then, I will no longer be here.' An image of SIA, seen from behind, seated in the same passage but facing the opposite direction, fades out to a video of the same place, as words, superimposed, fade in: Sven-Ingvar Andersson 1927–2007.

This ground-level view, looking down the garden's main passage, with grasses and leaves moving softly, opens and closes the twenty-minute film. To capture a magical quality, I had placed the camera on the ground: a 'Thumbelina' perspective, as in Danish Hans Christian Andersen's fairy tale, where plants are larger than life. At the beginning of the film, this shot signals that Marnas is no ordinary garden. At the end, when the still image of SIA fades out into the video, the moving leaves have an emotional impact that a still photograph would not. SIA is gone, but his spirit remains. Life is eternal. The haunting notes of Edward Elgar's piano music intensify the mood.

Multimedia affords the opportunity to bring image, word, and sound together in ways that animate the text and bring to life SIA's dialogues between ideas and place. Words convey ideas; images show how those ideas are embodied in the features and spatial experiences he created at Marnas. Sounds lend immediacy and intimacy. Multimedia is ideal for animating photographs to enhance a sense of passage through space and time. The seventeen short videos, some less than a minute in duration, the longest just over two minutes, offer diverse journeys through space, time, and ideas. The pace is varied, with most images on the screen for just a few seconds, to hold the attention of Internet users.

In 'Hawthorns' (01:27), that species appears as clipped hedges, topiary hens, and a freely growing grove. 'Hawthorns permit great variation. But not beyond those limits which lie in being a hawthorn.' 'Always proceed with a respect for their existence as living organisms,' SIA advises, and they 'can adapt to whatever circumstances are introduced'. And don't forget that 'humans, like hawthorns, are very adaptable, but still bound to their genetic structure. In our genetically fixed pattern, there is need for stability in our physical surroundings, as well as the need for freedom, diversity, and individual development. The task, then, is to find the balance.' A succession of photographs shows how he did that.

SIA reflects on sources of inspiration. He tells how 'the entrance to my house is completely hidden, so you need a clue for how to find it.' He was inspired by the villas of Palladio in Vicenza to mark the entrance with four willow trees, their trunks like columns, to form 'a loggia'. A distant view of the Villa Rotonda and the loggia's columns follows an image of the four newly planted trees ('The Willows', 01:20). 'Eventually, the willows' large branches threatened the roof,' so he 'cut off the side branches'. The idea to prune the willows like that was inspired by Meindert Hobbema's painting *The Avenue at Middelharnis* (1689). Where did the idea for the hawthorn hens come from? SIA says that he had long been fascinated by Max Walter Svanberg's erotic art. 'Birds with long necks. Majestic animals in a slow gliding movement. I imagined them wandering around in my garden. Looking out over the hedges': one sees drawings by

Svanberg morph into the long-necked hawthorn hens ('The Hens', 01:40). SIA also draws parallels between design and literature: 'I use poetry's methods. Like the juxtaposition of the small and the large. Daisies and pyramids' ('The Pyramid', 01:03). 'My garden is not a mystery novel,' he says, 'but there is a similarity in the way I organize experience. There are well-placed obstacles, detours and delays, which make you more alert' ('Mysteries', 01:38).

When 'ash saplings shot up and threatened to eliminate the view,' SIA 'cut off their tops. The shoots became trunks that pushed each other outward', forming the ideal foundation for a treehouse ('The Treehouse', 00:50). When snow drifted and pushed over a hedge, he adapted it. 'It became a living hedge, with alternately sharp and soft waves, the arbitrary distance between them determined by the snowstorm. A mistake opened up a path to renewal' ('A Path to Renewal', 01:03). When burning hedge clippings, 'fire burned away the grass, and a circular form arose on its own. With the help of string and pen, I ennobled the bare surface to a precise circle' ('Fire', 00:40); he then dampened the ashes and shaped them into a cone, like a cone of sand in a Japanese temple garden. For SIA, inspiration was often a response to circumstance.

'How much of life is determined by obscure subconscious intentions? What forces beyond the artist's own affect artistic creation?' SIA had set up a sculpture made of branches 'to mark the garden's longest sightline. The portal draws one forward.' 'It must mean something', people said. 'Those irregular branches look like bleached bones. Only then did I understand what I had done' ('The Portal of Death', 01:25). One of Edward Elgar's piano pieces 'explained' for SIA another part of the garden, a small circular pool with a dark bottom ('The Black Eye', 00:50). Multiple images of the pool flow into one other, ending at a video of the moving surface, as you hear Elgar's 'In Smyrna' for piano and SIA's reflection on how he knows the Black Eye is only five centimetres deep, but can imagine it as 'a bottomless well. A depth where secret thoughts and conflicting feelings are entwined like serpents, which rise and fall under the sunlit surface.' As SIA observed of Elgar's composition, 'although the music was gay and glittering, it had a magical depth.' Like the Black Eye. Like Marnas as a whole. Some meanings are intended from the beginning, others emerge over time.

The scripts for the videos are entirely in SIA's own words, drawn primarily from his book *Brev från min trädgård* (*Letters from My Garden*, 2008), a collection of short 'letters' to family, friends, colleagues, and mentors, living and dead, which was inspired by *Letters from My Windmill* by Alphonse Daudet. I selected and translated passages where SIA reflects on Marnas, its precedents, origins, elements, composition, plants, and meanings. I also looked for passages that reveal his personality (deliberate and intuitive, controlled and improvisational, whimsical and playful) and his love of literature, art, and music. I composed these excerpts into seventeen parts, which originally were intended to be viewed separately, encountered in the self-guided tour as the visitor wanders around the virtual garden. Ultimately, I realized that the short videos could be sequenced to create a story arc, beginning with the dream of a green den and ending in transformation.

Script in hand, it was then a matter of selecting and sequencing photographs, video, and sound to support and extend narrative and tone, establish rhythm, portray meaning, and evoke feeling. The viewer should feel as if SIA is showing them around his garden. 'Show, don't tell' is a principle of multimedia storytelling. Images give substance to words in the Marnas videos. Sometimes they punctuate the words, adding emphasis; sometimes they fill long pauses. In 'Portraits in Time', SIA says: 'My garden is a portrait of a person, a signal of the time, a mystical adventure, a gauntlet in time and space with a multitude of philosophical ideas, biographical notes, flashbacks and references in a constantly interrupted narrative.' The pace is slow, giving the viewer twenty-five seconds to absorb eight images, which convey each of these aspects of Marnas. We used music sparingly in the videos, preferring to add sounds, such as those of digging, clipping, and wind, which were implied or alluded to by the images or text. I found recordings of birdsong for six of the species that live in the garden.[16] Marnas is a paradise for birds, whom SIA addressed as 'winged friends'.

'Explore the garden': The self-guided tour

The self-guided tour begins at the entrance to the driveway, where four willow trees form a gateway (Figure 10.3a and b). Through trunks of trees, you can see Sven-Ingvar Andersson standing near his old half-timbered house. Apart from the photograph, which fills most of the screen, four features draw your attention. Each affords a choice. Touch (or click on) the symbol at the upper left, and it expands into a navigation bar, which leads to other parts of the website. Touch the glowing rectangle in the centre of the screen and see a short video, 'The Willows', which tells the story behind the threshold where you stand. Touch the symbol at the upper right to call up a plan of the garden, which notes your position. Small squares on the plan mark different locations; touch one to move there directly. The compass on the lower right tells you that the year is 1998. Its arrow points to the way ahead; touch the arrow to step forward. The view is now up the grassy driveway to the west. The year is 1992. A glowing square at the centre of the screen alerts you that the same spot can be seen at another time. Touch it and see that in 1979 the tall willows that lined the driveway in 1992 are saplings; the hedge at the garden's western boundary does not yet exist. The compass now offers three arrows, each pointing in a different direction: turn to the left, step back, or step forward. If you choose to step forward, you are in 1995, and the compass offers a choice of six directions: you can look left (south) alongside the house, turn south-west toward the entrance to the southern part of the garden, go straight ahead up the driveway, turn north-west toward the henyard, or turn right (north) into a leafy passage.

As you wander around the garden, sometimes the compass offers many ways ahead, at others only one. You progress, slipping in time, backward and forward across years. From time to time, you encounter a glowing rectangle or square, which provides either the same view at an earlier time or different season, or a video story about the ideas behind the place where you find yourself. The slippage in time, experienced in the interactive tour, mirrors the experience that I have at Marnas, and which SIA most certainly had. When I look down a passage or into a garden room, I see not only what is there before me in the present, but also remembered glimpses of past appearance. If you get lost, call up the plan in the upper right to see where you stand. Your perspective varies. You may look straight ahead in one image, then down at the ground in the next. Occasionally, you encounter SIA: planting, weeding, standing in the henyard, sitting in the gazebo, in his house. In self-guided tours, soundtracks become repetitive and distracting, so there is no sound.

Creating the tour posed challenges in image selection and sequencing, coding, and design. The first step was to select and sort SIA's photographs by garden section: house, guesthouse, patio, orchard, and gazebo to the south; driveway in the centre; henyard, flower garden, sun court, Beata's garden, place of fire, kitchen garden, compost, and seven north-south and east-west passages to the north. The next step was to take each section, one by one, to sequence photographs to create paths through it, and, finally, to connect these many pathways into a continuous network. There were gaps. Most were filled by my own photographs from 1990 to the 2000s. A few others remain, where the visitor encounters a blurred image and the announcement: 'You have entered a gap in space and time.' The whole process was complex and required many iterations.

While composing the tour, we also worked on a prototype for the code and design. For example, you should be able to 'walk' through the garden at your own pace, which means

Figure 10.3a, b Self-Guided Tour Screenshot, 2017. © Spirn, Anne Whiston.

that many images need to load quickly, in all possible directions from your current location, in preparation for whatever move you select next. The design of the compass went through many versions. As you move forward, to the left or the right, it rotates so that the direction you are facing is always at the top.

'Meet SIA' and 'Travel in Time'

After touring Marnas, the visitor may be curious to learn more about its designer/gardener, Sven-Ingvar Andersson. 'Meet SIA' is a gateway to his work, with links to selected writings, drawings, and professional projects. Andersson wrote hundreds of articles and monographs, which are difficult to find outside Scandinavia. With more essays and drawings, the website could become a digital archive of SIA's work. For the scholar or student who wants to study the garden's development systematically, another section of the website, 'Travel in Time', groups photographs in chronological order from the 1950s to the 2000s. 'Meet SIA' and 'Travel in Time' are organized like a typical archive. They provide access to the work, but do not bring it alive. However, there is potential to animate both in the future.

The Marnas digital archive, landscape architecture, and the digital humanities

The Marnas project employed multiple methods, both traditional and less conventional, from the scholarly and curatorial, to photography and the practice of website and video design as media of inquiry. As a scholar, I published a comprehensive, annotated bibliography of SIA's publications (nearly all in Swedish and Danish), wrote an essay on his ideas ('Texts, Landscapes, Life'), and translated his manifesto, 'A Letter from My Henyard', into English.[17] As a curator, I studied the thousands of photographs SIA took of Marnas and selected more than 600 images for this project. As a photographer, I documented changes in the garden, from my first visit in 1990 to the most recent in 2017, and sought to capture the ineffable, seeking appropriate light and significant details that carried metaphorical meanings, alluding to larger ideas.[18] As a landscape architect, I deciphered how SIA transformed ideas into the reality of three- and four-dimensional space, and recorded my field observations. My own photographs and fieldwork, along with contemporaneous journals, made it possible for me to determine the locations of SIA's photographs and to create the virtual paths through the garden. Previously, I had intended to write a book about Marnas that would portray SIA's ideas and their expression in his garden laboratory, but the book became instead multimedia videos and a multilayered website.

This project extends in significant ways my exploration of photography, multimedia, and website design as media of inquiry. In 1995, I was an early adopter of the Web as a creative medium for publishing and teaching. The website for my book, *The Eye Is a Door: Landscape, Photography, and the Art of Discovery*,[19] experimented in 2012 with interactive 'journeys'. In designing the Marnas website as a three-dimensional and four-dimensional spatial experience, innovative websites such as 'Highrise: The Towers in the World, World in the Towers' were an inspiration.[20]

The Marnas project exemplifies several aspects of the digital humanities, a field at the intersection of digital technology and humanities disciplines. The interactive experience of a landscape across time would be impossible without digital technology. Digital cameras, video editing software, and open-access online archives of images, music, and sounds facilitate production. The website itself is a digital archive, which draws from Dome, the MIT Libraries' open-access digital collections of images.[21] MIT Libraries collections in architecture, urban planning, and visual arts are also part of a larger online archive, SAHARA (Society of Architectural Historians

Architectural Resource Archive), 'a collection of over 100,000 images of architecture and landscapes contributed by [...] architects, scholars, photographers [...] and others who share an interest in the built environment [...] for all who teach, study, interpret, photograph, design, and preserve the built environment worldwide'.[22]

'To be fully felt and known landscapes must be experienced *in situ*; words, drawings, paintings, or photographs cannot replace the experience of the place itself, though they may enhance and intensify it.'[23] I wrote that in 1998, but that was before the technology for visual reality video was invented, when the capacity of the Internet to stream high-resolution photographs, videos, and audio recordings did not exist. It is no longer possible to visit the 20,000-year-old cave paintings at Lascaux—they are too fragile—but one now can move through the cave virtually and still feel wonder. The interactive video tour on the Lascaux website is no replacement for direct experience, but it is far better than looking at an assortment of individual still images.[24] The Marnas self-guided tour is no substitute for being in the place, but it permits a peek into gardens that once were, an experience afforded only through digital media. It is difficult now to imagine how fundamentally the invention of photography changed the experience and understanding of space and time. It suddenly became possible to see what until then could not be seen, to compare places and things distant in space and time, to stop time.[25] Today, multimedia and the Web have extended photography's power to enable new experience and understanding of landscape.

'Landscapes are a vast library of literature. [...] The library ranges from wild and vernacular landscape, tales shaped by everyday phenomena, to classic landscapes of artful expression, like the relationship of ordinary spoken language to great works of literature.'[26] We now have the tools to visit that library, virtually. The great potential that multimedia and the Web afford for the portrayal of place in space and time, and as an interplay between ideas and form, has barely been tapped.

Acknowledgements

This project would not have been possible without generous support of many individuals and institutions. Beata Engels Andersson, Steen Høyer, and the Royal Academy of Fine Arts made the photographs available to MIT Libraries, where Patsy Baudoin, Jolene de Verges, Christopher Donnelly, and Lorrie McAllister arranged for scanning and cataloguing. A grant from MIT's HASS Award funded two superb research assistants: John Moody, with his gift for multimedia video production, and Zhao Ma, with his talent for design, coding, and web development. Halla Moore designed and coded the original prototype for the self-guided tour.

Notes

1 Sven-Ingvar Andersson, 'En trädgård efter fem år', *Hem i Sverige* 1 (1965): 27.

2 Sven-Ingvar Andersson, 'Brev fra min hønsegård', *Arkitekten*, 1967; 'Letter from My Henyard', translated by Anne Whiston Spirn, http://web.mit.edu/spirn/www/newfront/henyard/henyard.htm.

3 Sven-Ingvar Andersson, personal communication, 29 May 2006.

4 Sven-Ingvar Andersson, 'Våra gamla trädgårdar och den nya tiden: bevara, förnya, förändra', *Lustgården*, 1958–9, 5–18; Andersson, 'Pastisch och antikvarisk restaurering', *Havekunst*, 1961, 33; Andersson, 'Principper for bevaring', *Landskab* 5–6 (1993): 112–21; Andersson, 'Uraniborgs renässansträdgård: restaureringsprojektets förutsättningar och hypoteser', *Stencil* 7 (1993): 13–20.

5 Andersson, 'Principper for bevaring', in *Sven-Ingvar Andersson*, ed. Steen Høyer (Arkitektens Forlag, 2002), 123.

6 Ibid., 122.

7 Sven-Ingvar Andersson, 'Söndagslanskap och måndagsstäder', *Havekunst* 7 (1966): 128.

8 I had published 'The Poetics of City and Nature: Toward a New Aesthetic of Urban Design' (*Landscape Journal* (1988)) and was writing *The West Philadelphia Landscape Plan: A Framework for Action* (Philadelphia: University of Pennsylvania, 1991). Marnas would later play a central role in one chapter of my book *The Language of Landscape* (New Haven: Yale University Press, 1998).
9 Andersson, 'Letter'; Sven-Ingvar Andersson, 'Häckar och höns: min torparträdgård', *Landskap* 57 (1976): 180–4; Sven-Ingvar Andersson, *Brev från min trädgård* (Malmö: Arena, 2008).
10 Andersson, 'Letter'.
11 Steen Eiler Rasmussen, Frans Lasson, Lisbeth Hertel, and Sven-Ingvar Andersson, *Karen Blixen's Flowers: Nature and Art in Rungstedlund* (Copenhagen: Christian Eilers, 1992), published originally in 1983. Blixen made bouquets with flowers from her garden, and Rasmussen photographed them. Many of these photographs of her bouquets, which resemble those of Andersson, are published in the book.
12 Andersson, *Brev från min trädgård*, 21.
13 'Marnas Garden', MIT Libraries, accessed 16 December 2017, https://dome.mit.edu/handle/1721.3/185535. The MIT archive consists of 736 photographs, of which 113 are my own and 623 from the collections of Beata Engels Andersson and the Royal Danish Academy of Fine Arts. In 2017, Beata found additional images, an album of black-and-white photographs from the 1950s and early 1960s, which are not yet part of the archive at MIT.
14 This approach is described in Anne Whiston Spirn, *The Eye Is a Door: Landscape, Photography, and the Art of Discovery* (Nahant, MA: Wolf Tree Press, 2014), published as an original e-book.
15 'Marnas', accessed 24 January 2018, http://www.marnasgarden.com/.
16 John Stearns Moody worked with me on multimedia development and was responsible for all aspects of production. Online sources of free music and sounds were an extraordinary resource. For example, xeno-canto, a website for sharing bird sounds from around the world, made it possible to locate sounds of specific bird species recorded within 100 kilometres of Marnas.
17 Anne Whiston Spirn, 'Texts, Landscapes, and Life' and 'Bibliography', in Høyer, *Sven-Ingvar Andersson*.
18 Spirn, *Eye Is a Door.*
19 'The Eye Is a Door', http://www.theeyeisadoor.com/.
20 Highrise, accessed 24 January 2018, http://highrise.nfb.ca. Google's 'Night Walk in Marseille' is another excellent example, which I discovered after producing the Marnas self-tour: Night Walk with Google, accessed 29 December 2017, https://nightwalk.withgoogle.com/en/home.
21 MIT Libraries Dome, accessed 10 December 2017, https://dome.mit.edu/.
22 'Sahara', Society of Architectural Historians, accessed 10 December 2017, http://www.sah.org/publications-and-research/sahara.
23 Spirn, *Language of Landscape*, 21
24 'Lascaux', Ministère de la Culture, accessed 10 December 2017, http://archeologie.culture.fr/lascaux/en/.
25 Spirn, 'Photography and the Art of Visual Thinking', in *Eye Is a Door.*
26 Spirn, *Language of Landscape*, 21.

Bibliography

Andersson, Sven-Ingvar, 'Våra gamla trädgårdar och den nya tiden: bevara, förnya, förändra', *Lustgården*, 1958–9, 5–18.

Andersson, Sven-Ingvar, 'Pastisch och antikvarisk restaurering', *Havekunst*, 1961, 33.

Andersson, Sven-Ingvar, 'Principper for bevaring', *Landskab* 5–6 (1993): 112–21.

Andersson, Sven-Ingvar, 'Uraniborgs renässansträdgård: restaureringsprojektets förutsättningar och hypoteser', *Stencil* 7 (1993): 13–20.

Andersson, Sven-Ingvar, 'En trädgård efter fem år', *Hem i Sverige* 1 (1965): 27.

Andersson, Sven-Ingvar, 'Brev fra min hønsegård,' *Arkitekten* (1967): 579–83. 'Letter from My Henyard', translated by Anne Whiston Spirn, http://web.mit.edu/spirn/www/newfront/henyard/henyard.htm

Andersson, Sven-Ingvar, 'Häckar och höns: min torparträdgård', Landskap 57 (1976): 180–4.

Andersson, Sven-Ingvar, *Brev från min trädgård* (Malmö: Arena, 2008).

Høyer, Steen, ed., *Sven-Ingvar Andersson* (Arkitektens Forlag, 1994).

Høyer, Steen, ed., *Sven-Ingvar Andersson* (Arkitektens Forlag, 2002).

Rasmussen, Steen Eiler, Frans Lasson, Lisbeth Hertel, and Sven-Ingvar Andersson, *Karen Blixen's Flowers: Nature and Art in Rungstedlund* (Copenhagen: Christian Eilers, 1992)

Spirn, Anne Whiston, 'The Poetics of City and Nature: Toward a New Aesthetic for Urban Design' *Landscape Journal* (Fall 1988).

Spirn, Anne Whiston, 'Bibliography,' in *Sven-Ingvar Andersson*, edited by Steen Høyer (Copenhagen: Arkitektens Forlag, 1994).
Spirn, Anne Whiston, 'Texts, Landscapes, and Life,' in *Sven-Ingvar Andersson*, edited by Steen Høyer (Copenhagen: Arkitektens Forlag, 1994).
Spirn, Anne Whiston, *The Language of Landscape* (New Haven, CT: Yale University Press, 1998).
Spirn, Anne Whiston, *The Eye Is a Door: Landscape, Photography, and the Art of Discovery* (Nahant, MA: Wolf Tree Press, 2014).

PART III

Urban stories from a green planet

Green stories from an urban planet

11

THE VERTICAL AND THE HORIZONTAL

Combining ethnographic and geographic methods in understanding landscape

Gareth Doherty

When I entered Rashid's office on my first day calling on the Ministry of Municipalities in Manama, on the desk on top of a pile of papers was a full-colour printed version of a report widely available on the Internet. This report, made soon after Google Earth became available in Bahrain, compared areas of the country with Bahraini royal properties and did not mince words. Comparing the densely populated capital city of Manama with the king's private island—of exactly the same size as the capital—the anonymous author asked: 'How many people live in Manama? And how many property owners are there? Who is allowed to enter the city? And what's the density of its population?' It goes on to say, 'Ask the same questions for this "Bahrain" island over here!'[1] The document, which was widely circulated by e-mail around the Gulf, contained screenshots made before Google Earth was banned in Bahrain (for a brief period, as it turned out). As one blogger put it, 'Pamphleteering doesn't get much more visceral than this (even if I have no easy way of verifying if it is true). If Bahrain's government wants to prevent the spread of this kind of information, it will have to ban email.'[2] This report's presence on a ministry desk indicated the seriousness with which online informal communication is taken.

I had expected that an affiliation with the ministry would allow me greater access to information and people. In general, this was not the case. While I made many good friends through the ministry, it took a few weeks for me to realize that being attached to the ministry included both obligation toward the ministry and a bias, perceived or actual. And then there was the issue that the ministry was not really that powerful.[3]

In general I found that those in authority, while sympathetic to my presence—and this includes a friend from the royal family—did not seem used to being questioned about government policies. They either didn't know the answers to my questions or did not want to tell me the answers, although no one ever actually said no. A couple of people who were especially sympathetic to my work explained that their willingness to open up to me was hampered by the fact that they feared giving out too much information might jeopardize their positions.

'Do you want the official figure or the real figure?' was a regular question asked in the various ministries in response to my queries, and hardly anyone batted an eyelash at the irony of such a question being posed. Obtaining any sort of quantitative information could be taxing and frustrating. In one instance, at an unnamed institution, a senior figure explained to me that a huge discrepancy between the published figures and the reality originated from a desire to please the rulers. Once the rulers became aware of a figure, exaggerated by a civil servant to make himself and his team look better in the eyes of the authorities, the organization became trapped in a set of double figures. There were those figures that were presented to the rulers and those that were used within the organization to try to catch up with what they were declaring. Other sets of double figures persist more pervasively.

When I enquired at a ministry in Manama about the square metres of green space in Bahrain, I was told that the official figure was both freely available and unreliable; meanwhile the real figure could only be obtained on my behalf through the written request of the president of my university to the minister. Furthermore, I was told that even if the president of Harvard wrote to the minister, she would probably not receive an answer. The complexities of the cultural conditions in Bahrain add to the complexities of gathering information. In general, I found that the majority Shiʿi[4] population was receptive to my presence and much more open with me than were the Sunni-controlled ministries. As the literature on the urbanism of landscape was so thin for Bahrain, it became clear that the broad range of data I needed for this research project could be obtained only from a long-term period of ethnographic fieldwork. I would need extended engagement with the location and its people to gather the data, qualitative and quantitative, that I needed.

Methods

Having secured funding for a year of fieldwork in Bahrain and the Gulf, I developed a schedule for the first few months. I wanted to confirm my plan's usefulness and asked some eminent professors for guidance on my methodology. The professors told me, 'just do what you need to do.' In other words, don't think too hard. It took me some time to realize and appreciate that this was not an evasion of the question but actually really good advice. I went to Bahrain and started doing. It was only when I was in the field, for instance, that I realized how important walking would be, as it increased the possibilities of serendipity that most anthropologists depend on and brought me into contact with many Bahrainis outside of my regular social orbit with whom I could engage.

Once in the field, I found that even the best-laid plans did not always work out as hoped. I constantly had to improvise, to abandon work I had been doing, to keep multiple lines of enquiry open, and hardest of all, to be patient. One of the biggest challenges I faced was that things took time, although being Irish I was already used to 'creative' schedules for things such as buses. When I asked an Irish bus driver once why the bus was always late, he quipped, 'If we left on time, everyone would miss it.' The expression *inshāllah* (a colloquial form of *in shāʾ Allāh*, meaning 'God willing') has many tonalities, each of which implies a different degree of commitment. Until I learned to appreciate and interpret the significance of these nuances, I found that the driver did not show up or promised meetings did not happen. The standard explanation for something not happening in the way that it was supposed to was that it couldn't have been God's will. God's intentions, it turned out, can be somewhat anticipated by noticing the different emphases between *inshāllah*, *inshāāllah*, and *inshāāāllah*. The longer the *ā*, the more likely something was to happen.

My expectations for my time in Bahrain were to live among Bahrainis, to meet locals as much as possible, and to speak Arabic on a daily basis—a plan that turned out to be incredibly difficult for many reasons, never mind the fact that there are so many local social groups and voices to be heard. Bahrain's polyvocality is evidenced by the plethora of accents in Bahrain; often they differ markedly even within families if children go to different schools or socialize with different social groups.[5] I had a romantic notion similar to that of Bronislaw Malinowski, one of the founders of ethnographic fieldwork, who advocates an immersive experience but describes an easy division of observer from observed, which is not always so easy in the field. Malinowski writes that the proper conditions for ethnographic fieldwork 'consist mainly in cutting oneself off from the company of other white men, and remaining in as close contact with the natives as possible, which can only be achieved by camping right in their villages'.[6]

Camping in Bahrain's villages would be useful only if the villagers camped too. Most Bahrainis live behind high walls, and it is unusual for expatriates such as myself to be invited inside. The lack of interaction between Bahrainis and expatriates is often attributed to issues surrounding women and modesty, but clearly there is more to it than that. In any case, I was surprised at how isolated my life was from what I expected local life to be. It was not at all the experience I was expecting; indeed my day-to-day life was in general not that different from my existence in the United States. This was due both to the impenetrability of Bahraini society to expatriates and to the fact that the Bahraini lifestyle is in many respects quite westernized. So when I did get invited behind the high walls, I was sometimes surprised at how life within was not that different.

When I met my Bahraini friends for lunch or dinner, I usually wanted to eat traditional Bahraini street food such as the most delicious local *tikka*—finely chopped cubes of lamb marinated overnight in dried lemon, which gives them a black colour, and grilled on a skewer. It is eaten with fresh *khubz* (bread) and raw green onion leaves. My friend Isa, the US-educated civil servant with a deep love for Bahrain and Bahraini traditions, introduced me to '*wāḥid dīnār tikka*' (*wāḥid* means 'one'; one dinar is about US$2.65). Except for Isa, my Bahraini friends and informants steered me toward fancy restaurants such as Monsoon, which they would describe as 'the best Thai restaurant in Bahrain'. Indeed, on one occasion when I returned to Bahrain for a short trip, all I wanted was to go find *tikka*, whereas my friends wanted to go to a posh restaurant with 'international cuisine', which also included pork and alcohol. They probably thought I would enjoy that. They preferred to showcase their cosmopolitan bona fides, but all I wanted was Bahraini food. One of the few restaurants near my home was Dairy Queen, the US fast-food restaurant, and although I initially tried to avoid eating there, I wound up being a regular customer known to the mostly Filipino staff. It had a largely student clientele, due to the proximity of the campus of AMA University. Ali, a younger brother of my friend Isa, said he often went to Dairy Queen in the early hours of the morning with his friends.

I did get invited into a few Bahraini homes during the course of the year, sometimes to a *majlis* and sometimes to the living area, depending on how traditional the family was. I was very fortunate in that Isa's family regularly had me over for meals, including Friday lunch. Following Friday prayers, this was the main family meal of the week. Soon after I arrived in Bahrain, Umm Isa invited me over for Friday lunch. *Umm* means 'mother of'. She told me to treat their home like my own. 'This is your house', she told me. 'Come over whenever you feel like it. It is your home.' Umm Isa's hospitality, I later learned, was not the norm, and it provided me with a welcome degree of security and stability in a sometimes harsh environment. Perhaps because they were of Persian descent, Friday lunch almost always included rice, yellow from saffron, mixed with pomegranates and crusted from oil on the bottom of the pan, as well as local fish such as *hāmūr* (grouper), *chanʿad* (mackerel), or *ṣāfy* (rabbitfish). Isa's family home became a second home for me, and I spent many hours there, although the distance—on the other side of the island out of walking range from my lodgings—always made it a little bit hard to get to.

My work became a multilayered ethnography, an ethnography based on seemingly disparate interviews and casual encounters, walking, photography, formal analysis of built projects, and some archival research. I studied green on a daily basis in public spaces, gardens, observing religious practices, government ministries, politics, and so forth. I would meet my core group of friends and informants on a regular basis, but they lacked social engagement with one another, and this led to a narrow, affiliation-based contact with the people I intended to study. In other words, I did not have access to any one pre-established community or wider extended family or group of friends in Bahrain; instead, I constructed my own. I interacted with a diversity of people and sites dispersed across the city, connected by green as discussed and practised on a daily basis. I came to call this interaction multilayered ethnography.[7]

In due course, I decided to accept the unpredictability of life and to treat it as a positive thing rather than an inconvenience. It reminds me of one relative who says she makes no plans because when she does, they never turn out as planned, so she has learned not to make any plans. I learned not just to deal with chance and serendipity as they arose, responding passively to whatever was thrown at me, but to actively engineer the likelihood of chance occurrences. Here, walking became critically important. My whole year was informed by walking and by the routes I took. Walking for long distances, I should add, is not a practice many people do in Bahrain during the daytime, due to the hot sun and a lack of pathways. I became a familiar sight to some interlocutors, who often told me that I was the only foreigner they saw walking around the island. I do not want to overly romanticize the power of walking, although it became an essential part of my method, since I do not drive and public transport can be confusing in Bahrain. Daytime walking in Bahrain is delightful for more than six months of the year, from October to March or April; the oppressive summer heat can make it very difficult, though. That's why I did many of my walks after dark.

My fieldwork carried with it several methodological points relating to the gathering of data, interviews, and my own access and sense of belonging. For example, I was not expecting to have much interaction with expatriates, but they were an important constituency for engagement.

Alfie was a destitute British-born expatriate who somehow fell through a gap in the system that encourages expatriates to leave Bahrain once they reach the age of retirement. Alfie found himself in his eighties, out of work and not wanting to die in Bahrain. I met him at Nirvana, an Indian restaurant in the suq where I liked to eat my lunch, usually green pea masala, extra spicy. The clientele included mostly Indian expatriate businessmen. The price of about 900 fils (US$2.50) for a lunch was high, but it seemed clean. That is why I chose to eat there the first time, and I returned often. Alfie was the only other white person I met in Nirvana. Seated at an adjacent table, he joined me and talked at length. Alfie had one son in Berlin and another in Bangkok, and was trying to decide where to spend his final years. He had previously worked in the film industry, including one stint in Hong Kong working with Sean Connery. We talked about language and accents. Alfie told me that the worst English accent he ever heard was among housewives in coffee shops in Saar: what he described as a mix of a bland expatriate twang and a frontal Essex accent from outside London. Saar is an area to the west of Bahrain where expatriates live in villas on land that was, until recently, desert or date palm groves. A couple of weeks later, I told Isa, who lived in Saar, about my conversation with Alfie. While Isa agreed about the accents, he casually asked if I had written about this encounter in my fieldnotes. I was surprised to realize that I had not. 'Why?' asked my friend. 'Is it because he is English?' I had to agree. Alfie did not matter to me because he was English. After my meeting with Alfie, I listened much more closely to what non-Bahrainis told me. Bahrain's polyvocality was not something I was prepared for.

Language also presented a problem. The Arabic classes I had taken equipped me with skills in written and spoken Modern Standard Arabic, useful for eyeing documents at the ministries or skimming the newspapers. Arabic helped me among the Baharna, a social group that I really only started meeting later on, during the Ashura religious festival. The governing classes were my primary interlocutors at first. Bahrain's business language is English, most people speak English, and the people I encountered in my everyday life—such as the laundryman, the tailor, and

Mr Kareem from the corner store (called *barrāda*, 'cold store' in Bahrain), who hailed from Kerala, which he said was very green—usually spoke English to me, their native language being Urdu, Hindi, Malayalam, or Tagalog. So, Arabic was in fact of limited use, and those who spoke Arabic would find my Arabic 'charming' and insist on speaking English to me for anything meaningful.

I lived and conducted fieldwork in Bahrain primarily during 2007 and 2008 and made several field visits before and after this time.[8] A few books I was reading at the time were especially influential in my thinking. First, I incorporated methods such as found in Farha Ghannam's *Remaking the Modern: Space, Relocation, and the Politics of Identity in a Global Cairo*, an ethnography about a new housing development constructed on the edge of Cairo.[9] Like Ghannam, I had the luxury of a year of fieldwork in one place. I also embraced the multilayered strategy of Diane Singerman and Paul Amar's *Cairo Cosmopolitan*, a collection of essays that I came to regard as almost a 'multi-authored' ethnography of Cairo.[10] I was also inspired by the combination of written and visual narrative in Orhan Pamuk's *Istanbul*, in that Pamuk's text is punctuated with melancholy historical photographs that, rather than illustrating facts, add to the mood and atmosphere of the book. I also closely consulted Steven Caton's *Yemen Chronicle*, an 'ethnomemoir' of his fieldwork in Yemen.[11] Caton's book, written twenty years after his original fieldwork, prepared me for some of the challenges of the field and what some might call the fieldwork blues.[12]

It was comforting to know of someone else's struggles, emotional and practical, with being in the field. I tried, for instance, to emulate Caton and maintain both analytic fieldnotes and a personal journal but was unable to maintain this duality, focusing instead on fieldnotes. Caton finds his journal a useful comparative tool to read in conjunction with his notes. I have found this more personal aspect represented in e-mails I sent to friends from the field, which became an important source for me as I analysed my time in Bahrain.

Predominantly, I follow the more traditional fieldwork of ethnographers like Ghannam, inasmuch as I was based in one site, and my notes are informed not just by what people said to me but also by the place as well as my personal bias.[13] Yet my research differs from Ghannam's significantly in two ways. The scale of the location is bigger: Bahrain is a nation-state of some 700-plus square kilometres (Figure 11.1), much larger than a single housing development on the fringes of Cairo.[14] But of more significance is that Ghannam had intimate contact with a family and social group with whom she spent most of her time and interacted almost daily on a deep personal level. Ghannam spent her time embedded within a pre-existing community that intimately informed her ethnography. If I had done this, my study would have been very different, perhaps an ethnography of green among one social group—say, the Baharna—or among the ʿAjam, or Keralites, or Western expatriates, rather than an ethnography based on a colour across the whole city-state of Bahrain.

The anthropologist Clifford Geertz, in *The Interpretation of Cultures*, suggests that loci cannot be taken as objects of ethnographic research, stating that 'anthropologists don't study villages (tribes, towns, neighborhoods); they study in villages.'[15] This line of argument ignores the fact that objects and things may have a social 'life' too, and as anthropologists move beyond the study of people to the study of the relationality of things, an extensive body of literature emerges on the social life of non-humans such as trees, movies, and everyday objects.[16] Geertz tells us: 'Man is an animal suspended in webs of significance he himself has spun.'[17] In a similar vein, this research began as a study of green in Bahrain, where the colour is an object, and evolved into an ethnographic study of landscape in Bahrain through the lens of colour. Critically speaking, landscape comprises land and the people who inhabit the land, as well as the relationships between them.[18] While it might seem strange that an assemblage of relationships could be

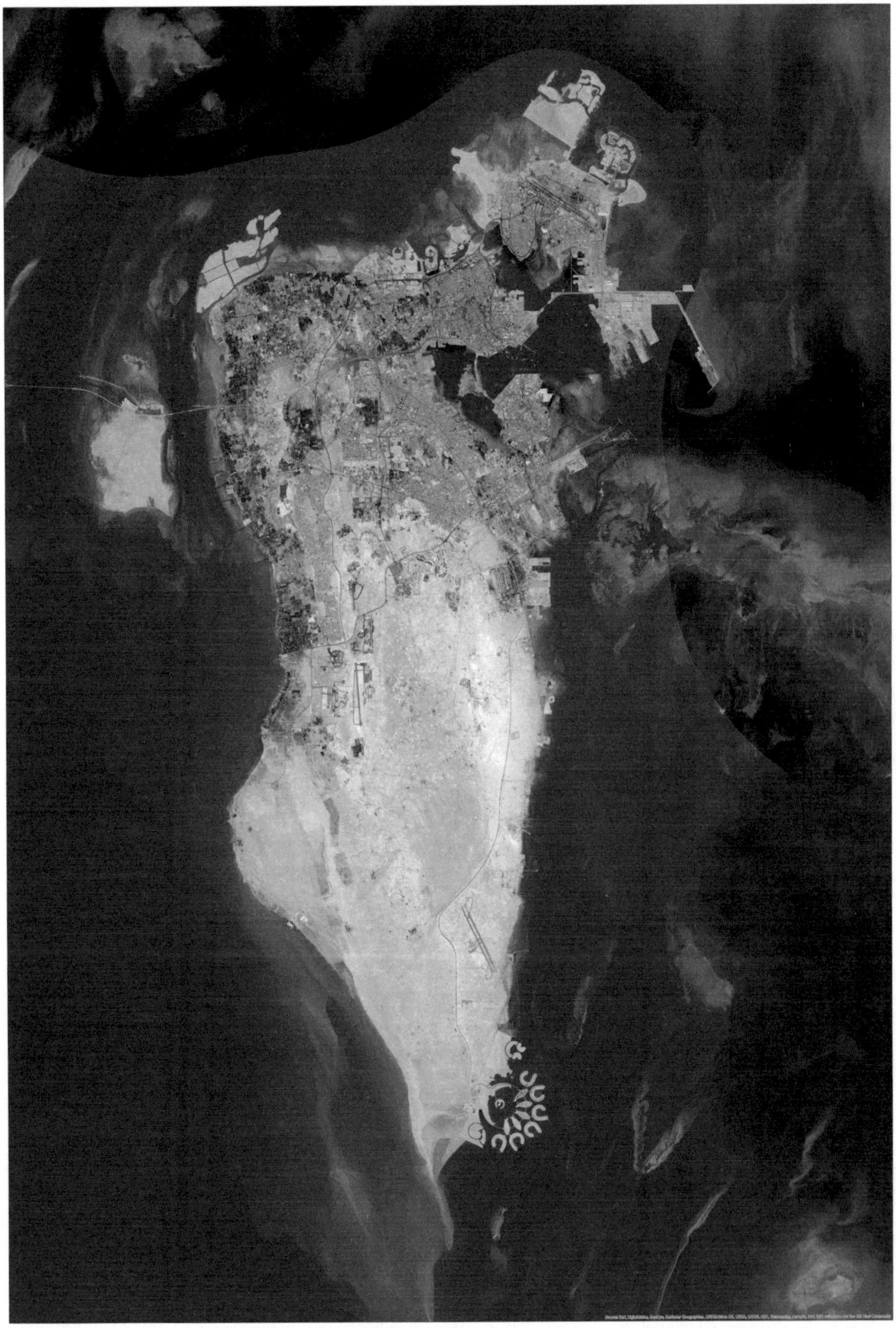

Figure 11.1 ESRI image of Bahrain, 2015. © Esri, Harvard University.

considered an interlocutor, others might agree that the landscape has the loudest voice of all. One of the ways that landscape speaks to us is through colour. For example, a brown, dried-out lawn could be saying it is thirsty, a grey-green, thriving date palm grove that it is content, and an evening red sky that a good day will follow.[19]

Walking

In the year of my fieldwork in Bahrain, my goal was to walk everywhere I needed to go. If landscape is the object, it is obvious that one needs to be *in* it. When one is inside or in a car, one is removed from the tactility of the landscape. In addition, when one walks, one has chance encounters in a way that does not happen in a car. Walking increases the opportunities for meeting people, and also for finding the unexpected and the strange. Ultimately, my goal was to make the strange familiar, to borrow from the oft-quoted phrase.[20] It is a phrase often used to explain anthropology: in making the strange familiar, and the familiar strange, we come to new knowledge regarding social patterns and relationships.

Once I had decided on the necessity of walking, my next and more difficult question was where to walk. I struggled with the routes for my walks. Does one just take the lines as drawn on the map? Does one use the seemingly arbitrary lines reflecting roads, laneways, and property boundaries? Or does one draw one's own line and follow that? Either way, I was thinking in lines. I was looking for patterns of green and landscape, and I thought the line would help to give structure to my walks. Inspired by the land artist Richard Long, I first drew ten straight lines running from the east to the west of Bahrain and determined that I would walk these straight lines, one per week.[21] The lines would cross various types of properties: public, private, wasteland, agricultural land, private housing, schools, and cemeteries, many of which are criss-crossed by walls, roads, and paths.

Walking these lines would not be intuitive or even necessarily legal; the aim was to enter diverse places that I might not normally encounter or might avoid. I expected to be introduced to people, landscapes, and green that I would not otherwise have access to. I would create something new in the process, connecting disparate parts via the lines I would draw on the map and walk by foot. While walking, I would record visually and with fieldnotes my interactions with green along the way. Another schema for conducting my fieldwork could have been to throw beans on a map and to go to the points on which the beans fell. A method inspired by Raoul Bunschoten and CHORA, the London-based design practice, this would at least in theory introduce a random sampling into the equation, also bringing me to places I might not normally visit. I did not follow either of these plans, not just for reasons of practicality and to avoid drawing too much attention to myself; my focus on green rendered these map-based schemata superfluous. The line was a distraction from the green.

Instead, I followed green. I had a number of methods for choosing the routes I would walk, and the aerial image coupled with intuition played the biggest part in selecting my routes. Basically, I walked wherever I needed to go. If I was going to a ministry in Manama to discuss green, I would walk for thirty minutes through the suq to get there. I recorded my interactions with green along the way, taking photographs and writing scratch notes in my pocket-size notebook and methodically noting details not just of the green but people's reactions to it as well. This meant starting conversations with strangers and being receptive to strangers starting conversations with me. The suq is not the greenest location, but still I would note the green vegetables being sold, the green winter *thiyāb* for sale in the tailor's, the green hubs on the car parked along the pavement, the green weeds poking through the pavement, the *mashmūm* (sweet basil) for sale at the cemetery on a Thursday evening. If I had to go to the ministry the following day, I would take a slightly different route.

Other times, I had to go farther than I could walk. On these occasions I would take a public bus, or Illias, my landlord's driver, would drive me. I followed the patches of green and the corridors I used to get there. I used the aerial image to help myself find the green areas, and I would set off to see those areas. In many ways my routes became a complex interplay between Google Earth, which guided me, (as seen in Figure 11.2) and the beholding of green scenery as seen from eye level (Figure 11.3). Rarely did my walks turn out as expected.

My study is based on a year of living and *walking* in Bahrain. The method of walking is fundamental to my understanding of Bahrain and intended to supplement rather than replace aerial reconnaissance, which was also an important tool in the research.[22] Peripatetics such as Francesco Careri, founder of the Italian Stalker/Osservatorio Nomade urban art workshop, describe walking 'as a primary act in the symbolic transformation of the territory'.[23] For me, the approach was much more pragmatic and designed to gather ethnographic data through the encounters I had with people, land, and colour. I met many of my friends and interlocutors through walking.

Abstraction and aerial

In their epic *The Landscape of Man: The History of Landscape Architecture from Prehistory to the Present Day*, Geoffrey and Susan Jellicoe position together one of Ian McHarg's maps of Philadelphia with a Jackson Pollock painting.[24] McHarg, a professor of landscape architecture and regional planning at the University of Pennsylvania, was known for his large-scale layered maps, which identified and separated various aspects of landscape such as soils,

Figure 11.2 Satellite image of northern Bahrain showing Manama to the right, 2016. © Google Earth Image © 2016 DigitalGlobe.

Figure 11.3 Villas in the greenbelt, 2008. © Doherty, Gareth.

hydrology, and so forth. Although the Jellicoes do not directly explain the relationship between the two images, the juxtaposition implies that spatial designers should take inspiration from the abstract arts (and vice versa), although the Jellicoes are careful to acknowledge that such interpretation is personal to every author. Concerned that humankind is moving into a phase where traditional understandings of space and time are no longer valid or in fact relevant, the Jellicoes advise their readers to look to the arts for a vision of the future, 'gaining confidence in the knowledge that the abstract art that lies beyond all art lives a life of its own independent of time and space'.[25]

This particular juxtaposition of McHarg and Pollock, the colours and application of the paint on the canvas informing how we read and live in the contemporary city, was the underlying inspiration for this research. Reading the aerial image of Bahrain as a painting, I was able to abstract the land use from the particular geography, infrastructure, or element of landscape. Then, in addressing the colour, I got a different view of the city than had I focused on the land use itself. The aerial view shows landscape in a certain nakedness.

Stretching back to Le Corbusier's fascination with the aerial image, since the rise of mass availability of aerial photography and indeed more recently Google Maps, Google Earth, and NASA images, designers have become used to reading the city as a collection of physical artefacts. There are many attempts at classification of the components of the city, such as Kevin Lynch's 'paths, nodes, districts, points and landmarks' and Stan Allen's 'points and lines'.[26] Richard T.T. Forman has a particular classification of landscape from the point of view of landscape ecology. For Forman, if people drop from a helicopter onto a city (again the aerial image is invoked), they will land on one of three components: a patch, a corridor, or the matrix. The matrix is the space that holds the patches and corridors together.[27] Thus, a city might be read as a collection of recognizable patches like houses, parks, and gardens; connected by corridors of infrastructures of roads, streets, and pavements; and all organized in a matrix, which are the leftover spaces that hold everything together in one (arguably) larger ecological system.[28]

The fascination with aerial photography has rightly permeated the design and ecological disciplines and offers an incredible tool to attempt to understand escalating urbanization and emergent urban, suburban, and exurban forms. Ian McHarg was of course an early pioneer, in *Design with Nature*.[29] McHarg's layered maps were a radical departure in their day and were primarily rational and scientific in their evaluations.[30] More recently, Charles Waldheim has been a vocal advocate of the importance of the aerial image. For Waldheim, aerial images make possible a new understanding of urbanism as a 'flatbed terrain' and 'horizontal surface'. Waldheim writes: 'New audiences and sites for work also offer the possibility of new formulations of landscape, recasting its image from green scenery beheld vertically to a flatbed infrastructure that includes both natural and urban environments.'[31] Waldheim challenges us to change our perspective on landscape from the traditional notion of green scenery seen at eye level toward one of landscape as a horizontal surface that makes no distinction between urban and rural, landscape and urbanism.

This research began as an attempt to describe the urbanism of landscape—as green scenery beheld vertically—in large part in response to the above quotation. In demonstrating the urban qualities of landscape, the aim is to take a slightly different approach to landscape urbanism, a disciplinary realignment including an influential body of literature and professional projects that have originated since the late 1990s. A basic tenet of landscape urbanism, again to quote Waldheim, is that 'landscape replaces architecture as the basic building block of contemporary urbanism.'[32]

Combining fieldwork—the vertical—with the aerial image (the horizontal) allows the opportunity for a 'thicker' reading of a landscape, and therefore is positioned to propose 'thicker' solutions that might ultimately be more successful.[33] (I should stress, however, that there is no linear relationship between fieldwork and proposition, as ethnography and design have very different epistemologies.) As the landscape ecologist Richard T.T. Forman affirms, 'Of course, dropping from the sky to examine the land closely is also essential.'[34]

For Forman, landscapes as read from an aeroplane are almost always a mosaic, and mosaics, as he points out, are coloured.[35] Yet as Walter Benjamin reminds us, the power of a country road—and presumably a city one too—is greater when walking along it than when one is looking at it from above.[36]

The vertical transect allows for a more intimate and nuanced reading of landscape than we can get from aerial images alone. 'Get out now. Not just outside, but beyond the trap of the programmed electronic age', urges John R. Stilgoe in *Outside Lies Magic*, a wonderful reading of the American landscape. Stilgoe urges the reader to 'Walk. Stroll. Saunter. Ride a bike, and coast along a lot. Explore.'[37] A city from above may look grey; from inside it can be dazzling in its colours: think of Times Square or Shinjuku. Indeed, when one is walking through the city, as the Urban Earth, a UK-based geography collective, do, it can be surprising how much green there is in the city outside the official category of green space. Why not behold landscape from both perspectives, from above and from eye level too? Large-scale geographies need to be understood ethnographically if we are not to lose touch with the people in those geographies.

Although the word *geography*, like *landscape*, is concerned with space and territory, geography differs from landscape in three significant ways. The first is inherent in the etymology of the word: *geo-graphy* means writing about the land as it is, recording its features and uses, whereas landscape overtly indicates a visual component. A second significant difference is the issue of scale: geography is not really tied to any one scale in the way a landscape is, or in the spatially hierarchical way the design disciplines are structured, from the broadest regional planning, to urban design, to landscape architecture, to architecture, to garden design, to interior design, and finally product design, all having a particular scalar focus.[38] We live in a multiscalar world—where the earth is not necessarily getting smaller, or bigger, just both—and geography liberates us from scale in a way that promises fresh insights into the study of that land. Aerial photography is a means to understanding that multiscalar end. Lastly, and significantly, geography inherently implies a social component. I suggest that designers need to rediscover people, and that ethnography offers a set of skills to engage with people. And doing so necessitates the beholding of green scenery vertically as well as horizontally.

Notes

1 'Bahrain and Google Earth', Ogle Earth, accessed 26 February 2016, http://ogleearth.com/BahrainandGoogleEarth.pdf.

2 Stefan Geens, 'Bahrain in Google Earth, Unplugged', *Ogle Earth* (blog), 21 September 2006, http://www.ogleearth.com/2006/09/bahrain_in_goog.html.

3 See, for example, Michael Lipsky, *Street Level Bureaucracy: Dilemmas of the Individual in Public Services* (New York: Russell Sage Foundation, 1980); Michael Herzfeld, *The Social Production of Indifference: Exploring the Symbolic Roots of Western Democracy* (Chicago: University of Chicago Press, 1993).

4 Arabic transliterations in this chapter are based on the conventions of the *International Journal of Middle Eastern Studies*.

5 See, for example, Mikhail Mikhailovich Bakhtin, *The Dialogic Imagination* (Austin: University of Texas Press, 1981).

6 Bronislaw Malinowski, 'Method and Scope of Anthropological Fieldwork', in *Ethnographic Fieldwork: An Anthropological Reader*, 2nd ed., eds. Antonius C.G.M. Robben and Jeffrey A. Sluka (Oxford: Wiley, 2012), 70.

7 See my review essay 'Review of *Cairo Cosmopolitan* by Diane Singerman and Paul Amar', *International Journal of Middle East Studies* 42, no. 4 (2010): 725–6. In it I describe the edited collection as a sort of multi-authored ethnography.

8 The research for the study included several trips around the Gulf to Kuwait, Iran, Oman, Qatar, Saudi Arabia, UAE, as well as Egypt. It is also based on several private visits to Bahrain extending back to my first visit in 2004. It was during the year of fieldwork that I decided to narrow my focus to Bahrain rather than studying the Gulf, as had been my original plan.

9 Farha Ghannam, *Remaking the Modern: Space, Relocation, and the Politics of Identity in a Global Cairo* (Berkeley: University of California Press, 2002).

10 See Doherty, 'Review of *Cairo Cosmopolitan*'.

11 Steven Caton, *Yemen Chronicle: An Anthropology of War and Mediation* (New York: Hill and Wang, 2005), 66.

12 Ibid., 14–15, 133.

13 I mean traditional in the sense that one spends an extended period of time living among so-called indigenous people.

14 Of course, there is a whole area of focus on the anthropology of the nation and nationality. See, for instance, Benedict Anderson, *Imagined Communities: Reflections on the Origin and Spread of Nationalism* (London: Verso, 1983); Michael Herzfeld, *Cultural Intimacy: Social Poetics in the Nation-State* (New York: Routledge, 1997).

15 Clifford Geertz, *The Interpretation of Cultures* (New York: Basic Books, 1973), 22.

16 See, for example, Arjun Appadurai, *The Social Life of Things: Commodities in Cultural Perspective* (Cambridge: Cambridge University Press, 1986); Laura Rival, ed., *The Social Life of Trees* (New York: Berg, 1998); Steven C. Caton, *Lawrence of Arabia: A Film's Anthropology* (Berkeley: University of California Press, 1999); Daniel Miller, *Stuff* (Cambridge: Polity Press, 2010). These are all important texts as they emphasize the anthropological study of things.

17 Geertz, *Interpretation of Cultures*, 5.

18 For more on the meaning of landscape, see Gareth Doherty and Charles Waldheim, eds., *Is Landscape . . . ?* Essays on the Identity of Landscape. (New York: Routledge, 2016); John R. Stilgoe, *What Is Landscape?* (Cambridge, MA: MIT Press, 2016).

19 See, for example, Eduardo Kohn, *How Forests Think: Toward an Anthropology Beyond the Human* (Berkeley: University of California Press, 2013).

20 T.S. Eliot, 'Andrew Marvell', *Times Literary Supplement*, 31 March 1921.

21 See, for example, Richard Long, *Walking the Line* (New York: Thames and Hudson, 2002).

22 See Charles Waldheim, 'Aerial Representation and the Recovery of Landscape', in *Recovering Landscape*, ed. James Corner (New York: Princeton Architectural Press, 1999).
23 Francesco Careri, *Walkscapes: Walking as an Aesthetic Practice* (Barcelona: Editorial Gustavo Gili, 2002), back cover.
24 See Geoffrey Alan Jellicoe and Susan Jellicoe, *The Landscape of Man: Shaping the Environment from Prehistory to the Present Day* (London: Thames and Hudson, 1987), 340, 343. In one of a series of unpublished interviews with the author in 1994–5, Jellicoe suggested that McHarg's maps could be considered art, 'if done well'.
25 Jellicoe and Jellicoe, *Landscape of Man*, 374.
26 See Kevin Lynch, *The Image of the City* (Cambridge, MA: MIT Press, 1960); Stan Allen, *Points + Lines: Diagrams and Projects for the City* (New York: Princeton Architectural Press, 1999).
27 Richard T.T. Forman, *Land Mosaics: The Ecology of Landscapes and Regions* (Cambridge: Cambridge University Press, 1995), 3–7.
28 One significant difference between Forman's and Lynch's approaches is that Lynch also includes the visual component (landmarks), perhaps because landmarks are not considered important for, or appreciated by, wildlife.
29 Ian L. McHarg, *Design with Nature* (Garden City, NY: Natural History Press, 1969).
30 McHarg also extensively engaged with anthropologists, seemingly preferring anthropologists to sociologists.
31 Waldheim, 'Aerial Representation', 136.
32 Charles Waldheim, *The Landscape Urbanism Reader* (New York: Princeton Architectural Press, 2006), 11.
33 See Clifford Geertz, 'Thick Description: Toward an Interpretative Theory of Culture', in *Interpretation of Cultures*, 3–30.
34 Forman, *Land Mosaics*, 3.
35 Ibid., 3–7.
36 Benjamin, *One-Way Street* (New York: Harcourt Brace Jovanovich, 1978), 28–9.
37 John R. Stilgoe, *Outside Lies Magic: Regaining History and Awareness in Everyday Places* (New York: Walker, 1998), 1.
38 For more on this, see Richard T.T. Forman's diagram outlining the scalar distinctions between an ecosystem, landscape, region, continent, and planet, in *Land Mosaics*, 12.

Bibliography

Allen, Stan. *Points + Lines: Diagrams and Projects for the City*. New York: Princeton Architectural Press, 1999.

Anderson, Benedict. *Imagined Communities: Reflections on the Origin and Spread of Nationalism*. London: Verso, 1983.

Appadurai, Arjun. *The Social Life of Things: Commodities in Cultural Perspective*. Cambridge: Cambridge University Press, 1986.

'Bahrain and Google Earth', Ogle Earth, accessed 26 February 2016, http://ogleearth.com/BahrainandGoogleEarth.pdf.

Bakhtin, Mikhail Mikhailovich. *The Dialogic Imagination*. Austin: University of Texas Press, 1981.

Benjamin, Walter. *One-Way Street*. New York: Harcourt Brace Jovanovich, 1978.

Caton, Steven. *Lawrence of Arabia: A Film's Anthropology*. Berkeley: University of California Press, 1999.

Caton, Steven. *Yemen Chronicle: An Anthropology of War and Mediation*. New York: Hill and Wang, 2005.

Careri, Francesco. *Walkscapes: Walking as an Aesthetic Practice*. Barcelona: Editorial Gustavo Gili, 2002.

Doherty, Gareth. 'Review of *Cairo Cosmopolitan* by Diane Singerman and Paul Amar'. In *International Journal of Middle East Studies* 42, no. 4 (2010): 725–6.

Doherty, Gareth and Waldheim, Charles eds., *Essays on the Identity of Landscape*. New York: Routledge, 2016.

Eliot, T.S. 'Andrew Marvell'. In *Times Literary Supplement*, 31 March 1921.

Forman, Richard T.T. *Land Mosaics: The Ecology of Landscapes and Regions*. Cambridge: Cambridge University Press, 1995.

Geens, Stefan. 'Bahrain in Google Earth, Unplugged'. In *Ogle Earth* (blog), 21 September 2006. http://www.ogleearth.com/2006/09/bahrain_in_goog.html.

Geertz, Clifford. *The Interpretation of Cultures*. New York: Basic Books, 1973.

Ghannam, Farha. *Remaking the Modern: Space, Relocation, and the Politics of Identity in a Global Cairo*. Berkeley: University of California Press, 2002.

Herzfeld, Michael. *Cultural Intimacy: Social Poetics in the Nation-State*. New York: Routledge, 1997.

Herzfeld, Michael. *The Social Production of Indifference: Exploring the Symbolic Roots of Western Democracy*. Chicago: University of Chicago Press, 1993.

Jellicoe, Geoffrey Alan and Susan Jellicoe. *The Landscape of Man: Shaping the Environment from Prehistory to the Present Day*. London: Thames and Hudson, 1987.

Kohn, Eduardo. *How Forests Think: Toward an Anthropology Beyond the Human*. Berkeley: University of California Press, 2013.

Lipsky, Michael. *Street Level Bureaucracy: Dilemmas of the Individual in Public Services*. New York: Russell Sage Foundation, 1980.

Long, Richard. *Walking the Line*. New York: Thames and Hudson, 2002.

Lynch, Kevin. *The Image of the City*. Cambridge, MA: MIT Press, 1960.

Malinowski, Bronislaw. 'Method and Scope of Anthropological Fieldwork'. In *Ethnographic Fieldwork: An Anthropological Reader*, 2nd ed., eds. Antonius C.G.M. Robben and Jeffrey A. Sluka, 69–82. Oxford: Wiley, 2012.

McHarg, Ian L. *Design with Nature*. Garden City, NY: Natural History Press, 1969.

Miller, Daniel. *Stuff*. Cambridge: Polity Press, 2010.

Rival, Laura ed., *The Social Life of Trees*. New York: Berg, 1998.

Stilgoe, John R. *Outside Lies Magic: Regaining History and Awareness in Everyday Places*. New York: Walker, 1998.

Stilgoe, John R. *What Is Landscape?* Cambridge, MA: MIT Press, 2016.

Waldheim, Charles. 'Aerial Representation and the Recovery of Landscape'. In *Recovering Landscape*, ed. James Corner, 121–139. New York: Princeton Architectural Press, 1999.

Waldheim, Charles. *The Landscape Urbanism Reader*. New York: Princeton Architectural Press, 2006.

12

TOWARD A SOMATOLOGY OF LANDSCAPE

Anthropological multinaturalism and the 'natural' world

Tao DuFour

In this chapter, we approach the theme of landscape from two theoretical perspectives: an anthropology of nature, and a phenomenology of the 'natural' world. The theme of landscape is of increasing interest for research in the social sciences[1] as well as among phenomenologists.[2] Anthropologist Philippe Descola, for example, situates the origin of landscape painting in relation to a modern mode of being that he terms 'naturalism'. Descola points to the motif of the 'Flemish window', dating to the first half of the fifteenth century, in which a profane landscape is isolated in the background of a religiously themed painting; to arrive at a landscape, '[a]ll that was then needed was to increase the size of the window to the dimensions of an entire canvas so that the picture within a picture became the actual subject of the representation'.[3] The relation between the perspectival geometric structure that underlies the representation of distant horizons, freed from religious narrative and symbolism, and the objectified landscape as a consequence stands as an emblem for an emancipated subjectivity that determines the order of the natural world through mathematical rules. The history of landscape is thus closely tied to an anthropology of modernity, and modernity is ontologically defined by a certain conception and experience of nature. Anthropology, as a modern science, is of course involved in this landscaping of nature, but because its interests are traditionally oriented toward a sociocultural 'other', it is constantly confronted with the studied societies' resistances to the project of naturalism as constitutive of the project of modernity. By taking these resistances seriously, and allowing them to inform and transform anthropology's research methods, anthropologists have crossed traditional epistemic frontiers concerning the concept of nature. I believe that some of these methodological and conceptual insights can be of value to research in landscape architecture, and I propose to explore one strand of anthropological thinking about nature, developed in the Amazonian anthropology of Brazilian anthropologist Eduardo Viveiros de Castro. I will enquire into Viveiros de Castro's concept of 'multinaturalism' and relate this to the phenomenology of nature in the work of Czech philosopher Jan Patočka. One of the earliest attempts at the systematic elaboration of the concept of nature in the phenomenological tradition is Patočka's habilitation thesis of 1936, *Přirozený svět jako filosofický problém* (*The Natural World as a Philosophical Problem*). In this work, Patočka develops a phenomenological analysis of the relationship between the body, language and the 'natural' world. This chapter therefore explores problems in the conception of landscape at the intersections of anthropology and phenomenology on the topics of 'nature' and the 'body' through Eduardo Viveiros de Castro's articulation of the concept of multinaturalism and the theme of the 'natural' world in Patočka.

Nature/culture

One of the fundamental dualisms that has traditionally plagued Western thought is that of 'nature' versus 'culture' or 'society'.[4] In relation to this problem, Descola's categorical divisions are instructive. Descola's category of 'naturalism', to which we first referred, points to the nature/culture dualism as ontologically definitive for modernity in general.[5] It is an inheritance of the Enlightenment of eighteenth- and nineteenth-century Europe, anticipated in the seventeenth century as a function of the Cartesian split of thinking subject and extended world. The nineteenth-century attempt at a resolution of the problem lay in a methodological division of the sciences between natural and human sciences (*Naturwissenschaften* and *Geisteswissenschaften*).[6] The critical contribution of philosophy and anthropology to this problem was to disclose this division itself as historically and culturally situated. This hermeneutical insight may have tempered the philosophical critique in part, but it raised again the spectre of 'culture' for anthropology, splitting the discipline into two camps: forms of 'naturalist reductionism'—biological, evolutionary, etc.—and forms of 'semiological idealism' or cultural relativism.[7] The problem is the transference of the earlier natural science/human science divide into the human sciences, and the splitting of human experience into 'natural' versus 'social' phenomena. The result, as Descola points out, is the continued acceptance of the premise 'that human experience must be understood as resulting from the coexistence of two fields of phenomena governed by distinct principles'.[8] Recent work in anthropology associated with what has been termed 'the ontological turn' has emphasized the fact that the nature/culture divide does not exist in many other ethnographic contexts in the form of an ontological dualism, which suggests that the problem is epistemic on the part of the anthropologist.[9] It is the anthropologist who lacks concepts appropriate to the ethnographic situations encountered, and perhaps the conceptual apparatus is inherent in the situations themselves. Furthermore, the phenomena being studied have their own power in turn to produce effects on the anthropologist that transfix and transform the concrete experience and conditions of ethnography, and by extension its theoretical implications.[10] The integrated theories of perspectivism and multinaturalism, which we will consider in the following sections, are Viveiros de Castro's attempt to place on equal epistemological ground the ontological constitution of indigenous communities as worlds that produce reciprocal effects on anthropology.

Perspectivism, multinaturalism

Viveiros de Castro presents his thesis of Amerindian multinatural perspectivism in an article first translated into English in 1998, titled 'Cosmological Deixis and Amerindian Perspectivism'. In this text, Viveiros de Castro is already projecting his ethnographic work on Amazonian societies toward the region of more general theoretical speculation. He would push this theoretical orientation further in a later 'metatheoretical' reflection on his own thesis in an article published in English in 2013, titled 'The Relative Native'. Martin Holbraad, in a brief preamble to this article, qualifies Viveiros de Castro's theory of multinaturalism as recasting 'the idea of nature as the manner in which people conceive it [...] provided "conception" here is understood in sharp contrast to representation, as an irreducibly ontological operation'.[11] The idea of 'nature' here is intended to recursively uproot the concept of 'concept' itself, as a conflation of epistemological and ontological categories.

This interpretation is in keeping with Holbraad's attempts to theorize the experience of 'things' as 'concepts' by giving priority to ethnographic situations in which these categories are blurred.[12] It would seem to me, therefore, that it is the ethnography proper which lies closest to this region of inflection where 'concept' and 'nature' fold into themselves, implying a metatheory in advance. I therefore turn to Viveiros de Castro's now classic ethnography of the Araweté of Amazonia, *From the Enemy's Point of View*, as a point of departure.[13]

In the closing chapter of his ethnography, titled 'The Anti-Narcissus', Viveiros de Castro shifts the focus of his inquiry from his own fieldwork on the contemporary Araweté to the ethnographic archive on the Tupinamba in general, and specifically the question of Tupinamba anthropophagy. Viveiros de Castro has elevated this motif of 'cannibalism' to the status of a 'philosophical enactment' as a metaphysics.[14] The figure of anthropophagy has a complex place in the development of Brazilian modernism, most notably in the influence of the 1928 polemic by Oswald de Andrade, 'Manifesto Antropófago'.[15] As Leslie Bary notes, 'Oswald's anthropophagist [...] neither apes nor rejects European culture, but "devours" it, adapting its strengths and incorporating them into the native self.'[16] Oswald draws on the motif's power to evoke the image of ritual cannibalism—an image accessible only through historical records—as a figure in the construction of a postcolonial paradigm that aims to deracinate the colonial archive.[17] He asserts: 'Cannibalism alone unites us. Socially. Economically. Philosophically'; and, foreshadowing Lévi-Strauss, 'We are concretists.'[18] It is this paradox of a flesh-eating theory that Viveiros de Castro stretches into an anthropophagic philosophy. One of the fundamental aspects of this 'cannibal metaphysics' is its thematization of the ontological status of the 'body'. The body for Viveiros de Castro is not merely a material 'thing'; rather, for the Amerindian societies that he studies, and in terms of the ethnographic archive in general, the animate body is the condition of possibility for the incorporation of a perspective, which is at the same time the constitution of a world. The body is a structure that facilitates the incorporation of a point of view as ontologically transformative, in the sense that the perspective is not a view on the world—a 'worldview'—but the body itself expresses the incorporated world. Viveiros de Castro presents ritual anthropophagy as the incorporation of the enemy—who is married to his killer's kin and resides in the community as a guest of honour—into the society as a transgression:

> In short, *the enemy was the center of the society*. Isn't this what was expressed in the solemn execution in the central plaza of the village, where the victim stood resplendent, superbly feathered as if he were the guest of honor? If [...] the relation of alliance is logically superior to that of filiation, since it rescues the latter from its continuity with nature and institutes society, here we have the same thing. The relationship to the enemy is anterior and superior to society's relationship to itself, rescuing it from an indifferent and natural self-identity....[19]

The implications of the practice of the incorporation of the 'enemy' are profound: it indicates the genuine *possibility* of the corporeal constitution of *another world*. The enemy as an 'Other' is here a structure: '*the Other is the expression of a possible world*.'[20] The condition of this possibility lies in the ontological status of the body as:

> [A]n ensemble of ways or modes of being that constitutes a *habitus*, ethos, or ethogram. Lying between the formal subjectivity of souls and the substantial materiality of organisms is a middle, axial plane that is the body qua bundle of affects and capacities, and that is the origin of perspectivism. Far from being the spiritual essentialism of relativism, perspectivism is a corporeal mannerism.[21]

'Perspectivism' is the name that Viveiros de Castro gives to a generally observed and recorded Amerindian animistic tendency to understand the differentiated nature of the world as a function of the specific differences of the corporeal constitution of all entities similarly ensouled as 'persons', which is to be understood as a formal category of intentionality.[22] Through specific ritual and instrumental means, including the appropriation of the 'clothing' of other species, certain persons—namely shamans—are able to transgress corporeal limits and exist as interspecific beings—simultaneously human and non-human—thereby gaining the 'perspective' of others, including animals, divine beings and the dead. This potential for interspecificity means that the shaman can inhabit multiple points of view on 'nature', and therefore multiple natures, in so far as the body expresses nature as a world-constitution. Multinaturalism thus follows logically and ontologically from perspectivism, such that in the continuous flux of inhabiting multiple perspectives as a vocation, the shaman is a being of *becoming*. This facility is a possibility not only for the shaman, but also for the killer, who paradoxically becomes his enemy by *not* participating in the consumption of the flesh of the victim. In becoming the enemy, the killer ensures his victim's humanity through discourse prior to the killing; hence the semiophagic nature of the rite, as the killer assumes a new name after fulfilling the act of vengeance and remains inside the house in solitude without speaking, with the victim's lips worn around his wrist.[23] The multinatural perspectivist potential is also a capacity of the gods who consume the dead, as described in Viveiros de Castro's ethnography of the contemporary Araweté.[24] Its most articulate level of presentation, however, is in terms of the conceptuality of the archive of myth, a topic to which we will return.

'Nature', therefore, is not an objective category over against subjective 'culture', but is the concrete, phenomenological experience of the structure: *world*. The body as 'clothing' is an expression of a world. Although Viveiros de Castro's identification of multinaturalism is specific to the context of Amerindian ethnography, I propose that there are general implications. Viveiros de Castro himself acknowledges this in his 'metatheoretical' reflections.[25] In order to draw out the wider philosophical implications of a theory of multinatural perspectivism for conceptions of landscape, I propose to bring this theoretical position into contact with the concept of the natural world in the phenomenology of Jan Patočka.

The natural world

The concept of the natural world entered architectural philosophy through the work of the Czech architect and theorist Dalibor Vesely, who initiated a distinct theoretical innovation informed by phenomenology and hermeneutics.[26] Vesely's impact on landscape architecture has been indirect, through the work of colleagues and students reflecting the influence of his phenomenological theses.[27] The natural world is an important heuristic concept for Vesely: it appears prominently in his major programmatic work, *Architecture in the Age of Divided Representation*, and is elaborated upon in the second chapter, titled 'The Nature of Communicative Space'.[28] Vesely employs three related notions in his attempt to articulate the significance of the motif of 'nature' for conceptions of space: the 'natural world', the 'lived world' and the 'latent world'. The terms imply levels of foundation as to the muteness of the world: from the more articulate lived world, to the natural world below, and ultimately to the lowest level of embodiment in the form of the latent world. As a mediating region between the latent world and the lived world, the natural world seems to be the domain of the meeting of 'natural conditions' and 'culture'. The second chapter of Vesely's work has a general, theoretical character that speaks to the broad theme of the perception of the spatiality of the world, which would include landscape. Vesely's generalization of the concept suggests that the natural world could be interpreted formally, in the manner of a transcendental philosophical claim. For the most part, the parallel notion of 'culture' is identifiable with a European cultural tradition, understood in terms of the motif of continuity, although when framed in relation to the natural world, 'culture' too takes on a certain formal character, reinforcing the sense that the phenomenon of the natural world itself seems to call for explanation at the level of the transcendental. The paradox is that the natural world in Vesely is both a relative condition, historically situated, and a universal ontological horizon and as such transcendental. The specificity of historical horizons is a fundamental ontological principle for Vesely, so we can safely assume that European culture is *one* horizon of 'nature' in the relative form of *a* natural world. At the level of 'culture', therefore, we are dealing with the paradox of a plurality of 'natural' worlds, understood ontologically in relation to *the* natural world. The problem parallels Viveiros de Castro's theory of multinaturalism, in that the multiplicity of 'natures' is an ontological plurality of worlds that do not constitute a relativism, in the sense that we are not dealing with mere subjective or collective 'representations'.

Vesely does not systematically elaborate the meaning of the concept of the natural world as a properly philosophical concern—one, for example, that would suggest the problem of an ontology of 'nature'. The notion of the natural world as a concept indicating an embodied spatial background remains, philosophically, a background concept. Its juxtaposition with the seemingly derivative notion of the 'latent world', which at times is characterized as muter than the natural world and at other times is conflated with it, points to an unresolved tension in the concept of the natural world in Vesely's work. The sense of a latent formalism or transcendental residue in Vesely's characterization of the natural world is neither accident nor mere oversight. It is rather a direct consequence of the influence of the philosophy of Jan Patočka, Vesely's former mentor, and his dyadic conception of the structure of the 'natural' world. The quotation marks around the term 'natural' in this instance are introduced by Patočka's biographer and translator Erazim Kohák to indicate the significative difference between the Czech term used by Patočka, *přirozený svět*—'natural' world—and the more general sense of the theme of the natural implied in the Czech *přírodní svět*—natural world associated with conventional notions of the idea of nature in the broadest terms.[29] The 'natural' world is a concept in Patočka's philosophy that is explicitly concerned with an elaboration and development of the notion of the life-world (*Lebenswelt*) in the philosophy of Edmund Husserl.[30] It is in relation to this specific philosophical background that the notion of the natural world in Vesely's work should be understood.[31]

In his essay of 1967, 'The "Natural" World and Phenomenology',[32] Patočka outlines his philosophical projection of the problem of the natural world beyond Husserl's elaboration of the life-world.[33] Patočka proposes that the problem of the natural world be characterized in relation to two central themes: the first is mereological, having to do with the consciousness of the 'world' as a whole in relation to its constitutive parts; the second concerns the *corporeity* of the natural world as foundational.[34] The theme of the relation between a part and the whole in Patočka is tied to the problem of 'ideality'. Patočka demonstrates it with the basic example of the visual and kinaesthetic perception of an object such as a table, in which the front is actually perceived *with* the non-actually perceived back and the entire material object as a whole as 'an idea that is constantly anticipated'.[35] The example is a simplification aimed at a phenomenological problem that has to do with the structure of the appearance of the world to a perceiving subject. Patočka's concern is to articulate the structure of the appearing as such, so that the world as a whole has the character of a structure rather than an aggregate of things. Steven Crowell, in his analysis of Patočka's conception of the human being as an 'ideality of nature', points to Patočka's concept of the 'autonomy' of the 'structures of appearing as such', that is, the world as an 'autonomous' structure.[36] Patočka's philosophical anthropological tendency is evident in the priority he places on the adult human being and its 'movements' as the locus of the philosophical inquiry into the question of Being, a particular prejudice he shares with and inherited from Heidegger's fundamental ontology.[37] Patočka's insistence, however, on the autonomy of the structure of the appearance of the world, which furthermore he proposes is given over to a 'formal', transcendental subjectivity as an 'empty position', subverts this anthropocentrism.[38] The notion of an autonomous structure suggests that it is purely the possibility of the *point of view* that is facilitated in this 'empty position', with relation to the appearance of the world, which is constitutive of subjectivity. The 'subject' is therefore an anti-subject, in the sense that its intentionality is a function of its position in the autonomous appearing of the world, rather than its own activity. Patočka does not follow through the ontological consequences of his asubjective phenomenological thesis to their limit, because he has determined in advance that the locus of the ontological problem is the human. Nevertheless, his conception of the natural world in terms of a 'formal transcendentalism of appearing as such', I argue, is already an opening onto the non-human, and an indication of the structure of the autonomous appearing of the landscape.[39] Its relation to the other structural feature of the natural world, its corporeity, must now be considered.

Situated perspectivism, multinatural situation

The motif of corporeity as a methodological principle in phenomenological description, that is, a method of grounding analysis based on a description of what is intuitively given to perception, finds its analogue in anthropology in the method of ethnographic description. The 'ethnographic moment', to use a term from Marilyn Strathern, characterizes the immediacy of the concrete experience 'in person' of the ethnographer to the subjects that she studies, be they Amerindian societies or scientists in a biological laboratory.[40] This methodological grounding does not, however, naively presuppose the absence of a discursive background or archive; even Marx in his late writings would come to recognize the significance of the ethnographic archive in relation to the question of history.[41] If there is a 'naivety' to ethnography, it is itself methodologically strategic, for example in the anthropology of science.[42] In Viveiros de Castro's work, the ethnographic ground is mixed not only with the ethnographic archive, but also with the discursive contexts of the societies, and specifically in the Amerindian case with that of myth: 'Myth, the universal point of flight of perspectivism, speaks of a state of being where bodies and names, souls and actions, egos and others are interpenetrated'. . . .[43]

In the perspectivist understanding, intentionality is a function of a corporeal schema, and as the 'soul' is 'formally identical in all species, it can only see the same things everywhere—the difference is given in the specificity of bodies'.[44] The 'body' is not a naturalistic object, in the parallel sense that the 'soul' is not a psychological subject. Rather, the body is an 'assemblage of affects or ways of being that constitute a *habitus*'.[45] The 'soul' has the character of 'personhood'. In the Amerindian context, the 'person' is a formal category, attributable to all species—human, animals and plants—meteorological phenomena, gods, the dead, artefacts, and the landscape as a whole as 'a center of intentionality' and 'equipped with the same general ensemble of perceptive, appetitative, and cognitive dispositions'.[46] It is the body as a corporeal, affective habitus with its dispositions and capacities which is the 'origin of perspectives', and therefore Amerindian perspectivism is a 'somatic perspectivism'.[47] The ability of shamans to metamorphose into other species is instrumentally a function of their appropriation of the bodies of other species as 'clothes', 'akin to diving equipment, or space suits, and not to carnival masks', which endow the shaman with the affects and capacities of the other species.[48] This immanent somatism, therefore, is the ontological basis of the multiplicity of nature as a constitution of worlds: 'all beings see ('represent') the world *in the same way*; what changes is *the world*

(a)

(b)

Figure 12.1a,b,c Swidden and anthropogenic forest in Amazonia, Floresta Nacional do Tapajós, Pará, Brazil, 2016. Extending Viveiros de Castro's ontological theme into the domain of the forest as an epistemological horizon, Eduardo Kohn proposes a thesis on the capacity for the constitution of "mean-ing" in species beyond the human: "all beings, including those that are nonhuman, are constitutively semiotic." Kohn's thesis draws explicitly on the philosophy of Charles Sanders Peirce, but I would suggest that the problem of "constitution" as a semiotic motif is anticipated in Patočka's early work, and its interrogation of problems first laid out by Husserl in his *Logical Investigations*. See Eduardo Kohn, *How Forests Think: Toward an Anthropology Beyond the Human* (Berkeley: University of California Press, 2013): 16. © DuFour, Tao.

(c)

Figure 12.1a,b,c Continued.

they see'—hence multinaturalism.[49] 'In the same way' here refers not to a biomechanism of vision, but to a structure of representation, that is, the constitution of meaning. The question remains, however, as to the temporality of this somatic world-constitution: does not a habitus imply some form of 'historical' being, and would not therefore all species, the landscape and even the Earth itself, as corporeal 'expressions' of habitus, be ontologically 'historical'?[50] The conceptual homology between Viveiros de Castro's description of the Amerindian conception of the 'soul' as a formal identity and Patočka's anti-subject as an 'empty position' is striking to me. For Viveiros de Castro, Amerindian 'souls' express worlds in terms of 'a plane of immanence', a notion he derives from Deleuze and Guattari.[51] On the other hand, and in spite of the transcendental formalism of Patočka's asubjectivity, the embodied subject is historically *situated*; the perspective is a historically situated perspective:

> [P]erception and the entire original givenness of the content of the natural world are essentially linked with corporeal life; life in the 'natural' world is a corporeal life [...T]he 'natural' world is essentially an oriented, perspectival, situational world.[52]

Viveiros de Castro's Amerindian 'soul' is somehow both immanent and formal, and Patočka's 'empty position' is transcendent yet situated. In the former we encounter the paradox of an immanent formalism, and in the latter that of a transcendental immanence. Through a coupling of these perspectives, Viveiros de Castro's 'formally identical' souls would be invested with historicity,[53] and Patočka's historically situated 'empty positions' would be indicative of not merely humans, but non-human entities and the landscape itself. We would as a consequence have the equivocal notions 'situated perspectivism' and 'multinatural situation', in which the concept of subjectivity would extend beyond the human to indicate a *multispecific* phenomenon (Figures 12.1a,b,c).[54]

Toward a somatology of landscape

With regard to corporeity, the critical question for Viveiros de Castro's explication of perspectivism and multinaturalism is *how* the body enables the constitution of perspectives and the ontological determination of multiple 'natures'. For an extended description of the formation of these concepts in specific reference to the problem of the body, I turn to the series of lectures given at the University of Cambridge's Department of Social Anthropology in 1998, published under the title 'Cosmological Perspectivism in Amazonia and Elsewhere'.[55] These lectures, particularly the third lecture published as 'Nature: The World as Affect and Perspective', indicate clearly the link between the concept of 'nature' and that of the 'body', and provide a descriptive framework for the theses of perspectivism and multinaturalism in terms of this relation.[56] I compare this with the chapter in Patočka's habilitation thesis titled simply 'The Natural World', where he outlines his philosophical position.[57]

Viveiros de Castro succinctly describes the relationship between the concept of the 'body' and that of naturalism as follows:

> [W]hat I call 'body' [...] is an assemblage of affects or ways of being that constitute a habitus. Between the formal subjectivity of souls and the substantial materiality of organisms, there is thus an intermediate plane which is occupied by the body as a bundle of affects and capacities and which is the origin of perspectives. The common, transpecific spirit has access to the same percepts, but species-specific bodies are endowed with different affects—and that is why we have multinaturalism.[58]

Precisely what is meant by 'bundle of affects and capacities', 'habitus', 'assemblage of affects' and so forth is what remains to be explored and clarified. Viveiros de Castro's approach to this problem is to draw on the idea of 'clothing' and the notion of 'equipment', as we saw, which he grounds ethnographically in terms of ritual performance and the bodily articulation, dress and masking that characterize it. Ritual involves 'clothing' the human in animal masks, which function to metamorphose humans into those animals. The transformation is neither 'representational' nor merely symbolic, but *affective*, in the sense that the capacities and powers realizable in potential as a function of the animals' corporeal constitution become available to the human *as* animal-body. Although Viveiros de Castro uses the notion of the body as 'tool', his concept of metamorphosis as performative is not merely instrumental. The human-cum-animal is a subjectivity with a particularized point of view because the transformation of the body is at the same time an ontological transformation of the world. I propose here to linger on this theme of the body as a 'bundle of affects and capacities' that function in terms of the potentiality inherent in the body's physiognomic appearance and its 'particular phenomenological quality'.[59]

It would seem to me that in speaking of 'affects and capabilities', Viveiros de Castro is implying the body's—human and non-human—inherent aesthetic and kinaesthetic structuring, and the relation of this to some environmental horizon.[60] The body's aesthetic-kinaesthetic dispositions imply particular capabilities and potentialities as positioned in relation to some context. This positionality and power to perform kinaesthetically is indicative, if not of a habitus in the full sense that implies a personal history and memory, at the very least of a habituality or 'tendency toward' such and such. Patočka describes the condition as follows:

> A tendency can be satisfied only by a subject, through the direct, active contact with its surroundings that is the agency of all our activities. These are our kinestheses, the

> movements that subsequently provide the ground for the apperception of our own bodiliness. *An animal subject is a priori possible only as body*. [...] The life in kinesthesis [...] is among the most important moments of the constitution of the objective world.[61]

In terms of a descriptive understanding of the body as apperceptive-kinaesthetic, Patočka posits a concept of subjectivity that opens the possibility for the constitution of the *objective* world, or properly speaking *an* objective world. Patočka does not, of course, inquire into the problem of interspecific metamorphosis, and he raises the question of the world-constitution of the non-human animal only tangentially. In spite of this apparent inattention, however, he does introduce the problem of the spatiality and temporality characteristic of bodily animality that is coincident with corporeal-human subjectivity, without reducing this formal identity of corporeal-subjectivity to a naturalistic, empirical concept of the body. Patočka describes the apperceptive-kinaesthetic animal-body as:

> [T]hat of an I governed by affective tendencies and the appertaining external stimuli; nonetheless, there is already a certain preliminary degree of freedom in this apprehension of the explicability of the unexplicated, in the understanding of the background as incessantly repeatable possibilities of progression and transition. For it lies in the very nature of space as horizon that it is infinite; its infinity stems from the *temporal* horizon in which the *process* of orientation is played out.[62]

Unfortunately, Patočka could not escape the pull of the philosophical anthropological tendencies of his time. Nonetheless, the passage quoted demonstrates that Patočka understood the animal-body-subject to be a phenomenological quality determined by a spatiality and temporality that is part of a formally universal structure of subjectivity inclusive, as formal, of human subjectivity. Patočka's attempt to propose an asubjective phenomenology and the concept of the 'empty position' arose later in the evolution of his thinking, but the germ could already be detected in his habilitation thesis. That the affective tendencies of animals and their spatially oriented bodily-kinaesthetic capacities are structured such that they participate in a universal temporal horizon would imply that the temporality and spatiality of animals, given in their concrete environmental specificities, could be understood in terms of *generational* categories.[63] As Husserl himself proposed, the problem could be extended to the vegetal world and even 'single-cell organisms'.[64] If the concept of the body from the point of view of a somatic phenomenology projected to its logical and phenomenological limits reveals that animal and vegetal worlds participate in a 'formal', transcendental subjectivity, and in their concreteness are determined by spatial and temporal specificities that would be indicative of a historicity, then *nature is*—ontologically—*subjective*. Such a phenomenology of the natural world, as a phenomenological somatology, would seem to possess an affinity with multinaturalism as articulated by Viveiros de Castro. It calls for an extended interrogation of the concept of the natural world, which would suggest an expansion of the horizon of our conception and experience of landscape. Understood as a somatic subject, the landscape could no longer be seen as merely a passive object of design and instrumental appropriation. The unity of the corporeity and subjectivity of the landscape introduces an ethical dimension to the problem of design in landscape architecture, one that is not limited to human consequences and thus requires what Aldo Leopold termed a *land ethic*.[65] The coupling of the concepts of multinaturalism and the natural world that I have proposed in this chapter function heuristically, in order to introduce the idea of a somatology of landscape and its implications for an ethics of landscape.

Notes

1 See, e.g., Philippe Descola, *Beyond Nature and Culture* (Chicago: University of Chicago Press), 2013; Anna Lowenhaupt Tsing, *The Mushroom at the End of the World: On the Possibility of Life in Capitalist Ruins* (Princeton: Princeton University Press, 2015); Tim Ingold, *The Perception of the Environment: Essays on Livelihood, Dwelling, and Skill* (London: Routledge, 2000).
2 See, e.g., Edward Casey, *Earth-Mapping: Artists Reshaping Landscape* (Minneapolis: University of Minnesota Press, 2005); Kuan-min Huang, 'Toward a Phenomenological Reading of Landscape: Bachelard, Merleau-Ponty, and Zong Bing', in *Phenomenology and Human Experience*, eds. Chung-chi Yu and Kwok-ying Lau (Nordhausen: Verlag Traugott Bautz, 2012); Jeff Malpas, ed., *The Place of Landscape: Concepts, Contexts, Studies* (Cambridge, MA: MIT Press, 2011).
3 Descola, *Beyond Nature and Culture*, 58–9.
4 Philippe Descola and Gísli Pálsson, eds., *Nature and Society: Anthropological Perspectives* (London: Routledge, 1996).
5 Descola, *Beyond Nature and Culture*, 172–200.
6 Hans-Georg Gadamer, *Truth and Method* (New York: Continuum, 2000), 173–264.
7 Philippe Descola, *The Ecology of Others* (Chicago: Prickly Paradigm Press, 2013), 27–8.
8 Ibid., 4.
9 Martin Holbraad and Morten Axel Pedersen, *The Ontological Turn: An Anthropological Exposition* (Cambridge: Cambridge University Press, 2017).
10 On the effective potency of situations encountered in ethnographic fieldwork, see Marilyn Strathern, *Property, Substance and Effect* (London: Athlone Press, 1999), 1–26.
11 Martin Holbraad, 'Turning a Corner: Preamble for "The Relative Native" by Eduardo Viveiros de Castro', *HAU: Journal of Ethnographic Theory* 3, no. 3 (2013): 470. Holbraad and Pedersen situate Viveiros de Castro's work in the context of anthropology's 'ontological turn' in 'Natural Relativism: Viveiros de Castro's Perspectivism and Multinaturalism', in *The Ontological Turn*, 157–98.
12 See for example Martin Holbraad, 'The Power of Powder: Multiplicity and Motion in the Divinatory Cosmology of Cuban Ifá (or *Mana*, Again)', in *Thinking Through Things: Theorizing Artefacts Ethnographically*, eds. Amiria Henare, Martin Holbraad and Sari Wastell (London: Routledge, 2007).
13 Eduardo Viveiros de Castro, *From the Enemy's Point of View: Humanity and Divinity in an Amazonian Society* (Chicago: University of Chicago Press, 1992).
14 Ibid., 303.
15 Oswald de Andrade, 'Cannibalist Manifesto', *Latin American Literary Review* 19, no. 38 (1991), first published as 'Manifesto Antropófago', *Revista de Antropofagia* 1, no. 1 (1928).
16 Leslie Bary, 'Oswald de Andrade's "Cannibalist Manifesto"', *Latin American Literary Review* 19, no. 38 (1991).
17 See Luis Madureira, *Cannibal Modernities: Postcoloniality and the Avant-garde in Caribbean and Brazilian Literature* (Charlottesville: University of Virginia Press, 2005), 35–51.
18 Andrade, 'Cannibalist Manifesto', 43. I refer here to Lévi-Strauss's 'science of the concrete' in *The Savage Mind* (Chicago: University of Chicago Press, 1962), 1–33. For an incisive critique of 'cannibalism' as a motif originating with the European conquest of the Americas see Silvia Federici, 'Colonization and Christianization: Caliban and Witches in the New World', in *Caliban and the Witch* (Brooklyn, NY: Autonomedia, 2014). On the motif of 'cannibalism' in the history of ideas see Cătălin Avramescu, *An Intellectual History of Cannibalism* (Princeton: Princeton University Press), 2011.
19 Viveiros de Castro, *Enemy's Point of View*, 301.
20 Eduardo Viveiros de Castro 'The Relative Native', *HAU: Journal of Ethnographic Theory* 3, no. 3 (2013): 478–9.
21 Viveiros de Castro, *Cannibal Metaphysics* (Minneapolis: Univocal, 2014), 72–3.
22 Viveiros de Castro draws the philosophical (as opposed to ethnographic) basis of the concept of perspectivism from Deleuze, who in turn traces a genealogy from Leibniz to Nietzsche. See Gilles Deleuze, *The Fold: Leibniz and the Baroque* (London: Athlone Press, 2001), 19–22.
23 Viveiros de Castro, *Enemy's Point of View*, 292–3.
24 Ibid., 252–72.
25 Viveiros de Castro, 'Relative Native'.
26 Vesely's contribution to architectural phenomenology is the topic of Joseph Bedford's doctoral dissertation, 'Creativity's Shadow: Dalibor Vesely, Phenomenology and Architectural Education (1968–1989)' (Princeton 2018). See also *Phenomenologies of the City: Studies in the History and Philosophy of*

Architecture, eds. Henriette Steiner and Maximilian Sternberg (London: Routledge, 2015) and *The Living Tradition of Architecture*, ed. José de Paiva (London: Routledge, 2017).

27 See for example David Leatherbarrow, *Topographical Stories: Studies in Landscape and Architecture* (Philadelphia: University of Pennsylvania Press, 2004).

28 Dalibor Vesely, *Architecture in the Age of Divided Representation: The Question of Creativity in the Shadow of Production* (Cambridge, MA: MIT Press, 2004), 43–107.

29 Erazim Kohák, ed., *Jan Patočka: Philosophy and Selected Writings* (Chicago: University of Chicago Press, 1989), 22–3.

30 Edmund Husserl, 'The Way into Phenomenological Transcendental Philosophy by Inquiring back from the Pregiven Life-World', in *The Crisis of European Sciences and Transcendental Phenomenology: An Introduction to Phenomenological Philosophy* (Evanston, IL: Northwestern University Press, 1970), 103–89.

31 It would be cumbersome to repeat the qualifying quotation marks throughout this text, so I follow Vesely's editorial decision and for the most part refer simply to the natural world, without quotation marks.

32 Jan Patočka, 'The "Natural" World and Phenomenology,' in *Jan Patočka: Philosophy and Selected Writings*, ed. Erazim Kohák (Chicago: University of Chicago Press, 1989), 239–73.

33 For a compact yet in-depth account of Husserl's phenomenology of the life-world, see Klaus Held, 'Husserl's Phenomenology of the Life-World', in *The New Husserl: A Critical Reader*, ed. Donn Welton (Bloomington: Indiana University Press, 2003). Held links this notion introduced in Husserl's late writings to central concepts developed early in Husserl's philosophy concerned with phenomenological method. See also David Carr, 'Ambiguities in the Concept of the Life-World', in *Phenomenology and the Problem of History: A Study of Husserl's Transcendental Philosophy* (Evanston, IL: Northwestern University Press, 1974), 190–211.

34 Patočka, '"Natural" World', 250–1.

35 Ibid., 252–3.

36 Steven Crowell, '"Idealities of Nature": Jan Patočka on Reflection and the Three Movements of Human Life', in *Jan Patočka and the Heritage of Phenomenology: Centenary Papers*, eds. Ivan Chvatík and Erika Abrams (Dordrecht: Springer, 2011), 14.

37 Patočka was not unaware of this as a problem, even directing criticism at Heidegger's thought and questioning whether it was not a 'transcendental philosophy trying to step over its own shadow' (quoted in Crowell, 'Idealities of Nature', 13). See Jan Patočka, *Body, Language, Community, World* (Chicago: Open Court Publishing, 1998), 101. Patočka's later writings on history tend to overshadow and perhaps even run counter to his phenomenological philosophy proper. My interpretation of Patočka's concept of the 'natural' world remains close to his phenomenology of perception and corporeity. It is in terms of this Husserlian orientation that I address the themes of historicity and habitus implied in Patočka's work. For Patočka's later writings on history see his *Heretical Essays in the Philosophy of History* (Chicago: Open Court, 1996), and Jacques Derrida, 'The Gift of Death', in *The Gift of Death and Literature in Secret* (Chicago: University of Chicago Press, 2008).

38 Crowell, 'Idealities of Nature', 18–19.

39 Quoted in ibid., 7.

40 Strathern, *Property, Substance and Effect*, 3–6.

41 I thank Bruno Bosteels for directing me to Marx's use of ethnographic sources in his late work, specifically the *Grundrisse*. See Karl Marx, *The Ethnological Notebooks of Karl Marx* (Assen: Van Goorcum and Co., 1974).

42 Bruno Latour and Steve Woolgar, *Laboratory Life: The Construction of Scientific Facts* (Princeton: Princeton University Press, 1986).

43 Viveiros de Castro, *Cannibal Metaphysics*, 68.

44 Eduardo Viveiros de Castro, 'Cosmological Deixis and Amerindian Perspectivism', *Journal of the Royal Anthropological Institute* 4, no. 3 (1998): 478.

45 Ibid.

46 Viveiros de Castro, *Cannibal Metaphysics*, 56, 58.

47 Viveiros de Castro, 'Cosmological Deixis', 478, 480.

48 Ibid., 482.

49 Viveiros de Castro, *Cannibal Metaphysics*, 71.

50 For a survey of the concept of habitus, see Dermot Moran, 'Edmund Husserl's Phenomenology of Habituality and Habitus', *Journal of the British Society for Phenomenology* 42, no. 1 (2011).

51 Gilles Deleuze and Félix Guattari, *What Is Philosophy?* (New York: Columbia University Press, 1994).

52 Patočka, '"Natural" World', 251. The concept of situation in Patočka implies both a world of perceived objects and corporeal spatiality along the lines of Merleau-Ponty's phenomenology of perception, as

well as the properly hermeneutical concept of historical situation as developed by Heidegger. See Patočka, *Body, Language, Community, World*, 127–34. On the concept of situation in Heidegger, see Martin Heidegger, *Being and Time* (Oxford: Blackwell, 2002), 341–8.

53 Viveiros de Castro, on the question of personal or individual history, draws reference to an Amerindian dual conception of the human soul which 'distinguishes between the soul (or souls) of the body, reified register of an individual's history, site of memory and affect, and a "true soul," pure, formal subjective singularity, the abstract mark of a person.' See Eduardo Viveiros de Castro, *The Relative Native: Essays on Indigenous Conceptual Worlds* (Chicago: HAU Books, 2015), 268.

54 I am pointing loosely here toward Viveiros de Castro's notion of *equivocation* as a conceptual framework for addressing the problem of comparison as 'translation' in anthropology: 'Good translation succeeds at allowing foreign concepts to deform and subvert the conceptual apparatus of the translator such that the *intentio* of the original language can be expressed through and thus transform that of the destination.' Equivocation also implicates the position of the subject of study as that of 'translator' as much as it does the anthropologist. See *Cannibal Metaphysics*, 87. Donna Haraway eminently develops the problematic of subjectivity as a multispecific concept, including the question of naturalcultural situation; in her discussion of the work of Barbara Smuts on baboons - with reference to Jacques Derrida's meditation on animals - Haraway points, for example, to the possibility 'to place baboons and humans together in situated histories, situated naturecultures, in which all the actors become who they are *in the dance of relating*, not from scratch, not ex nihilo, but full of the patterns of their sometimes-joined, sometimes-separate heritages both before and lateral to *this* encounter.' See Donna Haraway, *When Species Meet* (Minneapolis: University of Minnesota Press, 2008), 25.

55 Viveiros de Castro, *The Relative Native*, 191–324.

56 Ibid., 249–72.

57 Jan Patočka, *The Natural World as a Philosophical Problem* (Evanston, IL: Northwestern University Press, 2016), 52–84.

58 Viveiros de Castro, *The Relative Native*, 257.

59 Viveiros de Castro, speaking of the problem of the encounter with the other—the 'you'—implicated in 'becoming other', describes the relational context of this becoming as having 'a particular phenomenological quality'. It is in this context, effectively of metamorphosis *as* the exchange of perspectives, that he qualifies the meaning of the concept of 'supernature'. See Viveiros de Castro, *The Relative Native*, 290.

60 Viveiros de Castro's concept of 'affect' is perhaps derived in part from Deleuze and Guattari, considering the importance of their philosophy for his thought. See Deleuze and Guattari, *What is Philosophy?*, 163–200. For a critique of Deleuze and Guattari's concept of affect in their polemic on 'becoming-animal' see Donna Haraway, 'Becoming-Animal or Setting out the Twenty-Third Bowl?' in *When Species Meet* (Minneapolis: University of Minnesota Press, 2008), 27–35.

61 Patočka, *Natural World*, 73, my emphasis.

62 Ibid., 75.

63 The phenomenological analysis of generational structures has been explored by Anthony Steinbock with the concept of *generativity*. Steinbock hesitates to project beyond the human, but his work indicates that the germ for a generative phenomenology of the non-human is already present and even developed in Husserl. See Anthony J. Steinbock, *Home and Beyond: Generative Phenomenology After Husserl* (Evanston: Northwestern University Press, 1995). Such a generative phenomenological understanding of non-human life is antithetical to conceptions of animality projected onto colonized bodies as 'eminently mechanical, almost physical things', critically exposed in the work of philosophers such as Achille Mbembe and Valentin Mudimbe. See Achille Mbembe, *On the Postcolony* (California: University of California Press, 2001), 236.

64 Edmund Husserl, *Zur Phänomenologie der Intersubjektivität: Texte aus dem Nachlaß Dritter Teil: 1929–1935* (The Hague: Martinus Nijhoff, 1973), 173. In the full passage from which this quotation is drawn, Husserl considers animal and plant 'generativity', indicating clearly the anthropological neutrality of his concept of subjectivity. I was directed to this in Steinbock, *Home and Beyond*, 280, note 13. The extent to which Husserl's phenomenology is open to non-human intentionality and world constitution is a topic that calls for further research. See Tao DuFour, *Husserl and Spatiality: Toward a Phenomenological Ethnography of Space* (London: Routledge, 2019 forthcoming).

65 Aldo Leopold, *A Sand County Almanac* (Oxford: Oxford University Press, 1968), 201–6. The urgency of the matter in the face of anthropogenic climate change is taken up with polemical force by Viveiros de Castro in the book that he co-authored with philosopher Déborah Danowski, *The Ends of the World* (Cambridge: Polity Press, 2017).

Bibliography

Andrade, Oswald de. 'Cannibalist Manifesto'. *Latin American Literary Review* 19, no. 38 (1991): 38–47.

Avramescu, Cătălin. *An Intellectual History of Cannibalism*. Princeton: Princeton University Press, 2011.

Bary, Leslie. 'Oswald de Andrade's "Cannibalist Manifesto"'. *Latin American Literary Review* 19, no. 38 (1991): 35–6.

Bedford, Joseph. 'Creativity's shadow: Dalibor Vesely, Phenomenology and Architectural Education (1968–1989)'. PhD dissertation. Princeton University, 2018.

Carr, David. *Phenomenology and the Problem of History: A Study of Husserl's Transcendental Philosophy*. Evanston, IL: Northwestern University Press, 1974.

Casey, Edward. *Earth-Mapping: Artists Reshaping Landscape*. Minneapolis: University of Minnesota Press, 2005.

Crowell, Steven. '"Idealities of Nature": Jan Patočka on Reflection and the Three Movements of Human Life'. In *Jan Patočka and the Heritage of Phenomenology: Centenary Papers*, edited by Ivan Chvatík and Erika Abrams, 7–22. Dordrecht: Springer, 2011.

Danowski, Déborah, and Eduardo Viveiros de Castro. *The Ends of the World*. Cambridge: Polity Press, 2017.

Deleuze, Gilles. *The Fold: Leibniz and the Baroque*. London: Athlone Press, 2001.

Deleuze, Gilles, and Félix Guattari. *What Is Philosophy?* New York: Columbia University Press, 1994.

Derrida, Jacques. 'The Gift of Death.' In *The Gift of Death and Literature in Secret*, 1–116. Chicago: University of Chicago Press.

Descola, Philippe. *Beyond Nature and Culture*. Chicago: University of Chicago, 2013.

Descola, Philippe. *The Ecology of Others*. Chicago: Prickly Paradigm Press, 2013.

Descola, Philippe, and Gísli Pálsson, eds. *Nature and Society: Anthropological Perspectives*. London: Routledge, 1996.

DuFour, Tao. *Husserl and Spatiality: Toward a Phenomenological Ethnography of Space*. London: Routledge, 2019 forthcoming.

Federici, Silvia. *Caliban and the Witch: Women, the Body and Primitive Accumulation*. Brooklyn, NY: Autonomedia, 2014.

Gadamer, Hans-Georg. *Truth and Method*. New York: Continuum, 2000.

Haraway, Donna. *When Species Meet*. Minneapolis: University of Minnesota Press, 2008.

Heidegger, Martin. *Being and Time*. Oxford: Blackwell, 2002.

Held, Klaus. 'Husserl's Phenomenology of the Life-World.' In *The New Husserl: A Critical Reader*, edited by Donn Welton, 32–62. Bloomington: Indiana University Press, 2003.

Henare, Amiria, Martin Holbraad, and Sari Wastell, eds. *Thinking Through Things: Theorizing Artefacts Ethnographically*. London: Routledge, 2007.

Holbraad, Martin. 'The Power of Powder: Multiplicity and Motion in the Divinatory Cosmology of Cuban Ifá (or *Mana*, Again)'. In *Thinking Through Things: Theorizing Artefacts Ethnographically*, edited by Amiria Henare, Martin Holbraad and Sari Wastell, 189–225. London: Routledge, 2007.

Holbraad, Martin. 'Turning a Corner: Preamble for "The Relative Native" by Eduardo Viveiros de Castro'. *HAU: Journal of Ethnographic Theory* 3, no. 3 (2013): 469–71.

Holbraad, Martin, and Morten Axel Pedersen. *The Ontological Turn: An Anthropological Exposition*. Cambridge: Cambridge University Press, 2017.

Huang, Kuan-min. 'Toward a Phenomenological Reading of Landscape: Bachelard, Merleau-Ponty, and Zong Bing.' In *Phenomenology and Human Experience*, edited by Chung-chi Yu and Kwok-ying Lau, 45–63. Nordhausen: Verlag Traugott Bautz, 2012.

Husserl, Edmund. *The Crisis of European Sciences and Transcendental Phenomenology: An Introduction to Phenomenological Philosophy*. Evanston, IL: Northwestern University Press, 1970.

Husserl, Edmund. *Zur Phänomenologie der Intersubjektivität: Texte aus dem Nachlaß Dritter Teil: 1929–1935*. The Hague: Martinus Nijhoff, 1973.

Ingold, Tim. *The Perception of the Environment: Essays on Livelihood, Dwelling, and Skill*. London: Routledge, 2000.

Kohák, Erazim. *Jan Patočka: Philosophy and Selected Writings*. Chicago: University of Chicago Press, 1989.

Kohn, Eduardo. *How Forests Think: Toward an Anthropology Beyond the Human*. Berkeley: University of California Press, 2013.

Latour, Bruno, and Steve Woolgar. *Laboratory Life: The Construction of Scientific Facts*. Princeton: Princeton University Press, 1986.

Leatherbarrow, David. *Topographical Stories: Studies in Landscape and Architecture*. Philadelphia: University of Pennsylvania Press, 2004.

Leopold, Aldo. *A Sand County Almanac*. Oxford: Oxford University Press, 1968.
Lévi-Strauss, Claude. *The Savage Mind*. Chicago: University of Chicago Press, 1966.
Madureira, Luis. *Cannibal Modernities: Postcoloniality and the Avant-garde in Caribbean and Brazilian Literature*. Charlottesville: University of Virginia Press, 2005.
Malpas, Jeff, ed. *The Place of Landscape: Concepts, Contexts, Studies*. Cambridge, MA: MIT Press, 2011.
Marx, Karl. *The Ethnological Notebooks of Karl Marx*. Assen: Van Goorcum and Co., 1974.
Mbembe, Achille. *On the Postcolony*. California: University of California Press, 2001.
Moran, Dermot. 'Edmund Husserl's Phenomenology of Habituality and Habitus'. *Journal of the British Society for Phenomenology* 42, no. 1 (2011): 53–76.
Paiva, José de, ed. *The Living Tradition of Architecture*. London: Routledge, 2017.
Patočka, Jan. 'The "Natural World" and Phenomenology'. In *Jan Patočka: Philosophy and Selected Writings*, edited by Erazim Kohák. Chicago: University of Chicago Press, 1989.
Patočka, Jan. *Heretical Essays in the Philosophy of History*. Chicago: Open Court, 1996.
Patočka, Jan. *Body, Community, Language, World*. Chicago: Open Court, 1998.
Patočka, Jan. *The Natural World as a Philosophical Problem*. Evanston, IL: Northwestern University Press, 2016.
Steinbock, Anthony J. *Home and Beyond: Generative Phenomenology After Husserl*. Evanston, IL: Northwestern University Press, 1995.
Steiner, Henriette, and Maximilian Sternberg, eds. *Phenomenologies of the City: Studies in the History and Philosophy of Architecture*. Surrey: Ashgate, 2015.
Strathern, Marilyn. *Property, Substance and Effect*. London: Athlone Press, 1999.
Tsing, Anna Lowenhaupt. *The Mushroom at the End of the World: On the Possibility of Life in Capitalist Ruins*. Princeton: Princeton University Press, 2015.
Vesely, Dalibor. *Architecture in the Age of Divided Representation: The Question of Creativity in the Shadow of Production*. Cambridge, MA: MIT Press, 2004.
Viveiros de Castro, Eduardo. *From the Enemy's Point of View: Humanity and Divinity in an Amazonian Society*. Chicago: University of Chicago Press, 1992.
Viveiros de Castro, Eduardo. 'Cosmological Deixis and Amerindian Perspectivism'. *Journal of the Royal Anthropological Institute* 4, no. 3 (1998): 469–88.
Viveiros de Castro, Eduardo. 'The Relative Native'. *HAU: Journal of Ethnographic Theory* 3, no. 3 (2013): 473–502.
Viveiros de Castro, Eduardo. *Cannibal Metaphysics*. Minneapolis: Univocal, 2014.
Viveiros de Castro, Eduardo. *The Relative Native: Essays on Indigenous Conceptual Worlds*. Chicago: HAU Books, 2015.

13 DESIGNING LANDSCAPES OF ENTANGLEMENT

Martin Prominski

The philosophy of landscape architecture consists of many strands, including ethics, ontology and aesthetics, which constantly have to be reinterpreted in a rapidly changing world. This chapter addresses one of these strands, the philosophy of nature, and reflects upon contemporary changes in how it is interpreted and the relevance of this for landscape architectural research and practice.

Why nature? Because, arguably, nature is currently the most seriously challenged key concept within landscape architecture—a concept that needs to be reassessed philosophically. The strongest argument for such a reassessment comes from the field of geology, which has announced a new geological epoch, following the Holocene: the Anthropocene. This term was coined by Paul Crutzen and Eugene Stoermer in 2000, and designates a completely new geological situation in which humanity now influences every square metre of the earth's surface, and its atmosphere too, for example through carbon or nitrogen emissions. Geology is able to trace human impact in the earth's sediments, and the impact is so strong that a return to the Holocene seems to be impossible. Consequently, there is no nature left in the classical philosophical sense. According to this classical notion—a concept that has dominated the Western world for centuries—there is a dichotomy between nature and (human) culture, and nature is a powerful agent in its own right, with its own inherent value, independent of human influence.

The philosophical challenges posed by the concept of the Anthropocene are considerable. The introductory statement by the curators of the 'Anthropocene' Project, a transdisciplinary international project over several years currently running at the Haus der Kulturen der Welt in Berlin, expresses it aptly: 'Nature, as we know it, is a concept that belongs to the past. No longer a force separate from and ambivalent to human activity, nature is not an obstacle nor a harmonious other. Humanity forms nature. Humanity and nature are one, embedded from within the recent geological record'.[1] Acknowledging the Anthropocene amounts to an unmistakable call to transcend the West's dualistic conception of nature versus culture and move towards a non-dualistic, unitary philosophy of nature. In other words, one may say that while unitary concepts of nature and culture are nothing new, especially in non-Western cultures, acknowledging the Anthropocene increases their importance enormously. Thus, we can build upon existing attempts to define non-dualistic philosophies of nature in order to develop landscape architectural approaches to research and design in unitary, synthetic ways. To develop such a foundation for landscape architectural theory and practice, this chapter starts by discussing the philosophies of Philippe Descola, a French anthropologist, and Bruno Latour, a French sociologist. For them, the Anthropocene is a productive category for developing concepts of the human and non-human that transcend the nature-culture dichotomy.

Dethroning naturalism: Descola's relative universalism

In a recent lecture, French anthropologist Philippe Descola characterized the Anthropocene as a new era in which humankind has become a natural force.[2] He distinguishes clearly between the Anthropocene and Anthropization. The latter is a process which has been happening for 200,000 years, a co-evolution of humans and non-humans affecting most parts of the earth. Not even the Amazon rainforest is untouched nature; it is a largely anthropogenic ecosystem. This sounds similar to the characterization of human effects in the Anthropocene—yet Descola sees a difference between the two. Compared to the rather local co-evolutionary effects of Anthropization, human impact today has reached a global and systemic scale, leading to cumulative and accelerated climate change, ocean acidity and biodiversity loss. Descola explains this radical development with the term 'naturalism'. By naturalism, he means the specifically Western type of relationship between humans and non-humans, according to which the privilege of possessing a mind and soul is bestowed only on humans, while non-humans are just physical matter. One of the main motivations for his work is to explain that this notion—with all its destructive consequences—is just one of four possible ways in which humans can relate to non-humans, and that we need to modify Western naturalism in order to escape from its 'contemporary tyranny'.[3] In his magnum opus, *Beyond Nature and Culture* (2013), Descola develops his framework of four ontologies regarding the relationship between humans and non-humans. In addition to naturalism, by analysing a huge number of ethnographic examples from all over the world he also identifies animism, totemism and analogism as such ontologies. The latter three all operate without the dichotomy of nature and culture.[4]

When Descola addresses nature conservation as an example of the potential positive effects of acknowledging multiple ontologies, his thinking becomes immediately relevant for landscape architecture. He criticizes the fact that international nature conservation policy is closely linked to the cosmology of naturalism, which has dominated European thought for at least two centuries. This relatively young cosmology is certainly not shared by all others on the planet. What, then, would a more universal ethics of nature look like? In terms of nature conservation, it would mean that measures proposed by naturalism, such as the preservation of biodiversity, ecosystem services and carbon storage, might be of secondary importance for proponents of animism. According to the latter, humans have intersubjective relationships with non-humans. Animals are treated like humans, yet hunting them is allowed if done respectfully and carefully. To illustrate this ontology of animism, which sounds utopian to Western ears, Descola cites a document which the indigenous community of Sarayaku in the Ecuadorian Amazon presented to the Climate Summit in Paris in December 2015, and which calls for a new legal category of protected territories called *Kawsak Sacha* (Living Forest):

> Whereas the western world treats nature as an undemanding source of raw materials destined exclusively for human use, *Kawsak Sacha* recognizes that the forest is made up entirely of living selves and the communicative relations they have with each other. [...] These selves, from the smallest plants to the supreme beings who protect the forest, are persons (*runa*).
>
> *Kawsak Sacha* is where [we] interrelate with the supreme beings of the forest in order to receive the guidance that leads [us] along the path of *Sumak Kawsay* (Good

> Living). This continuous relation that we [...] have with the beings of the forest is central, for on it depends the continuity of the Living Forest, which, in turn permits a harmony of life among many kinds of beings, as well as the possibility that we all can continue to live into the future.[5]

This approach expresses a diversification of conservation strategies beyond the dominant dualistic Western concept of nature versus culture. Descola aims to define an ecology of relationships, in which the various relationships between human and non-human are analysed and developed in a differentiated way. The value of Descola's conception for the Anthropocene, as well as for landscape architecture, is that it offers consistent alternatives to predominant Western naturalism. Reflecting on these alternatives in an integrated manner, and applying them specifically in each design context, could lead to less destructive human actions in relation to the earth.

In the Anthropocene, everything becomes a matter of design: Bruno Latour's philosophy of design

Another eminent French figure in the fields of anthropology, philosophy and sociology who has reflected on the consequences of the Anthropocene for our understanding of nature is Bruno Latour. Since his book *We Have Never Been Modern* (1993), he has been grappling with the modern distinction between nature and culture. For him, the Anthropocene is another reason to refocus on new entanglements between former adversaries:

> 'Tomorrow,' those who have stopped being resolutely modern murmur, 'we're going to have to take into account even more entanglements involving beings that will conflate the order of Nature with the order of Society; tomorrow even more than yesterday we're going to feel ourselves bound by an even greater number of constraints imposed by ever more numerous and more diverse beings'.[6]

From this starting point—the abandonment of the distinction between nature and society—Latour develops a nexus of thought which is of particular interest for landscape architecture. If everything on earth (and beyond) is steeped in human activity and meaning, then there are inevitably countless relations between humans and non-humans in an 'entangled pluriverse'.[7] We cannot reflect on these relations passively, from a distance, as matters of fact; we are called on to work actively on these relational matters of concern. Latour calls this active work 'composition' and goes as far as writing a 'Compositionist Manifesto'. It includes a comprehensive reflection on the meaning of composition:

> Even though the word 'composition' is a bit too long and windy, what is nice is that it underlines that things have to be put together (Latin *componere*) while retaining their heterogeneity. Also, it is connected with composure; it has clear roots in art, painting, music, theater, dance, and thus is associated with choreography and scenography; it is not too far from 'compromise' and 'compromising', retaining a certain diplomatic and prudential flavor. Speaking of flavor, it carries with it the pungent but ecologically correct smell of 'compost', itself due to the active 'de-composition' of many invisible agents. [...] Above all, a composition can fail and thus retains what is most important in the notion of constructivism (a label which I could have used as well, had it not been already taken by art history). It thus draws attention away from the irrelevant difference

between what is constructed and what is not constructed, toward the crucial difference between what is well or badly constructed, well or badly composed.[8]

For a landscape architect, this description of composition has a familiar ring: putting together heterogeneous assemblies on the basis of creativity, compromised by the site's conditions and the client's brief, and acknowledging that there is no perfect solution, only good ones or bad ones. Thus, it is no large step from composition to design, which is landscape architecture's mode of action. In Latour's reflections on design,[9] he explains that it had a rather limited meaning until very recently, especially in his native France: it meant putting a cosmetic sheen on things invented by serious engineers or scientists. But today, he asserts, design is more than just the surface; it is part of the very substance of production processes, and 'has been extended from the details of daily objects to cities, landscapes, nations, cultures, bodies, genes and to nature itself'.[10] That shaping a landscape is a design task will raise no eyebrows among landscape architects—but Latour's ambition goes much further, and the manner in which he includes nature and genes makes the radicalism of his arguments apparent. He concludes that everything is designed today, and quotes '*Dasein ist Design*'[11]—the marvellous German pun coined by the Dutchman Henk Osterling, which means 'being is designing'. Latour draws similar conclusions to Descola regarding nature conservation:

> Not only has nature disappeared as the outside of human action (this has become common wisdom by now); not only has 'natural' become a synonym of 'carefully managed', 'skilfully staged', 'artificially maintained', 'cleverly designed' (this is true especially of so called 'natural' parks or 'organic foods'); but the very idea that to bring the knowledge of scientists and engineers to bear on a question is to necessarily resort to the unquestionable laws of nature, is also becoming obsolete. Bringing in scientists and engineers is quickly becoming another way of asking: 'How can it be better redesigned?' The bricolage and tinkering elements always associated with design have taken over nature.[12]

To sum up, the Anthropocene is for Latour a strong final indicator that any hope of differentiating between science and politics, facts and values, nature and culture, has died.[13] The entanglements between humans and non-humans are entanglements of composition, an issue of design. Arguing in this way, he enlarges enormously the field of activity as well as the responsibility borne by designers, including landscape architects.

The Anthropocene and landscape architecture

Once the Working Group on the 'Anthropocene', which is part of the Subcommission on Quaternary Stratigraphy of the International Commission on Stratigraphy, has formalized the Anthropocene as a geological epoch (this ongoing procedure is complex and includes many questions, such as when the Anthropocene started),[14] it will be in all school textbooks and influential at the very base of society. Such a far-reaching concept needs constant critique regarding its meaning, its scope or even its name.[15] Yet even sharp critics such as Donna Haraway conclude that 'we will continue to need the term *Anthropocene*. I will use it too, sparingly; what and whom the Anthropocene collects in its refurbished netbag might prove potent for living in the ruins and even for modest terrain recuperation'.[16] Thus, it seems productive to discuss the qualities and deficits of the concept by using the term itself instead of inventing new ones. It is a quality of the concept of the Anthropocene that it is broad enough to include high-tech freaks as well as 'five-minutes-past-midnight pessimists'. Personally, I find these extremes misleading, and would rather argue for multilayered, intermediate approaches such as relative universalism or compositionism, as discussed above, or for conviviality,[17] *mésologiques*,[18] andscapes[19] or 'sociality among all living things'.[20]

The ideas put forward by Descola and Latour have shown that the Anthropocene is a strong motivator for developing new concepts of the relations between non-humans and humans. This motivation is accompanied by the conviction that the Anthropocene is not simply a neutral description of the enormous consequences of human impact; it is also a call to change and halt negative developments. According to Jan Zalasiewicz, professor of palaeobiology and convenor of the Working Group on the 'Anthropocene', 'much of this global change will be to the detriment of humans. Not all of it (Greenland, for example, is currently greening and booming), but the present and likely future course of environmental change seems set to create substantially more losers, globally, than winners'.[21] Thus, the new concepts should help to guide global change in a positive, sustainable direction. In the following, I will ask what role landscape architecture can play in this process. I will use 'entangling' as a keyword, since it summarizes the core of Descola's and Latour's ideas (and has also been used by many others),[22] and I will reflect on landscape architectural strategies to entangle non-humans, humans and time.

Entangling non-humans

Landscape architecture is the design discipline which—like no other—has the privilege of dealing with non-human living things. There are hardly any projects in which plants or soil are not addressed. So is it even necessary to raise this issue here? Well, there is a difference between *addressing* non-humans in a design and *entangling* them in it. If we as humans do not relate, via our senses, to non-humans, it will not be an entangling design. For example, ironically, in those tasks which address non-humans most intensively—i.e. in nature conservation areas—a separating approach prevails, based on naturalism in Descola's sense of the word. Entanglements between humans and non-humans are avoided and sometimes even forbidden by nature protection laws, for example in the European Union. Jon Hoekstra, chief scientist for the World Wildlife Fund, calls this 'fortress conservation' which 'sets nature apart from people' and 'forces a mutually exclusive trade-off between conserving biodiversity and meeting human needs'.[23] What are needed are designs which take care of plants and animals as well as human users in open spaces and allow them to relate to one another. A good example of this is the Buchholzer Bogen (Buchholz Arc) in Hannover. When the decision was made in 1995 to widen the Mittelland Canal in the built-up area of Hannover, the destruction of valuable habitats along the existing canal bank had to be compensated for under German environmental impact assessment law. The landscape architects at Nagel, Schonhoff and Partner took the opportunity to initiate a new habitat while at the same time enhancing the means of experiencing and accessing the canal landscape, which plays an important role in Hannover's open-space system. For the Buchholz Arc, they proposed that a small bulge in the canal bank should be constructed, thus creating an unusual waterland locality on the otherwise straight canal edge. The multilayered structure of indigenous plants attracts a lot of insects and birds, and the site is characterized by above-average biodiversity, so this is an entanglement of non-humans. Instead of being protected from humans, this new habitat was carefully integrated into the linear open-space system along the canal with the installation of a walkable sculpture. The Japanese artist Tadashi Kawamata designed a raised wooden boardwalk, which spans the waterbody and serves as a pathway as well as a platform for observing flora and fauna below (Figure 13.1). Thus, the Buchholz Arc fulfils the strict demands of nature conservation by means of a completely human-made design, and offers rich experiences of the interplay between humans and non-humans. It is an example of how to overcome the divide between nature and culture, or naturalism as characterized by Descola. However, projects which entangle humans and non-humans in this way are by no means the norm in landscape architecture. How can the discipline live up to Latour's prediction, cited above, that in the Anthropocene 'we're going to have to take into account even more entanglements involving beings that will conflate the order of Nature with the order of Society'?[24] In his 'Compositionist Manifesto', Latour suggests one possible option: 'tackl[ing] the tricky question of animism anew'.[25] To consider animism as a productive ontology for designing entanglements of non-humans and humans, we would have to acknowledge that other entities, such as animals, plants or minerals, have an 'interiority'. Interiority is associated with attributes usually assigned to the soul, mind or consciousness, such as intentionality, subjectivity, reflexivity or emotion.[26] The problem with such an approach has already been identified by Latour himself: 'It immediately gives a sort of New Age flavor to any such efforts, as if the default position were the idea of the inanimate and the bizarre innovation were the animate'.[27] However, there are some recent scientific findings which should convince even the most stubborn proponents of naturalism that plants can see, feel and remember.[28] This knowledge increases our understanding of how humans and plants are entangled, but the implications for landscape architecture of this new understanding have not yet been researched and still wait to be addressed. For entanglements between animals and

Figure 13.1 A bulge or small bay has been created on the edge of the Mittelland Canal (in the background) at the Buchholz Arc in Hannover, 2016. The bulge, a habitat with over-average biodiversity, is spanned by a walkable sculpture, which is connected to the local path network. © Prominski, Martin.

humans, there are already some research findings relevant to design. The Animal Aided Design project[29] researches how one can design using the life cycles of six exemplary species to enhance urban wildlife. One might complain that here one has almost lost sight of the human role in entanglement. To sum up, in terms of the entanglement of non-humans and humans, landscape architectural practice and research is waiting to be re-animated.

Entangling time

As Donna Haraway pointed out so precisely, the Anthropocene (or in her terminology, the Chthulucene) is characterized by 'dynamic, ongoing [...] forces and powers of which humans are a part, within which ongoingness is at stake'.[30] It is about 'real and possible timespaces',[31] and she proposes 'chipping and shredding and layering like a mad gardener, [to] make a much hotter compost pile for still possible pasts, presents, and futures'.[32] For landscape architecture, this entangling of time fits very well into a dynamic, open understanding of ecosystem design, which has been discussed and developed in landscape architecture for decades[33] and was recently well summarized in *Projective Ecologies*, edited by Chris Reed and Nina-Marie Lister (2014).

River landscapes are an appropriate example to illustrate the challenges of time-based design. The classical engineering-based approach has tried to control river processes and limit interactions by confining rivers within channels. Recent landscape architectural designs propose to break up these channels and enable unpredictable future entanglements of water, sediments, plants, animals and people. Many good examples have been realized, such as the River Ebro in Zaragoza[34] or the River Isar in Munich;[35] I would like to illustrate the approach in more detail by using the example of the River Aire, close to Geneva.[36] This design, by George Descombes and the Atelier Descombes and Rampini SA, is remarkable for two reasons. First, it avoids introducing any 'naturalistic' aesthetics into the morphodynamic processes of erosion and sedimentation which the river generates. A completely new, eighty-metre-wide riverbed has been designed by converting agricultural land to provide more space for floodwater. The riverbed's initial form consisted of precisely shaped lozenges, through and over which the river has been allowed to flow, slowly transforming the geometric shapes into the organic shape of a braided river over the years (Figure 13.2). Here one sees the artistic interplay of human and non-human agents, with an unpredictable outcome. Second, the project not only choreographs future processes, but also entangles the past. The historic canal running parallel to the new riverbed was retained as a cultural artefact by transforming it into a linear series of gardens positioned along and partially above it. The juxtaposition of and contrast between the historic, linear canal structure and the newly introduced dynamic river space running next to it provokes reflection on the river's state before and after, and on the ecological and cultural aspects of the current river. I do not know of any other landscape architecture project which entangles past, present and future in such a creative, multidimensional way.

In such cases, is it appropriate to say that landscape architectural research and practice is already doing well at entangling time? I would say not. Beyond river projects, it is difficult to find inspiring examples. There are many fascinating competition entries, such as the proposals by OMA for La Villette from 1982 or Field Operations for Downsview Park from 2001, but they have not been realized. It seems that clients—and often designers as well—prefer fixed, controlled appearances which are not intended to change over time. Thus, there is a lot of potential for future landscape architectural research to intensify the focus on process aesthetics[37] as well as process strategies[38] in order to increase time-based entanglements of humans and non-humans.

Figure 13.2 The two juxtaposed river courses of the River Aire near Geneva: the retained canal to the left with a linear series of gardens, the new riverbed to the right with partly eroded lozenges, 2014. © Chironi, Fabio.

Entangling humans

Landscape architectural projects are usually used by people; thus a focus on humans goes without saying in the discipline. However, this focus is often abstract rather than concrete. For example, when landscape architects sitting in their offices imagine a public space, they conjure up images of future use and users in their creative minds, but they are not in direct physical contact with the users or the site. A true entangling of humans calls for more levels of design interaction. For Bruno Latour, the design of a thing—e.g. a park or public square—is a 'gathering' (he follows Heidegger here) and thus automatically a 'collaborative design'.[39] What options are there for a complex collaboration or entanglement of designers and site users? Gleisdreieck Park in Berlin[40] is an excellent case study for examining this question, which addresses issues that have been discussed intensively since the 1960s. One major and widely applied path towards entangling humans in the design process is public participation. At Gleisdreieck Park, the participation was a mixture of established and new methods.[41] The first phase started in 2005 with a questionnaire which the administration and a social research institute had developed together with focus groups. It was sent to 1,600 randomly chosen households within a twenty-minute walk of the park. The 400 answers revealed that fifty per cent did not even know that a park was going to be built in their neighbourhood; therefore the administration decided to offer guided tours of the site. There were 2,200 participants on the thirty-four tours. This was followed by workshops with thirty-two groups of up to thirty people, who developed recommendations for the competition brief. Parallel to this, there was a moderated Internet forum. It received 70,000 visits by 7,800 users, and in a moderated phase a text with recommendations for the competition brief was written jointly by 200 participants. These recommendations submitted by the public played a crucial role in the brief for the first phase of the park competition, in which eighty-six landscape architecture firms participated. The nine jurors, including one representative from the community initiatives, chose eleven contributions to proceed to the second phase of the competition. With these eleven designs, a 'planning weekend' was held at which the public were invited to discuss the plans directly with the designers and jurors. Six hundred people attended this weekend. The results of these dialogues were integrated into the recommendations for the eleven teams as to how they should improve their designs in the second, decisive phase of the competition. According to one juror, this weekend had a significant impact on the development of the eleven designs, as well as on the decision criteria for the jury.[42] This type of intense public participation, which in this case was highly successful, still leaves the subsequent decisions about shaping and transforming the space to the experts, i.e. landscape architects and administration.

A less widely applied but even more intense means of entangling humans is 'commoning', whereby people transform the space themselves. According to David Bollier, commoning is a social practice characterized by 'acts of mutual support, conflict, negotiation, communication and experimentation that are needed to create systems to manage shared resources. This process blends production (self provisioning), governance, culture, and personal interests into one integrated system'.[43] At Gleisdreieck Park, this approach was significant for several reasons. First of all, today's park was only able to become reality because many local initiatives saw the derelict railway site as their own urban common land and had been fighting since the 1970s against ideas about changing it into a road, an amusement park or a residential area. Later, while designing the park, community groups managed to claim some areas as their 'commons'. Here they were able to realize their own ideas, such as intercultural gardens, allotments or a space for experiencing nature (Figure 13.3). This process was not without conflict—the landscape architects, and even more so the community groups, had to compromise—but in the end the park was seen as a

Figure 13.3 The 'Nature Experience Space' in the eastern part of Gleisdreieck Park in Berlin, which has been designed and built by a community initiative, 2014. © Prominski, Martin.

great success,[44] and the administration has dubbed it 'The Park of 1,000 Voices'.[45] Furthermore, the entanglement will continue in the future, because in November 2014 a group of ten civic representatives was elected by residents, park users and interested parties to decide, together with other stakeholders, on future developments in the park.[46]

This example expresses the enormous potential of landscape architecture for entangling humans. The toolbox of public participation is already well equipped. However, intensive strategies of entangling become tricky: they demand that the designer take a back seat, because it is the people who are realizing their own ideas and transforming the space, rather than the designer. In an era when commons are gaining more and more momentum,[47] the relationship between designing and commoning is one of the most challenging topics of landscape architectural research.

Conclusion and outlook

Charles Darwin started the final paragraph of *On the Origin of Species* with a poetic image:

> It is interesting to contemplate an *entangled* bank, clothed with many plants of many kinds, with birds singing on the bushes, with various insects flitting about, and with worms crawling through the damp earth, and to reflect that these elaborately constructed forms, so different from each other, and dependent on each other in so complex a manner, have all been produced by laws acting around us [My emphasis].[48]

His contemplation of entanglements operates from a distance and expresses the modernist philosophy of nature, with a separation of objects and subjects. In the Anthropocene, we need a new perspective on the entangled bank. A distanced contemplation is impossible if non-humans and humans are inextricably intertwined. As humans are an active part of any entangled bank, we can conclude that these complex entanglements are always a matter of concern and an issue of design.

This leads to consequences for landscape architecture. I interpret this new perspective as a call to focus on entanglements at different levels. I have categorized three types, i.e. the entanglements of non-humans, humans and time, and it is important to add that these types should ideally overlap on each project. I see the integration of all three categories as an indicator of excellence in landscape architectural projects—a project such as Gleisdreieck Park serves as an example of this. I have also developed ideas and questions for landscape architectural research that arise from the three types of entanglements, and I have hinted at new directions, such as re-animation, time-based aesthetics and design strategies, and designing commons.

To summarize, landscape architecture has already experimented with answers to questions raised by the Anthropocene. Landscape architectural theory and practice—as can be seen when reflecting on the three projects above (and there are many more)—is already able to operate from a non-dualistic perspective, and is currently developing complex entanglements between non-humans and humans in space and time. Is there any other discipline that works so creatively at the vibrant interface between humans and non-humans? The door is open now for landscape architectural research to leave its underdog position, and to articulate its unique qualities in the context of transformative science and the Anthropocene.

Notes

1 Bernd Scherer and Katrin Klingan, 'The "Anthropocene" Project: An Opening', Haus der Kulturen der Welt, January 2013, http://hkw.de/en/programm/projekte/2013/anthropozaen_eine_eroeffnung/start_anthropozaen_eine_eroeffnung.php.

2 Philippe Descola, 'Relativer Universalismus', *Lettre International* 112 (2016): 107–12.
3 Ibid., 112.
4 For a short summary, see Philippe Descola, 'Who Owns Nature?' *Books & Ideas*, 21 January 2008, http://www.booksandideas.net/Who-owns-nature.html.
5 Eduardo Kohn, 'Ecopolitics', *Cultural Anthropology*, 21 January 2016, https://culanth.org/fieldsights/796-ecopolitics.
6 Bruno Latour, *An Inquiry into Modes of Existence: An Anthropology of the Moderns* (Cambridge, MA: Harvard University Press, 2013), 10.
7 Bruno Latour, 'An Attempt at a "Compositionist Manifesto"', *New Literary History* 41 (2010): 471–90.
8 Ibid., 473.
9 Bruno Latour, 'A Cautious Prometheus? A Few Steps Toward a Philosophy of Design (with Special Attention to Peter Sloterdijk)', Bruno Latour website, 2008, http://www.bruno-latour.fr/node/69.
10 Ibid.
11 Ibid.
12 Ibid.
13 Latour, *Inquiry*, 10.
14 Working Group on the 'Anthropocene', 'What is the Anthropocene? Current Definition and Status', accessed 18 February 2017, https://quaternary.stratigraphy.org/workinggroups/anthropocene/; Colin Waters et al., 'The Anthropocene Is Functionally and Stratigraphically Distinct from the Holocene', *Science* 351 (2016).
15 E.g. Donna Haraway, *Staying with the Trouble: Making Kin in the Chthulucene* (Durham, NC: Duke University Press, 2016).
16 Ibid., 47.
17 Steve Hinchliffe and Sarah Whatmore, 'Living Cities: Towards a Politics of Conviviality', *Science as Culture* 15, no. 2 (2006): 123–38.
18 Augustin Berque, 'Mésologiques', *Mesologiques: Études des milieux*, 29 October 2011, http://ecoumene.blogspot.de/p/argument.html.
19 Martin Prominski, 'Andscapes: Concepts of Nature and Culture for Landscape Architecture in the "Anthropocene"', *Journal of Landscape Architecture* 9, no. 1 (2014): 6–19.
20 Kinji Imanishi, *A Japanese View of Nature: The World of Living Things* (London: Routledge Curzon, 2002).
21 Jan Zalasiewicz et al., 'The New World of the Anthropocene', *Environmental Science and Technology* 44, no. 7 (2010): 2228–31.
22 See for example Anne Tietjen, *Towards an Urbanism of Entanglement: Site Explorations in Polarised Danish Urban Landscapes* (Aarhus: Arkitektens Forlag, 2011); Christopher Hight, 'Designing Ecologies', in *Projective Ecologies*, eds. Chris Reed and Nina-Marie Lister (New York: Actar, 2014), 101; Elisabeth Meyer, 'Beyond Sustaining Beauty: Musings on a Manifesto', in *Values in Landscape Architecture and Environmental Design: Finding Center in Practice and Theory*, ed. Elen Deming (Baton Rouge: LSU Press, 2015), 30–53.
23 Jon Hoekstra, 'Conservation 3.0: Protecting Life on a Changing Planet', *Live Science*, 26 July 2013, https://www.livescience.com/38481-new-approach-to-conservation.html.
24 Latour, *Inquiry*, 10.
25 Latour, 'Compositionist Manifesto', 481.
26 Descola, 'Relativer Universalismus', 109.
27 Latour, 'Compositionist Manifesto', 481.
28 Daniel Chamovitz, *What a Plant Knows* (New York: Scientific American, 2013).
29 Thomas Hauck and Wolfgang Weisser, *AAD: Animal Aided Design* (Freising: Technische Universität München, 2015).
30 Haraway, *Staying with the Trouble*, 101.
31 Ibid.
32 Ibid., 57.
33 Anne Spirn, *The Granite Garden: Urban Nature and Human Design* (New York: Basic Books, 1984); James Corner, 'Ecology and Landscape as Agents of Creativity', in *Ecological Design and Planning*, ed. George Thompson and Frederick Steiner (New York: Wiley, 1997), 81–108; Martin Prominski, *Landschaft entwerfen* (Berlin: Reimer, 2004).
34 Martin Prominski et al., *River. Space. Design* (Basel: Birkhäuser, 2017), 198.
35 Martin Prominski, 'Orchestrating the Agencies of Landscape', *GAM Architecture Magazine 07: Zero Landscape* (2011): 192.

36 Cf. Prominski et al., *River. Space. Design*, 286–9.
37 E.g. Elisabeth Meyer, 'Sustaining Beauty: The Performance of Appearance', *Journal of Landscape Architecture* 3, no. 1 (2008): 6–23.
38 Reed and Lister, *Projective Ecologies*.
39 Latour, 'Cautious Prometheus?'
40 Landscape architect Atelier Loidl: competition 2006, completion 2013. See, e.g. Andrea Lichtenstein and Flavia Alice Mameli, eds. *Gleisdreieck/Parklife* (Bielefeld: Transcript Verlag, 2015).
41 Senatsverwaltung für Stadtentwicklung und Umwelt, ed., *Der Park am Gleisdreieck: Idee, Geschichte, Entwicklung und Umsetzung* (Berlin: Medialis, 2013).
42 Ibid., 55.
43 David Bollier, 'Commoning as a Transformative Social Paradigm', *The Next System Project*, 28 April 2016, http://thenextsystem.org/commoning-as-a-transformative-social-paradigm/.
44 Agnes Müller, 'From Urban Commons to Urban Planning—or Vice Versa? "Planning" the Contested Gleisdreieck Territory', in *Urban Commons: Moving Beyond State and Market*, eds. Mary Dellenbaugh et al. (Basel: Birkhäuser Verlag, 2016), 148–63.
45 Senatsverwaltung für Stadtentwicklung und Umwelt, *Der Park*, 6.
46 Müller, 'Urban Commons to Urban Planning', 155.
47 Bollier, 'Commoning'.
48 Charles Darwin, *On the Origin of Species* (London: John Murray, 1859), 489f.

Bibliography

Berque, Augustin. 'Mésologiques'. *Mesologiques: Études des milieux*, 29 October 2011. http://ecoumene.blogspot.de/p/argument.html.

Bollier, David. 'Commoning as a Transformative Social Paradigm'. *The Next System Project*, 28 April 2016. http://thenextsystem.org/commoning-as-a-transformative-social-paradigm/.

Chamovitz, Daniel. *What a Plant Knows*. New York: Scientific American, 2013.

Corner, James. 'Ecology and Landscape as Agents of Creativity'. In *Ecological Design and Planning*, edited by George Thompson and Frederick Steiner, 81–108. New York: Wiley, 1997.

Crutzen, Paul. 'Geology of Mankind'. *Nature* 415 (2002): 23.

Crutzen, Paul, and Eugene Stoermer. 'The "Anthropocene"'. *E. F. IGBP Newsletter* 41 (2000): 17–18.

Darwin, Charles. *On the Origin of Species*. London: John Murray, 1859.

Descola, Philippe. 'Who Owns Nature?' *Books & Ideas*, 21 January 2008. http://www.booksandideas.net/Who-owns-nature.html.

Descola, Philippe. *Beyond Nature and Culture*. Chicago: Chicago University Press, 2013.

Descola, Philippe. 'Relativer Universalismus'. *Lettre International* 112 (2016): 107–12.

Haraway, Donna. *Staying with the Trouble: Making Kin in the Chthulucene*. Durham, NC: Duke University Press, 2016.

Hauck, Thomas, and Wolfgang Weisser. *AAD: Animal Aided Design*. Freising: Technische Universität München, 2015.

Hight, Christopher. 'Designing Ecologies'. In *Projective Ecologies*, edited by Chris Reed and Nina-Marie Lister, 86–105. New York: Actar, 2014.

Hinchliffe, Steve, and Sarah Whatmore. 'Living Cities: Towards a Politics of Conviviality'. *Science as Culture* 15, no. 2 (2006): 123–38.

Hoekstra, Jon. 'Conservation 3.0: Protecting Life on a Changing Planet'. *Live Science*. 26 July 2013. https://www.livescience.com/38481-new-approach-to-conservation.html.

Imanishi, Kinji. *A Japanese View of Nature: The World of Living Things*. London: Routledge Curzon, 2002.

Kohn, Eduardo. 'Ecopolitics'. *Cultural Anthropology*, 21 January 2016. https://culanth.org/fieldsights/796-ecopolitics.

Latour, Bruno. *We Have Never Been Modern*. Cambridge, MA: Harvard University Press, 1993.

Latour, Bruno. 'A Cautious Prometheus? A Few Steps Toward a Philosophy of Design (with Special Attention to Peter Sloterdijk)'. The website for Bruno Latour. 2008. http://www.bruno-latour.fr/node/69.

Latour, Bruno. 'An Attempt at a "Compositionist Manifesto"'. *New Literary History* 41 (2010): 471–90.

Latour, Bruno. *An Inquiry into Modes of Existence: An Anthropology of the Moderns*. Cambridge, MA: Harvard University Press, 2013.

Lichtenstein, Andrea, and Flavia Alice Mameli, eds. *Gleisdreieck/Parklife*. Bielefeld: Transcript Verlag, 2015.

Meyer, Elisabeth. 'Sustaining Beauty: The Performance of Appearance'. *Journal of Landscape Architecture* 3, no. 1 (2008): 6–23.

Meyer, Elisabeth. 'Beyond Sustaining Beauty: Musings on a Manifesto'. In *Values in Landscape Architecture and Environmental Design: Finding Center in Practice and Theory*, edited by Elen Deming, 30–53. Baton Rouge: LSU Press, 2015.

Müller, Agnes. 'From Urban Commons to Urban Planning—or Vice Versa? "Planning" the Contested Gleisdreieck Territory'. In *Urban Commons: Moving Beyond State and Market*, edited by Mary Dellenbaugh, Markus Kip, Majken Bieniok, Agnes Katharina Müller and Martin Schwegmann, 148–63. Basel: Birkhäuser Verlag, 2016.

Nowotny, Helga, Peter Scott and Michael Gibbons. *Re-Thinking Science*. Cambridge: Polity, 2001.

Prominski, Martin. *Landschaft entwerfen*. Berlin: Reimer, 2004.

Prominski, Martin. 'Orchestrating the Agencies of Landscape'. *GAM Architecture Magazine 07: Zero Landscape* (2011): 184–93.

Prominski, Martin. 'Andscapes: Concepts of Nature and Culture for Landscape Architecture in the "Anthropocene"'. *Journal of Landscape Architecture* 9, no. 1 (2014): 6–19.

Prominski, Martin. 'Research and Design in JoLA'. *Journal of Landscape Architecture* 11, no. 2 (2016): 26–9.

Prominski, Martin, Antje Stokman, Daniel Stirnberg, Hinnark Voermanek and Susanne Zeller. *River. Space. Design*. Basel: Birkhäuser, 2017.

Reed, Chris, and Nina-Marie Lister, eds. *Projective Ecologies*. New York: Actar, 2014.

Scherer, Bernd, and Katrin Klingan. 'The "Anthropocene" Project: An Opening'. Haus der Kulturen der Welt. January 2013. http://hkw.de/en/programm/projekte/2013/anthropozaen_eine_eroeffnung/start_anthropozaen_eine_eroeffnung.php.

Schneidewind, Uwe, and Mandy Singer-Brodowski. *Transformative Wissenschaft*. Marburg: Metropolis Verlag, 2013.

Schneidewind, Uwe, Mandy Singer-Brodowski, Karoline Augenstein and Franziska Stelzer. 'Pledge for a Transformative Science: A Conceptual Framework'. Wuppertal Paper 191, Wuppertal, Institute for Climate, Environment and Energy, 2016.

Seggern, Hille von, Julia Werner and Lucia Grosse-Bächle, eds. *Creating Knowledge: Innovation Strategies for Designing Urban Landscapes*. Berlin: Jovis, 2015.

Senatsverwaltung für Stadtentwicklung und Umwelt, ed. *Der Park am Gleisdreieck: Idee, Geschichte, Entwicklung und Umsetzung*. Berlin: Medialis, 2013.

Spirn, Anne Whistorn. *The Granite Garden: Urban Nature and Human Design*. New York: Basic Books, 1984.

Tietjen, Anne. *Towards an Urbanism of Entanglement: Site Explorations in Polarised Danish Urban Landscapes*. Aarhus: Arkitektens Forlag, 2011.

Waters, Colin, Jan Zalasiewicz, Colin Summerhayes, Anthony D. Barnosky, Clément Poirier and Agnieszka Gałuszka. 'The Anthropocene Is Functionally and Stratigraphically Distinct from the Holocene'. *Science* 351 (2016). doi:10.1126/science.aad2622.

Working Group on the 'Anthropocene'. 'What is the Anthropocene? Current Definition and Status'. Accessed 18 February 2017. https://quaternary.stratigraphy.org/workinggroups/anthropocene/.

Zalasiewicz, Jan, Mark Williams, Will Steffen and Paul Crutzen. 'The New World of the Anthropocene'. *Environmental Science and Technology* 44, no. 7 (2010): 2228–31.

14

ENLARGING THE URBAN ORCHESTRA

Re-thinking current approaches to landscape architecture

Matthew Gandy—interview by Henriette Steiner

Matthew Gandy is Professor of Geography at the University of Cambridge. He is a cultural, urban and environmental geographer with particular interests in landscape, infrastructure and more recently biodiversity. His publications include the award-winning books Concrete and Clay: Reworking Nature in New York City *(MIT Press, 2002) and* The Fabric of Space: Water, Modernity, and the Urban Imagination *(MIT Press, 2014). This contribution is a record of a conversation between Henriette Steiner and Matthew Gandy which took place in Cambridge in November 2016.*

HS: In your work on cities and landscape, you have dealt with the effects and legacy of modernity in its many guises. These are material, but also have to do with ideas and ideals. Today, it is commonly accepted that a hard distinction between urbanity and landscape no longer exists. From the point of view of landscape architecture, this means that the landscape architect needs to deal as much with the built as with the grown—with what has traditionally been called 'city' and what has been called 'nature'. In light of these challenges, could you outline your use of the different notions of 'the modern', including modernity and modernism, and the way you confront them in your work—discussing not just their meanings and definitions, but also their continued relevance?

MG: My starting point would be that we have multiple modernities, and these are often contradictory. We are already dealing with a very complex terrain. I find it quite useful to make a distinction between the modern as a particular experience of space, or a changing experience of space, and modernization, which often refers more to material transformations in the built environment, such as technological networks and things of that kind. The term 'modernism' I tend to use more sparingly, in relation to specific aesthetic or design-related developments, particularly in the early or middle decades of the twentieth century. In some of my writing, I have used the term 'late modern' as a loose epochal signifier, which for many reasons is greatly preferable to the brief focus on the postmodern that is now fading away. In terms of these different distinctions, and the development of landscape architecture in particular, there is an interesting set of debates about the spaces between buildings. I think this is where there is an interesting point of departure from some of the traditional architectural discourse—a more synoptic or integrated perspective on metropolitan areas, but at different scales. The other thing to mention is the increasing focus in landscape architecture on the proliferation of marginal zones from the 1970s onwards, especially related to post-industrial transitions. Clearly, geopolitical factors are involved in Berlin and other border zones, and I think that these empty or ambiguous spaces have been a particular focus of interest. I would emphasize that there is no consensus about what to do with these empty or marginal spaces. Something that has particularly interested me is the contrast between cultural and ecological responses, which is based to some degree on intrinsic values rather than more speculative appropriations. If you look at East London and developments surrounding the 2012 Olympics, a lot of very interesting spaces were erased by this vast speculative zone, often using the rhetoric of ecological restoration, biodiversity, and 'urban greening'.

HS: If we can therefore say that the question of multiple existing modernities challenges our understanding of landscape, may we ask to what extent the legacy of modernity hinders us from actually seeing and understanding, or even talking about, these vague spaces left over by the processes of modernization? And can we be clearer about defining what landscape as a category could be in light of this complex situation?

MG: I think many of the classic modern appropriations of landscape have been oriented towards a utilitarian ethos. What is especially interesting about recent debates is that they are trying to move away from that, attempting to acknowledge or include a wider range of voices or forms of cultural appropriation. We could perhaps argue that our understanding of landscape, and urban and industrial landscapes in particular, is now more complex and sophisticated than it may have been in the past. But there is another strand here which I think is important, which is that some of the classic modern landscapes were connected with the public realm, or a particular understanding of public culture. This presents a tension with the transition towards more neoliberal or speculative approaches to urban space. There is clearly a tension here, which you see exemplified in projects such as the High Line in Manhattan. On the one hand, it is presented as a very aesthetically and ecologically sophisticated appropriation of an abandoned space, and in another way, it is promoted as an enrichment of public culture. But it can also be read as a powerful agent for the speculative dynamics of gentrification in that particular neighbourhood. If you visit the site you have the feeling of stepping into the atrium of a museum, or even a corporate headquarters, with an extensive list of rules for how you should behave or respond to the space, along with its pervasive security apparatus. I understand that it is one of the most expensive public spaces in the world per square metre. The High Line provides a stark contrast with other attempts to appropriate abandoned transport infrastructure, such as the Parkland Walk in North London, which is very simple, and just comprises a means of access and a pathway. It is clearly a very different kind of structure. And in Paris, for instance, we have the Promenade Plantée, which is another simple example of an appropriated walkway that connects different spaces within the heart of the city.

HS: I have also looked at the High Line and found that it is extremely expensive to maintain. There is this feeling of something that is spontaneous in terms of the plants, but it actually requires a high level of maintenance—much more so than many other parks. At the same time, the spontaneous growth of plants that were at the site previously has been completely eliminated, along with a lot of urban life that was there. It is now no longer allowed or considered appropriate.

MG: I think that is a good point. I think that the High Line has involved elements of aesthetic and ecological erasure, as well as the loss of collective memories. And in terms of alternative social cultural histories of New York City, these kinds of spaces had a significant role, so the elimination or transformation of these spaces for a very specific kind of public is quite exclusionary in its wider dynamics.

HS: Definitely. It's so choreographed, almost as if it is a theatrical space. It's not very relaxing!

MG: No. And there are other examples in New York. The Central Park Conservancy was set up as part of a post-municipal park system, and it became a kind of cultural foundation with much stronger links to elite patterns of giving and wealthy benefactors. What was simply part of the municipal park system became incorporated into a much more hierarchical dynamic in the use and maintenance of public space. New patterns of philanthropic urbanism are now emerging. There is another project under consideration now, called the Low Line, which attempts to appropriate a series of underground spaces using a similar cultural and political dynamic, including crowdfunding for public space.

HS: That is interesting, because it all seems to come out of opposition to the neoliberal logics of urban space, which bears similarity to part of the original idea behind the High Line. Urban activist representatives of a homosexual community in the area were heavily involved in transforming the High Line into a public park, as an act of opposition which in many respects was born out of the contrast with some of the norms of mainstream urbanism, consumerism, etc. that now arguably dominate the area. It is as if the area was gentrified to an extreme point, so that it was completely transformed in a short period of time.

MG: The relationship between the transformation of public space and sexuality in the city is very significant. You could say that the dynamics of the High Line have ironically moved towards a 'de-queering' of public space, pulling marginal spaces into a more mainstream conception of sexuality within public space and culture.

HS: And it is steeped in all these notions of 'the wild' and 'the different', which have almost been transformed into a picture at the High Line, but as a deeper phenomenon they have simply gone from that space.

I'd now like to return to the question of scale, which you brought up a few moments ago, and also to questions about method and methodology. In contrast to other designers, who work at the level of an object, a building, or even an urban plan, landscape architects are in a special position, and they often need to look at scales that are much bigger. I guess you confront similar situations in your work when you look at large infrastructures. What landscape architects do is that they often make use of non-verbal forms of representation— that could be models, or forms of mapping, or drawing, and it can be analogue or digital. But I think there is often a problem with these methods, especially with mapping on this large scale, because they work in a Cartesian, abstract space. Even though these forms of representation can show something different, they are also dependent on the extreme form of abstraction that mapping needs. In disciplines such as human geography, this is a recognized problem. I wonder whether you see any parallels in your own considerations of how to deal with large-scale phenomena, in terms of finding a method and being reflective about it. Could there be possible crossovers between your experiences from geography and what happens in landscape architecture? (Figures 14.1, 14.2, 14.3.)

MG: I have always found it extremely useful to visit the spaces or places that I am writing about, and that is an integral part of my methodology. When I was writing about Parc Henri Matisse in Lille—which I studied because it was designed by Gilles Clément, who is a very interesting figure—it was important for me to go and spend time in the park at different times of the year, observing patterns of behaviour, small changes, and everyday practices such as maintenance. All of these things were important to me in building a richer sense of how that particular space functions. There was also the possibility to compare the material complexities of the site with some of the more abstract design sketches, plans, and writings of Clément himself. Bringing all of these things together forms part of my methodological approach.

Figure 14.1 Parc Henri Matisse, Lille (2011) © Matthew Gandy.

Figure 14.2 Parc Henri Matisse, Lille (2011) © Matthew Gandy.

Figure 14.3 Parc Henri Matisse, Lille (2011) © Matthew Gandy.

Another methodological development to mention is the growing technical sophistication of mapping such as the use of GIS, remote sensing, and other techniques. Some of these technologies of representation are now so sophisticated that people confuse the representation with reality, and I think this is potentially problematic, because we are still only looking at a cartographic or spatial representation. In some of the work I have done on urban biodiversity, I am struck by the impossibility of accurately mapping or representing the entire complexity of urban space, and I have been really fascinated by attempts to do that. For example, at the Institute of Ecology at the Technical University Berlin, their extremely sophisticated botanical maps of West Berlin from the 1980s are really remarkable cartographic and scientific achievements that nevertheless are by definition incomplete. They are representations of socioecological dynamics that were changing just as the map was completed. For me these botanical-cartographic exercises reveal some of the limits to urban ecology in terms of the numbers of people who can collaborate or take part in a sophisticated mapping project. How many people can accurately identify grasses, for example? More recently, there has been increasing emphasis on citizen science. Again, in realms such as environmental monitoring, biodiversity recording, and so on, we can produce remarkable cartographic documents that involve hundreds or even thousands of different contributors. But they are still nevertheless always by definition incomplete, and they are always created for a particular purpose, which may sometimes generate tensions around their production when data produced by the public becomes commodified. There is a degree of public suspicion over how data might be used. In the case of 'big-data' sets in the ecological sphere, for example, there are opportunities to sell data that have been collected by ordinary people to help with fields such as environmental consultancy, which can ironically contribute towards the destruction of really interesting sites. Here we see a series of tensions between the limits of knowledge, the development of citizen science, and the potential for commercial exploitation of different types of data.

HS: I wonder, when we look at research output in more traditionally academic forms, if we could also bring in this situated perspective in a positive way? Donna Haraway talks about what she calls 'feminist speculative fabulations in the scholarly mode' as one example of how to try out different anti-establishment forms of writing, but there could be many other ways of challenging typical forms of argument which could be useful in landscape architecture research.

MG: I have always found Donna Haraway's work very interesting. One of the things that is important about her work is the epistemological rigour that runs through it with the contextualization of specific debates and intellectual developments. On the one hand she is acutely aware of some of the larger-scale developments, such as Cold War science, the perturbations of global capital, and structural dimensions such as racism and gender inequality. But she is also sensitive to the local context, whether that be a laboratory, a scientific institution, a government department, or a field site. So there is this extremely interesting sensitivity to different levels or scales of causality.

HS: I would now like to return to your work on wastelands in light of the more general movement that is going on currently, including theoretically, in different areas of the social sciences and humanities, which we could call a 'material turn'. As material objects, wastelands are difficult to define. I wonder how we can understand their materiality without establishing new schools or inflexible methodologies, so that we can understand the claim of the concrete context in relation to both an individual as well as collective context.

MG: I should say that wastelands or marginal spaces have been a long-standing fascination for me. I have my own memories of empty spaces in London, with the profusion of particular kinds of nature. More recently I have been looking more closely at these sites, and I want to bring together a combination of perspectives. I want to take the scientific aspect seriously, and this is somewhat lacking in much of the landscape literature in terms of looking closely at how complex socioecological assemblages function. It is important to have a richer understanding of the material, socioecological complexities of these sites. I also want to bring other dimensions into the interpretive or analytical frame, such as collective memory, ethnographic insights into how people use such spaces and what they think about them, and engage with alternative cultural representations. The film *Fish Tank*, for example, directed by Andrea Arnold, has a remarkable landscape interlude that engages with one of these marginal spaces on the edge of London, which nevertheless serves as a vernacular public space characterized by spontaneous or accidental forms of nature. I am interested in trying to expand our conception of what a landscape is, and what it can be. I explore this in an essay entitled 'Unintentional Landscapes', where I try to think about spaces that are interesting, but not because they were made intentionally interesting in a narrow design sense, so that we encounter this tension between design and 'non-design', or at least have a richer understanding of the spontaneous dynamics of other-than-human nature, and the relationship between these different levels or forms of materiality and urban culture.

HS: If we begin to think about these in-between spaces as an epistemological problem in its own right, where do you see interesting things happening on the theoretical level to challenge binary thinking?

MG: I think that we can develop more nuanced theoretical perspectives on these types of urban spaces. The body of literature that is referred to as 'urban political ecology' has been extremely interesting and useful in advancing more historically and politically engaged perspectives on the production of urban space and the production of metropolitan nature, but I think there is still a lot of interesting work to be done. For example, in relation to new developments in ecological science itself, it strikes me that some of the urban political ecology literature doesn't really engage with the scientific literature much beyond the role of ecology as a kind of metaphorical marker or rhetorical signifier. It does not look closely enough at recent developments within the biophysical sciences, such as epigenetics and other key fields. Equally, in addition to the recent interest in the extension of urban frontiers and planetary dimensions to urban research, there is a whole series of meso or micro scales of analysis that are potentially very interesting. I also think that in terms of 'new materialities', the corporeal and epidemiological dimensions of the human body could be looked at in a more focused way. Particular developments such as the Zika virus require a much more complex analytical and theoretical framework.

In order to explore the theme of urban epidemiology, landscape design and landscape architecture we need to look at the specific hydrological dimensions of urban landscapes in a much more serious way that moves beyond generic readings of nature-culture relations. I am always intrigued by those architectonic projections with digitally rendered human figures, trees, and other elements of the landscape that are automatically generated from some obscure template, with almost no relation to the complexities of material or cultural conditions on the ground.

HS: The Anthropocene may be seen as one of the concepts that tries to grasp what could be called the planetary effect of industrial culture. Yet it also is a highly problematic concept, because it is almost a spatial metaphor that contains everything. It can be a 'wild projection plane' for various techno-optimist fantasies that connect with Paul Crutzen's original conceptualization, including an idea of geoengineering, and it involves a historicist logic that relies on typical 'modern' concepts such as progress, progression, and linear, epochal temporal understandings.

MG: I think the Anthropocene is an extremely difficult and complicated term to use in a rigorous way. The idea has been extensively appropriated by this geoengineering approach which you refer to. There are some new critiques emerging which are extremely important. I think that the neologism 'Capitalocene', which Jason Moore suggests, is a much more pertinent term to use. Of course, the term 'Capitalocene' would never gain approval within the geological sciences or the wider scientific community, but it actually carries more analytical weight than the Anthropocene.

HS: So if we work with a deeper background to culture, which will always be richer and will always be concrete—and which will also be a receiver or continuum of difference, rather than a perfect space that is homogenous—that begins to raise a lot of questions to do with ethics and political urgency. What do you see as the most important questions currently arising that challenge landscape architectural research and practice?

MG: The most pressing environmental challenges are clearly the interrelated threats of climate change and biodiversity loss. Of course, the rise of xenophobia, racism, and spiralling socioeconomic inequalities are also critically important, and impact on the purpose and meaning of public or shared spaces.

HS: My final question, therefore, concerns the role of the designer. Should we begin to think of a new kind of habitus, or a new kind of embodied knowledge that the designer should appropriate? And how might that become operationalized in this context?

MG: I think that urban designers—and within that category I would also include architects—have tended on the one hand to exaggerate their potential role, even though to a significant degree they are often subservient to their clients. On the other hand, however, they have tended to underplay their potential contribution to some of the smaller-scale possibilities of enriching the public realm and showing a greater sensitivity to social and ecological difference. There is much scope for design to be more sensitive and self-reflective in relation to the complexities of urban space. We should recall the work of the German urban sociologist Hartmut Häußermann, who consistently pointed out that making cities better must be achieved through social policy and cannot be advanced through a reliance on architectural design alone. In the political sphere, critical decisions affecting education, health, and other areas are ultimately more important than attempts to shape physical space.

HS: So, to return to the beginning of our discussion, this question of the design disciplines still being concerned with the modernist agenda to heal the city and get rid of all of the 'bad stuff'—the illnesses, and the differences.

MG: In a way we are still grappling with the Olmstedian legacy of the nineteenth century and a romantic perspective that 'beautiful places' can produce 'beautiful societies'. The full complexity of capitalist urbanization has not yet been engaged with sufficiently by the design profession.

HS: So your final point would be for designers to take a step back and take in those differences before they start to change the physical framework?

MG: I think designers in the twenty-first century need to have an understanding of a very wide range of topics and developments. They need to recognize their own limitations, and where appropriate they need to reach out to other sources of expertise, including local forms of vernacular knowledge as well as technical or scientific experts. I think of the German urban planner Martin Wagner's expression in relation to his experience of Weimar-era Berlin, where he referred to urban transformations being undertaken by an 'orchestra of experts'. Although this metaphor seems quite anachronistic, I do like this idea of people coming together to solve common challenges by using a huge range of expertise and insights.

HS: So perhaps if we really enlarge this orchestra to encompass things, and plants, and non-human participants…

MG: Yes, why not?

HS: …we could update the metaphor, maybe.

MG: Yes, exactly! We could update the metaphor to include a wider range of participants.

15

CITY, NATURE, INFRASTRUCTURE

A brief lexicon

Jane Wolff

Walking along the edge of almost any coastal city in North America, streets and buildings to one side, waves to the other, you might think that the landscape fell into tidy categories: land and water, culture and nature, metropolis and wilderness. Our observations are conditioned by our vocabularies, and everyday language points us toward oppositions between human forces and forces that people do not control. But cities test those divides. Living together in urban concentrations requires constant negotiation between people's intentions and the dynamic systems of the places they inhabit. Establishing and sustaining dense settlements means instigating reciprocal, cumulative engagements with the varied, plural processes of the environment. These back-and-forth exchanges are woven into the structures and practices of every urban landscape. The city is a hybrid, part culture and part nature.

Thirty years ago, Anne Whiston Spirn put forward the idea of the city as a granite garden, an ecosystem produced by:

> Complex interaction between the multiple purposes and activities of human beings and other living creatures and [...] the natural processes that govern the transfer of energy, the movement of air, the erosion of the earth, and the hydrologic cycle.[1]

She made the point that nature persists even in landscapes that seem entirely manufactured, in:

> Trees and gardens, and weeds in sidewalk cracks and vacant lots [...] the air we breathe, the earth we stand on [...] rain and the rushing sound of underground rivers buried in storm sewers [...] an evening breeze, the sun and the sky [...] dogs and cats, rats in the basement [...] falcons crouched on skyscrapers.[2]

She called for new urban design methods that considered non-human processes as powerful forces to be understood and cultivated rather than dismissed or controlled, and she offered strategies and tactics for remaking the city at scales from the region to the detail. Writing for both popular and professional audiences, she aimed to mobilize an effective response to the environmental crises of the moment.

A generation after *The Granite Garden*, designers and citizens faced with the escalating calamities of climate change and sea level rise are still working to come to terms with the hybridity of urban landscapes. Too often, language is an obstacle. Vocabulary shapes the way we know the world,[3] and the bluntness of ordinary definitions means that the city's complicated mix of geographical circumstances, cultural ambitions and environmental processes goes unrecognized—and unconsidered—in discussions about the future. This gap is both a technical and a political problem. Because it obscures the landscape's material characteristics, it impedes the development of strategies for physical transformation. Because it limits popular understanding of the forces that shape the landscape, it diminishes the content of public conversations about change. To move forward in ways that respond effectively to the complexity of what surrounds us, we have to address the distance between common language and common landscapes. We need a lexicon that speaks to the nuanced conditions of the city.

The word **infrastructure** makes a suitable test entry. Translated directly from its Latin components, the word means 'below structure'; it appeared in French in 1875 to describe the subgrade of a railway line. Imported shortly afterwards into English, it has come to designate the range of networks that support contemporary life: roads, bridges, telephone lines, the power grid, the water supply, the sewer system. In North America, where infrastructure spending is a significant subject of public debate and policy, conversations usually centre on concrete and cables. But *in situ*, in the context of particular places and cultures, infrastructure's operational definitions reveal much more than fixed constructions. Its manifestations—as objects, systems, practices and ideas—express and embody the complex, constant negotiation of people's plans for the city with environmental forces they cannot contain.

Registering and articulating the range of infrastructure's possible meanings means looking closely at actual landscapes, and the edge of San Francisco Bay offers useful raw material for examination. In many ways, it is a typical North American shoreline: the land we now take for granted was constructed to provide access to the deep water of a significant harbour; buildings sprang up to support commercial and industrial activities in and around its port; container shipping and offshore manufacturing left wharves and factories derelict; the water's edge emerged as an urban amenity; and now sea level rise points to radical change. It is also an unusually compelling landscape, and examining its version of typical conditions as a case study offers insights that pertain to places whose dramas are less vividly expressed.

Since colonization, the inhabitation of San Francisco Bay and its boundaries has involved intense interchanges between cultural intentions and natural phenomena. The bay has had simultaneous and different meanings: as the route to the mineral and agricultural wealth of California's interior; as the site of military fortifications that protected those resources; as the centre of a dispersed metropolitan region that grew around its harbours; and as the last chamber of the largest tidal estuary on the west coast of the Americas. It has been the subject of countless daydreams, individual and collective, gritty and romantic, about power, commerce, culture, nature and paradise. Those plural interpretations and the endeavours they engendered have produced an unending dialogue between people and the dynamic systems around them, at scales from the microscopic to the regional and beyond. A walk along the edge of the bay offers a careful observer the chance to see how that negotiation has shaped ideas, artefacts and practices fundamental to urban life[4] (Figure 15.1).

One of San Francisco's first negotiations was with the tides, and the physical constructions that ensued rested on the conceptual infrastructure of the **law**. Before the middle of the nineteenth century, large parts of what now comprises the edge of San Francisco Bay were marshes and mudflats. Flooded twice a day when the tides came in, sometimes wet and sometimes dry, the tidelands were disputed territory, neither land nor sea. Their status changed through legislation. In 1850, the United States Congress passed the Arkansas Swamp Lands Act, and American states gained title to swamp and overflowed lands. In 1851, the California Legislature followed

Figure 15.1 Seen as a whole and at a glance, San Francisco's shoreline looks fixed and unambiguous, 2012. © Wolff, Jane.

with the San Francisco Beach and Water Lots Act, which authorized the tidelands' sale on the condition that they were made dry.[5] The water's edge was redefined as property. The tidelands became a commodity, and speculators turned marshes and shallows into ground high enough to be developed and sold.

Negotiating the physical demands of ownership led to the construction of an infrastructural object, the **sea wall**. Built between 1860 and 1910 by the State Board of Harbor Commissioners,[6] the stone and concrete wall followed the mean low tideline across the bay's mudflats. Constructed so that its top was above high high tide,[7] the sea wall provided a stable boundary for rock and rubble that people piled on tidal flats to make new land. The level surface built between the deep water of the bay and the steep hills of the Coast Range supported (and still supports) warehouses, factories and office towers, and it provided the route for a **freeway** that connected the city centre to suburbs across the bay and down the peninsula. But the constructed surface looks more trustworthy than it is. Unconsolidated fill, saturated by water that migrates through the sea wall, lacks the structural integrity to withstand earthquakes. In big shocks, the ground loses the strength and stiffness to support what is above it. In 1906, blocks and blocks of buildings collapsed. In 1989, the freeway, which had looked big and tough enough to last forever, was damaged so badly that it had to be taken down piece by piece.

Just up from the water's edge, the **retaining walls** that turned San Francisco's hills into real estate represent another kind of negotiation between people and terrain: they make steep slopes into house lots. But their mediation has not produced a lasting treaty. Walls cannot stop change in the ground they have interrupted. Soil settles; water flows; tree roots grow; and earthquakes make everything move. The walls crack, and the cracks are colonized by plants that like dry, rocky soil. Untended, the walls would lose their hold on the hills. Retention means maintenance. Infrastructure consists of practices as well as objects (see Figure 15.2 a,b,c).

Inhabiting the bay landscape means contending with the sky, too, and the ritual of the **foghorns** turns infrastructural practices into a performance that echoes through the city. Its *mise en scène* is the dense, low fog that covers the bay and obscures the towers of the Golden Gate bridge. There are two actors: a foghorn operator and a ship's pilot. The operator watches the horizon from the deck of the bridge. When the fog approaches, he or she turns the foghorns on, and their coded sounds tell the pilot where the bridge's hidden towers stand. When the fog lifts, the pilot can navigate by sight. The operator turns the horns off, and the dialogue ends. The foghorn is a tool to negotiate the weather, but its efficacy depends on the vigilance of its operators and the skill of the pilots who guide commercial ships in and out of San Francisco Bay. The pilots, in turn, rely on an infrastructure of **information**: tide tables, charts of the bay's changing floor, signals and instructions from the Coast Guard's Vessel Traffic Service. Those data—and an understanding of physics—allow them to direct the weights and counterweights of tugboats that keep ships on a safe course to harbour.[8]

When ships arrive at **port**, they become part of an infrastructure system designed to negotiate the transfer of goods between land and sea. At Oakland, the bay's largest seaport, railways and highways converge at terminals where cranes move containers in and out of vessels more than a thousand feet long. Logistics and operations favour speed and volume, and the constant traffic dispatches and delivers manufactured goods and raw materials to and from all over the world.[9] But the port also brokers a second, less orderly exchange. A ship's ballast water can carry stowaways along with its registered cargo: plants and animals from foreign ports. Until the end of the twentieth century,[10] flora and fauna dumped with ballast established themselves in the bay and migrated up the estuary. Often without local predators, they wrought havoc on native species. Ten years ago, the Nature Conservancy found that the bay had a higher proportion of exotic species than any other aquatic region in the world.[11] Eradicating the invaders is impossible, because measures taken to kill them would harm natives as well. Negotiating globalized trade has created uncontrollable relationships among the bay's biota. The scale and volume of what comes and goes mean that San Francisco Bay is no longer just part of a regional ecosystem: it exists at the centre of a habitat map of the world.

Across the bay, transport infrastructure at Heron's Head has also had unexpected consequences for living systems: the **anchorage** of a bridge intended to mitigate the distance between San Francisco and the bay's east shore became the armature for an accidental wetland. In 1970, the Port of San Francisco began to build the landing for a southern bay crossing. Seven years later, a referendum stopped construction on the new bridge. The embankment sat abandoned, and waves and tides dropped sand and sediment on its southern edge. Over the course of two decades, plants and animals colonized the emerging soil, and a marsh grew out of the rubble. Then a legal settlement, restitution for the use of substandard fill, required the port authority to cultivate what had been wild. Port funds removed concrete, asphalt and debris from the site, and dug a channel to let the tides flow in. They still pay environmentalist gardeners to tend to the marsh. Fifty years after its construction, infrastructure meant for cars has made space for flora and fauna: the unfinished jetty anchors habitat rather than a southern bay crossing.

The design and construction of **bridges** represent a conversation not just with sediment and soil but with bedrock geology. The hills that surround and punctuate San Francisco Bay formed thirty million years ago, as the tectonic plate under the Pacific Ocean began to slide northward against the edge of the plate that supports North America. Folded and crumpled by the friction, the sea floor was pushed up into low mountains along the edge of the continent.

(a)

a. **retaining wall:** a wall designed to hold back earth.

b. **crack:** a linear fissure in a surface that has split but not yet broken into separate parts.

c. **volunteer:** a plant that emerges spontaneously where it finds soil and space.

d. **weephole:** a small hole that allows water to escape from the ground behind a retaining wall.

(b)

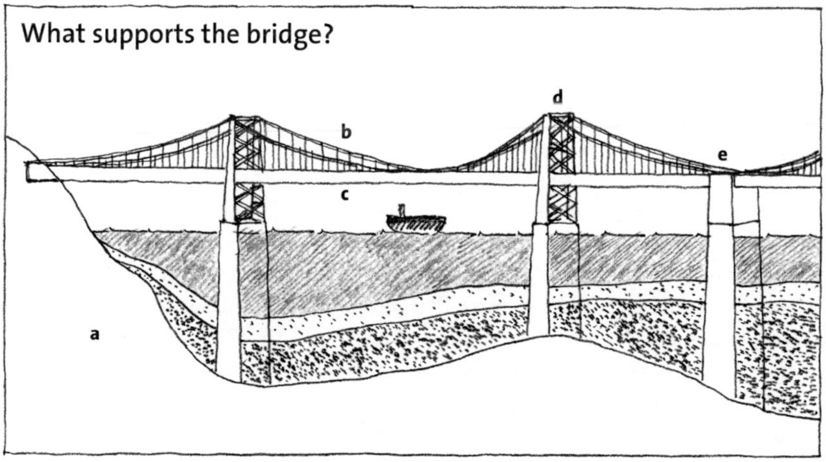

a. **bedrock:** the solid rock that lies under soil and sediment.

b. **cable:** a bundle of steel strands compressed in a steel casing.

c. **deck:** the travel surface of a bridge.

d. **tower:** the tall, slender structure that carries the cables of a suspension bridge.

e. **anchorage:** the structure to which suspension cables are fixed.

Figure 15.2a,b,c The places described by these drawings reveal that the city's infrastructure is a negotiation between cultural intentions and natural processes. © Wolff, Jane.

(c)

a. river: a natural watercourse, flowing in a line from higher to lower ground.

b. delta: a landscape created by rivers as they approach the sea. In the California Delta, the Sacramento and San Joaquin Rivers split into a series of winding, slow-moving channels that converge at the Carquinez Strait and flow into San Francisco Bay.

c. dam: a barrier built across a river to create a reservoir.

d. aqueduct: a pipe or channel engineered to bring water to the city from far away.

Figure 15.2a,b,c Continued.

Ten thousand years ago, when the last ice age ended, San Francisco Bay filled the valley between two chains of hills. One hundred and seventy years ago, give or take, San Francisco and Oakland were established on either side of the bay. Eighty years ago, the Bay Bridge created a thin strip of continuous ground across the watery centre of the metropolis. Though it defies land mass, it depends on the old, solid rock of the hills. The western half of the crossing comprises two suspension bridges. Each begins from an anchorage in bedrock—Rincon Hill on the west and Yerba Buena Island on the east—and they meet at a constructed anchorage halfway in between. Hung from concrete and steel towers and pulled tight at each anchorage, the cables that hold the bridge deck transfer its weight vertically to rock that lies below the shifting surface of the bay floor. At Yerba Buena, solidity demanded (and offered) two more rounds of geomorphological negotiation. A **tunnel** was blasted through the bedrock of the island to meet the bridge's eastern crossing. The rubble blasted out was used to build a sea wall around Yerba Buena Shoals, a high patch of bay floor just north of the island. Filled with sand and sediment dredged from the bay, the navigation hazard of the shoals became Treasure Island. Formed in historical time, it will likely as not be lost to the rising waters of the Anthropocene.

The **aqueduct** that brings drinking water to San Francisco mediates between people and what might be the largest and most ephemeral of environmental processes, the climate. In central California, summers are dry, and even rainy winters are not enough to provide sufficient water to support large populations. In the early twentieth century, San Francisco's public utility commission built a dam in the Sierra Nevada mountains, a few hundred kilometres inland, to capture the abundant snow and rain that fell there. Since then, pipes and reservoirs have brought the impounded water straight to town. The negotiation has meant better habitat for people and worse habitat for fish. The aqueduct carries water that would otherwise run down the west slope of the mountains through the Tuolumne River, join the San Joaquin as it crosses the Great Central Valley of California, and enter the vast inverted delta that feeds San Francisco Bay. What flows from the taps of the metropolis[12] is used at the expense of the largest estuary on the west coast of the Americas. For many decades that trade-off was not the subject of wide discussion. But the estuary is now in the late stages of collapse, and the public, whose concerns gave rise to the National Environmental Policy Act and the California Environmental Quality Act,[13] has registered the loss. The original, single negotiation between people and the environment, the negotiation that brought water to a big city in a dry region, has become a negotiation among people with plural values about where the water should go, to the city or to the fish. It will be more fraught, not less, as the metropolis keeps growing and the climate continues to change. The aqueduct has become a question about how people can live together (and with other living creatures) in intensity—and uncertainty.

Observing and imagining

This brief lexicon is meant as a manifesto, an argument that the observation and documentation of landscapes as we find them are necessary steps toward design.

The edge of San Francisco Bay is not unique. Inflected by the particulars of local geography, culture, institutions and mythologies, negotiations between people and a dynamic environment shape every urban landscape. The pressures of climate change mean that cities all over the developed world will be revisiting those conversations. The reprises need to take place in language that does justice to the complexity of the places they address. Landscapes exist not only as physical realities but also as ideas, and the vocabulary we use to describe the places we know now becomes the vocabulary we use to imagine what might be next. Our landscape lexicons constitute a framework for discussions—and decisions—about values, uses and forms. They set the terms for change.

So how and where might appropriate language emerge? The point of departure for this brief lexicon was the landscape as a person encounters it, and its definitions began as a series of drawings of artefacts, processes, places and phenomena that often go unnoticed. Documentary drawing is a powerful tool for delimiting observation.[14] The processes of framing and composition allow the exclusion of information that is not relevant to a situation, so a drawing can bring attention to what is most important. Adding analytical and diagrammatic conventions such as cross-section to a picture in perspective allows a drawing to reveal the hidden conditions and relationships that underlie what someone sees. Drawings that use familiar symbols (for instance, an old-fashioned figure of the wind or devices that measure time and distance) can depict invisible forces and dynamics (for instance, the energy that motivates waves or the gradual rising of the tide). When they include human figures, drawings show us our own place in the landscape; when they juxtapose different scales, they suggest the conjunction of places and moments that might seem remote from each other.

In addition to their inherent representational capacities, drawings can become the anchors for layers of annotation and commentary gathered through field study and library research. Naming and explaining the essential parts of a scene offers the chance to define subjects and vocabulary from the wide range of fields with a bearing on the hybrid circumstances of urban landscapes. Connecting technical terms and conditions to a drawing situates them in a context that different kinds of audiences can understand; it makes them more broadly legible and intelligible. And drawings can become springboards for open questions about the significance of what they show. The range of possible answers opens another avenue for landscape lexicons: to speak directly to the complex and often invisible relationships between natural forces and cultural intentions.

This way of working positions the definition of meaningful language in a reciprocal and intimate relationship to the habit of thoughtful observation—and to the practices of planning and design. If it became a standard part of the conceptual design phase of landscape architecture commissions, the development of a landscape lexicon would change the way designers, clients and constituents thought about context, programme, materials and form. In situations beyond the usual scope of landscape architecture, it could put design skills to work in a new way: developing a framework for technical and political discussions of the future. Broadly adopted as an analytical method, it would allow landscape architects to recast themselves as interpreters and translators of the constant negotiation between nature and culture, and as ecological diplomats whose job is not only to propose specific solutions for particular sites but also to engender public discussion about the range of possible futures.

Looking for a way forward in the hybrid conditions of the Anthropocene, we need to make ourselves literate about the places we live in. Learning to read and reimagine nature in the city means asking how the landscape has been transformed by patterns of inhabitation and because of ideas about meaning and value. It requires the consideration of local conditions in the context of their region. It depends on the ability to tie specific sites and moments to broad, complex forces and processes at work behind the scenes. It demands the examination of the present for traces of the past; it asks about the needs and desires we bring, as experts and as citizens, to our evolving negotiations with the environmental processes around us.

Vocabulary is a tool whose value depends on use. A landscape lexicon should offer verbal and visual language in the service of related causes: perception, the awareness of what is around us; observation, the act of careful watching; understanding, the recognition of significance; and speculation, the contemplation of possibility. It should bring together different kinds of information, at different scales, from different fields of expertise and points of view. It should ask people to consider the ways in which specific situations shape and reshape the meanings of everyday words, and it should give them the means to identify tangible patterns that point to abstract categories of meaning. It ought to be a technical document that examines how things work, a historical document that looks at how the landscape evolved (and is still evolving), a cultural document that brings values to light, and a political document that can be used to frame public discussion about how and why the landscape might respond to the pressures of the moment.[15]

Every dictionary is the product of its time and place; each one, in its 'wide-ranging engagement with words, meanings, artefacts, usage, and culture—and [...] the perspectives which might be variously taken on [them... reveals] how we, as speakers, understand the world and articulate what we perceive'.[16] Language is never fixed and never final, and neither are the hybrid landscapes of our cities. They vary over the course of days, seasons, years, decades and centuries, both as material artefacts and in our perceptions. And what we want from them and for them changes, too. A landscape lexicon can only be defined as a work in process.

Notes

1 Anne Whiston Spirn, *The Granite Garden: Urban Nature and Human Design* (New York: Basic Books, 1989), 4.

2 Ibid.

3 Linguists Edward Sapir and Benjamin Whorf argued that language simultaneously shapes and mirrors perception. Whorf wrote:

> The categories and types that we isolate from the world of phenomena we do not find there because they stare every observer in the face. On the contrary the world is presented in a kaleidoscopic flux of impressions which have to be organized in our minds. This means, largely, by the linguistic system in our minds.

Paul Kay and William Kempton, 'What is the Sapir-Whorf Hypothesis?' *American Anthropologist* 86 (1984): 66.

4 This essay belongs to a long-term research project that includes the exhibition 'Bay Lexicon' for the Bay Observatory gallery at the Exploratorium in San Francisco; a book manuscript of the same name, currently in process; 'Bay Lexicon: A Manifesto', a broadside prepared for the Exploratorium as a companion to the exhibition; and the essay 'Lexicon as Theory: Some Definitions at the Edge of San Francisco Bay', in *Pamphlet 20: Delta Dialogues*, eds. Christophe Girot, Susann Ahn, Isabelle Fehlmann and Lara Mehling (Zurich: GTA Verlag/ETH Zurich, 2017).

5 The Arkansas Swamp Lands Grant Act of 1850, which gave the states title to all swamp and overflowed lands, made tidelands available for reclamation. California received title to over two million acres; a series of state laws authorized their sale and reclamation. For a detailed account of this legislation and its consequences, see Matthew Booker, *Down by the Bay: San Francisco's History Between the Tides* (Berkeley: University of California Press, 2013) and Gerald Robert Dow, 'Bay Fill in San Francisco: A History of Change' (master's thesis, California State University, 1973), 4–18.

6 Dow, 'Bay Fill', 33–9.

7 San Francisco Bay's tides are semi-diurnal: there are two unequal high tides and two unequal low tides every day.

8 This information comes from an interview with Captain Drew Aune of the San Francisco Bar Pilots; the conversation took place aboard a pilot boat sailing from Pier 9 to the Golden Gate Bridge, 27 June 2012.

9 Approximately 1,750 ships travelled through the Port of Oakland in 2016. 'Port of Oakland May Get Second-Fewest Ship Visits in a Decade', Port of Oakland, last modified 17 August 2017, http://www.portofoakland.com/seaport/port-oakland-may-get-second-fewest-ship-visits-decade/.

10 Since 2000, the Ballast Water Management for Control of Nonindigenous Species Act (California Assembly Bill 703) has prohibited ships from releasing non-indigenous species into US territorial waters. To that end, the California State Lands Commission has established protocols for the management of ballast water. US Department of the Interior, Bureau of Land Management, *California Coastal National Monument, Draft Resource Management Plan/Draft Environmental Impact Statement* (Monterey: California State Office, Bureau of Land Management, 2004), 3.20–4.

11 Jennifer L. Molnar et al., 'Assessing the Global Threat of Invasive Species to Marine Biodiversity', *Frontiers in Ecology and the Environment* 6, no. 9 (2008).

12 The San Francisco Public Utility District's Hetch Hetchy reservoir and aqueduct supply water to approximately eighty-five per cent of the agency's 2.7 million residential, commercial and industrial customers in San Francisco and the suburban jurisdictions of Alameda, San Mateo and Santa Clara counties. San Francisco Water Power Sewer, 'Overview', accessed 1 December 2017, https://sfwater.org/index.aspx?page=355.

13 'National Environmental Policy Act', accessed 17 November 2017, https://ceq.doe.gov/index.html; California Natural Resources Agency, 'CEQA: The California Environmental Quality Act', accessed 17 November 2017, http://resources.ca.gov/ceqa.

14 I became convinced of the importance of looking carefully at the world—and of understanding the stories it offers—during my undergraduate education in documentary film-making. My thoughts about the capacity of drawing to communicate complex documentary information to diverse audiences emerged later through the development of projects on difficult landscapes, including *Delta Primer: A Field Guide to the California Delta* (San Francisco: William Stout Publishers, 2003) and the 2011 'Gutter to Gulf' initiative (co-authors Elise Shelley and Derek Hoeferlin), accessed 15 December 2017, http://www.guttertogulf.com.

15 This proposal for a landscape lexicon comes from my 'Lexicon as Theory', 17.

16 Lynda Mugglestone, *Dictionaries: A Very Short Introduction* (Oxford: Oxford University Press, 2011), 16.

Bibliography

Booker, Matthew. *Down by the Bay: San Francisco's History Between the Tides*. Berkeley: University of California Press, 2013.

California Natural Resources Agency. 'CEQA: The California Environmental Quality Act'. Accessed 17 November 2017. http://resources.ca.gov/ceqa.

Dow, Gerald Robert. 'Bay Fill in San Francisco: A History of Change'. Master's diss., California State University, 1973.

Kay, Paul, and William Kempton. 'What is the Sapir-Whorf Hypothesis?' *American Anthropologist* 86 (1984): 65–79.

Molnar, Jennifer L., Rebecca L. Gamboa, Carmen Revenga and Mark D. Spalding. 'Assessing the Global Threat of Invasive Species to Marine Biodiversity'. *Frontiers in Ecology and the Environment* 6, no. 9 (2008): 485–92.

Mugglestone, Lynda. *Dictionaries: A Very Short Introduction*. Oxford: Oxford University Press, 2011.

'National Environmental Policy Act'. Accessed 17 November 2017. https://ceq.doe.gov/index.html.

Port of Oakland. 'Port of Oakland May Get Second-Fewest Ship Visits in a Decade'. Accessed 1 December 2017. http://www.portofoakland.com/seaport/port-oakland-may-get-second-fewest-ship-visits-decade/.

San Francisco Water Power Sewer. 'Overview'. Accessed 1 December 2017. https://sfwater.org/index.aspx?page=355.

Spirn, Anne Whiston. *The Granite Garden: Urban Nature and Human Design*. New York: Basic Books, 1984.

US Department of the Interior, Bureau of Land Management. *California Coastal National Monument, Draft Resource Management Plan/Draft Environmental Impact Statement*. Monterey: California State Office, Bureau of Land Management, 2004.

Wolff, Jane. *Delta Primer: A Field Guide to the California Delta*. San Francisco: William Stout Publishers, 2003.

Wolff, Jane. 'Lexicon as Theory: Some Definitions at the Edge of San Francisco Bay'. In *Pamphlet 20: Delta Dialogues*, edited by Christophe Girot, Susann Ahn, Isabelle Fehlmann and Lara Mehling, 14–24. Zurich: GTA Verlag/ETH Zurich, 2017.

Wolff, Jane, Elise Shelley and Derek Hoeferlin. 'Gutter to Gulf'. Accessed 15 December 2017. http://www.guttertogulf.com.

PART IV

Designing with the past in the future

Politics, heritage, and sustainability

16

URGENT INTERVENTIONS NEEDED AT THE TERRITORIAL SCALE—NOW MORE THAN EVER

Kelly Shannon

In a view from the future, 2016–17 might well prove a critical turning point, at least from an Anglo-Saxon bias: for geopolitics, the environment, science, economics, activism and ethics. This chapter is an explicit response that was ignited by the precarious state of affairs which left so many around the world utterly bewildered; it is a way of traversing what seems like a genuine impasse, to find a possible 'means' through the workings of the discipline of landscape architecture. In June 2016, the UK voted to exit the European Union. Five months later, Donald Trump, the billionaire real-estate mogul and self-styled strongman with no political experience, who is contemptuous of allies, civic discourse and democratic convention, shocked the world in an electoral upset. He quickly proved business would not proceed as usual—with wildly inappropriate cabinet appointments (many of whom are vehemently at odds with the departments they run), compulsive tweeting, and departure from nearly all the conventions of a global leader. Numerous analyses blamed rising economic and geographical inequality as the primary reason for Trump's victory.[1] Continental Europe appears to have been emboldened by the American result, and since then elections in Germany, France, Italy and the Netherlands have further undermined fragile welfare states with far-right parties claiming victories. Governments are blamed for slow growth, rising unemployment, open immigration policies and continuing allegiance to the corporate, neoliberal international order over the common citizen.

Meanwhile, natural disasters continue to wreak havoc worldwide. In 2016, the El Niño weather event triggered more severe droughts, floods and landslides than usual, and more frequent cyclones across Africa, Asia and Latin America; it has been blamed for food and water shortages, heatwaves, forest fires and increased vulnerability to diseases, including the Zika virus. At the time of writing, as a string of devastating hurricanes, wildfires and mudslides from August to December 2017 has served as a wake-up call for many Americans to new environmental realities (Figure 16.1), while the American government remains steadfast in its refusal to discuss climate change; the issue remains sharply contested and entrenched along partisan lines. At the same time, for decades, science has evinced that the world has become significantly warmer, and experts have numbers to back their claims. According to America's National Oceanic and Atmospheric Administration and National Aeronautics and Space Administration, 2016 was the world's hottest year on record, with the temperature having risen 1.45 degrees Celsius above pre-industrial averages, precariously close to the 1.5 degree Celsius threshold that is considered a new international goal for limiting warming.[2] A report by the United Nations Office for

Figure 16.1 Hurricane Harvey, Houston, Texas, 2017. Entire residential districts of Houston, Texas were severely inundated during Hurricane Harvey in 2017. It should have been a wake-up call to new environmental realities in the US. © Sgt. Martinez, Daniel J. US National Guard.

Disaster Risk Reduction, *The Human Cost of Weather Related Disasters 1995–2015*, reports that weather-related disasters have killed over 600,000 people (roughly 30,000 per year) and injured or adversely impacted 4.1 billion global citizens.[3] Overall, it is estimated that annual economic losses from disasters stand at between US$250 billion and US$300 billion, extrapolating from a study of nationally reported disaster losses.[4]

At the same time, the world is witnessing the largest species migration ever. On the one hand, flora, fauna and humans are relocating to escape the perils of climate change. On the other hand, segments of the natural world (in the 'sixth extinction'[5]) and human world (including in war-torn regions, and among peasants and indigenous people who are prey to new 'markets' and ethnic cleansing) are facing extinction due to human-induced atrocities. In late 2017, the prospect of nuclear confrontation remains at its highest level since the crisis of the Bay of Pigs. Meanwhile, the consequences of ongoing human population growth, and its global economy driven by the digital revolution, are becoming more and more evident as time passes. Clearly, the expanding and wealthier populace demands more of the earth's finite resources; nearly globally, innovative technology is seen as *the* saviour. However, writers such as Bill McKibben, Naomi Klein, Edward O. Wilson, Naomi Oreskes, Erik Conway, Jedediah Purdy and even Pope Francis, amongst others, argue that there is an enormous risk attached to technical mega-fixes and a blind confidence in technical solutions and the manner in which market logics are grafted onto them without consideration of larger issues of the progress of humanity and history—inclusive of social development, ecology and ethics. According to these writers, we are now at a precipice, a critical tipping point, standing before the wholesale destruction of the planet. The never-ending, interlinked and adaptive cycles of growth, accumulation, restructuring and renewal occur in human and natural ecologies of nested sets of scales and time frames.

Figure 16.2 2 Freeway, Los Angeles, California, 2016. In the post-petrol age, LA's massive freeway network is envisioned to become social connector: a smog and carbon absorber, an important piece of sustainable hydrologic infrastructure, and a catalyst for civic, recreational, cultural and economic activities. © Stoss Landscape Urbanism: Reed, Chris, and Katherine Harvey, Scott Mitchell, Chris Reznich, Chloe Street and Tim Wilson of Stoss Landscape Urbanism, with Paul Kassabian of Simpson, Gumpertz and Heger.

Environmental writers and activists quite convincingly declare that there must be a reaffirmation of civic values, and that the worldwide momentum of protests which are forever springing up prove that there is a mounting awareness of climate change. The growth of new anti-establishment movements, in a wide and ever-broadening variety of contexts, challenges the neoliberal order. Grassroots organizations hold leaders accountable to the realities of science and the principles of justice. It was reported that more than 600,000 people from 175 countries worldwide took to the streets in November 2015 to call for a strong deal in France that would see a swift transition from fossil fuels to renewable energy.[6] Ultimately, on 22 April 2016, leaders from those same 175 countries signed the United Nations Paris Agreement, recognized as a huge victory for international collaboration. Yet it is a fundamentally weak and profoundly unjust pact, condemning future generations and front-line communities to bear the full force of climate catastrophe, with non-binding emissions reduction targets, lack of financial support, and no clear date to phase out fossil fuels. In another example of tempered success, the 4 December 2016 outcome of the Standing Rock Sioux Tribe (North Dakota) protest has temporarily blocked an oil pipeline being built near its reservation and invigorated the American environmental movement, but is perhaps threatened by the aggressive pro-oil polices of the Trump administration. The 'water protectors'—with their rallying cry '*Mni wiconi*!' ('water is life!')—upended their lives and those of sympathizers from around the globe to defend the claim that the tribe should have a say over its natural resources, specifically claiming the legal *rights* granted to them by the US government through treaties in the nineteenth century and federal laws in the twentieth century. At the time, it seemed that the US Army Corps of Engineers would have to search for alternative routes for the US$3.7 billion Dakota Access Pipeline (part of the nearly complete 1,170-mile pipeline, which crosses four states and is intended to transport as much as 550,000 barrels of oil a day from the oil fields of western North Dakota to a terminal in Illinois). However, all changed with the Trump administration, which has been forging a full-fledged assault on the environment. Ecological policies that were decades in the making, and their respective regulating agencies, are vanishing. On 1 June 2016, Trump announced that he would withdraw American commitment to the Paris Agreement on climate change mitigation. There was dissent. One hour after Trump's pronouncement, French president Emmanuel Macron addressed the American public from the Elysée Palace and offered refuge to American scientists studying climate change with the goal to 'make our planet great again'. At the national level, the governors of several US states formed the United States Climate Alliance[7] to, in a bi-partisan manner, continue to advance the Paris Agreement's objectives, despite federal withdrawal: businesses and mayors have also added commitments. As quickly as executive orders are issued, protests erupt (Figure 16.2).

It is heartening that opposition and social movements are taking root and beginning to spread far and wide, and that incredibly bold ideas about the environment and climate change are being initiated by leading intellectuals, policymakers, think tanks and entrepreneurs. Pope Francis' 2015 180-page encyclical on the environment reframed climate change with sweeping moral implications.[8] In subsequent speeches to the US Congress and United Nations, he bolstered his stance; at the UN, he said: 'It must be stated that a true "right of the environment" does exist.' An attack on the environment was an assault on the rights and living conditions of the most vulnerable, he said, warning that at its most extreme, environmental degradation threatened humanity's survival. 'Any harm done to the environment, therefore is harm done to humanity,' Francis said. 'The ecological crisis, and the large-scale destruction of biodiversity, can threaten the very existence of the human species.'[9] Naomi Klein, at once an advisor to the Pope and an activist writer, was also part of the group that penned Canada's *Leap Manifesto: A Call for a Canada Based on Caring for the Earth and One Another*,[10] which calls on the government to honour indigenous rights and move towards a clean economy fuelled by renewable energy. Academic, activist and prolific writer Bill McKibben's 350.org, begun in 2008, now operates in 188 countries as a grassroots organization to reduce the carbon dioxide in the atmosphere, from more than 400 particles per million to safe levels below 350 particles per million. Architects turned fierce political and environmental activists Arundhati Roy in India and Nnimmo Bassey in Nigeria are both prolific essayists and polemicists. Roy has been fighting dam-building in her country, while Bassey has led the charge against the destructive practices of oil and gas extraction not only in West Africa, but also more generally in the Global South.

Yet extremes of wealth, poverty and insecurity remain in perceptible levels of racial resentment, terror and simply the struggle to survive—in urban centres worldwide, and highlighted in the US by the Black Lives Matter campaign, in targeted urban warfare in Aleppo, and in the increasing demand for water, with 783 million people without access to clean and safe water, and nearly 2.4 billion without adequate sanitation.[11]

It is imperative for designers of the built environment to actively, not passively, situate themselves in the ever-evolving context. Landscape architects have a responsibility in these larger issues. They must be at the core of a categorical shift. Climate change is not pseudoscience, and social tensions are not conspiracy theories. Compromised summit declarations do not have a forceful impact. Landscape architecture must set the boundaries for policymakers and orient social movements. Bold and inspired projects must lead policy. It is now, more than ever, that landscape architecture is a necessity, since there is no assurance that policies will be directed in an intelligent trajectory.

As has been mentioned above, the contemporary world is at a tipping point—one that is disturbingly divided and environmentally devastated. Landscape architecture is the most powerful tool to marry social and ecological justice. It has the capacity to address the world's most pressing and fundamental problems, amongst which are myriad issues linked to climate change, deforestation, energy, water and food security. There are six broad strategies outlined below that address habitat preservation, new spatial and programmatic reconfigurations of territory, resource management and settlement morphologies, all of which require a crossing of disciplinary and scalar boundaries and out-of-the-box design thinking. It is partially the work of the landscape architect, who must work together with a myriad of other specialists.

Nature conservation to restore ecologies

The bold environmental prescription proposed by double Pulitzer Prize winner and biologist Edward O. Wilson in his impassioned 2016 book *Half-Earth: Our Planet's Fight for Life* is to set aside roughly half of the planet as a permanent natural preserve, undisturbed by humans. A large part of this consists of areas where nature—existing ecosystems with strong biodiversity—is already unique and intact (where, according to Wilson, habitats protected by governments and agencies already account for fifteen per cent of the earth's land area), including the Amazon region, the Congo Basin, New Guinea and Antarctica, amongst others. The oceans are also part of his fifty per cent; he suggests halting fishing in open seas and letting life recover. According to him, it is particularly in the industrialized world that nature can be restored through the linking of patches to create wildlife corridors.[12] At different scales, this can clearly still be achieved worldwide, although Wilson gives no clear practical propositions. This is largely the work of the landscape architect: to spatially envision the interconnectedness of systems, of flora and fauna, in the manner in which thinkers like Alexander von Humboldt once described the 'web of nature'.[13]

The skills of the landscape architect, to visualize and to create visions for territories, to understand where to build and, more importantly, where to *not* to build, are vital. In other areas, open space can be hybridized with other uses, including recreational with the creation of more parks, at the national, state and local levels. As has been commented by many others, geographic information systems, remote sensing, and satellite and aerial photography now afford a vast array of information to the field:[14] the challenge remains how to intelligently utilize, synthesize and interpret the data in order to develop spatial strategies that simultaneously create land mosaics for expanded ecologies and work with larger nested systems of the territory, including, for instance, the logics of integrated watershed management, afforestation and connection of habitat corridors, logical systems of requalified and densified areas of human settlement, etc.

Bold territorial design projects can lead to policy decisions that could include open space acquisition for an expanded public realm at one scale, while at another scale could lead to the expansion of federal lands for preservation and habitat regeneration as a foil to strategic growth and densification. The landscape architect can once again become leader of the regional project—to direct the development of large-scale territories. The landscape architect is the mediator of the urban and nature, the untamed and the civilized; the conservation of nature and restoration of ecologies is of imminent importance in the (re)development of the future of the earth.

Coastal afforestation to respond to sea level rise

Coasts, and particularly deltas, with their inevitable expanded gradient of wetness—ranging from sea and river water with sedimentation or sediments with a high degree of wetness, to soils with a low degree of wetness—have corresponding natural variations of fauna and flora. The total length of the earth's shoreline is 573,000 kilometres—almost exactly the distance from the earth to the moon. The coast's milieu of meiofauna (Chordata, Arthropoda, Mollusca, Annelida, Nematoda, Tardigrada and Rotifera)[15] and marine vegetated habitats (seagrasses, salt, macroalgae and mangroves) are less studied than their terrestrial counterparts, but most of its biodiversity persists. Coastal vegetation contributes significantly to carbon burial, and buffers the impacts of sea level rise and wave action that are associated with climate change. There is an enormous opportunity in the intelligent spatial reconfiguration of coastlines and their afforestation and reforestation, which combines new living, productive landscapes, recreational possibilities or simply mere habitat preservation.

Forests, specifically mangrove forests, naturally and organically infiltrate areas between water and land. They occupy brackish water zones along tropical and subtropical coasts, and they are highly productive ecosystems which support fish and crustacean habitats, trap sediment, recycle nutrients and protect shorelines from erosion. Tree planting on dykes, small or large, in a certain way imitates this land-making or stabilizing process. Throughout the world, the domestication and colonization of the landscape led to the shift from a gradual transition between water and land towards a categorical division between wet and dry. Over time, the gradient was largely eliminated through the introduction of linear elements—canals, dykes, ditches and so on; mangroves mediated between land and water, gradually 'making' land, while planted trees fixed higher elements. Territories transformed from dynamic mud plains with softly undulating topography and gradients of wetness, to assemblages of low/wet and dry/high components (where high remains a relative notion). The historic paradigms of delta and coastal land management are undergoing fundamental shifts, and living with floods and interconnected canal systems is no longer as evident as it once was.[16] Entire hydrological regimes are facing major disturbances due to excessive dam-building and extensive riparian re-engineering. Upstream damming disrupts fauna migration, upsets the natural flood cycle crucial to the ecosystem, and disturbs recession agriculture, while at the same time deltas are overexploited for basin resources.[17] All along coasts and deltas, manipulated water pressure and subsidence, coupled with sea level rise, makes saline intrusion a growing threat to enormous territories. The construction of large, protective sea dykes is creating gigantic concrete barriers between land and sea in regions that once thrived solely on natural processes of erosion and sedimentation.

The landscape architect can develop coastal afforestation and 'mangroving' strategies that could reverse the colonization and domestication of the wetness gradient. Afforestation of tropical and subtropical coastal necklaces with mangroves—and also other marine vegetated habitats for other landscape regions—can protect coasts, work *with* natural dynamics, heal salinated landscapes, rebalance water salinity, increase and diversify food production, and even be developed hand in hand with new settlement morphologies.[18] Afforestation in deltas can be an insulator, protecting adjacent lands against the turbulent effect of monsoons, tsunamis and spring tides while simultaneously delivering a renewable resource in terms of medicinal and wood products.

The landscape architect can work together with experts from marine science, architects and engineers to develop biodiverse coasts, which can enhance the food chain and create new environments for habitation and renewable energy. Organic aquaculture can be selectively interwoven into a designed forest mosaic that could also include new settlement typologies and wind and tidal energy mechanisms, and which would be strategically protected from the threats of inclement weather patterns.

Water urbanism to mitigate flood and drought

Traditionally, the workings of urban and rural regions are the result of a subtle and fragile balance between water and land, permeable and impermeable surfaces, organized by territorial hydraulic systems of water management and soil stabilization. The primitive logic of 'cut-and-fill' and differences in micro-topography were a powerful tool. Levels of inundation determined distinct land uses, and therefore the definition of wet/dry, productive/inhabited, and safe/unsafe component parts of the land mosaic was considered essential. Indigenous water management methods—often simultaneously addressing pragmatism, urbanism and symbolism, and with innovative engineering and an understanding of topography and seasonal weather patterns—provide invaluable lessons. At the same time, the definition of the city as shaped by water has been a cornerstone of urbanism.[19]

Water urbanism reflects upon the growing challenges of water in the city, infrastructural landscapes, and the reuniting of engineered and natural processes. The predicted consequences of climate change (particularly floods, but also droughts), new pressures of storm water and basin management, and ecological concerns create rich interdependencies of water and urbanism. Water urbanism requires first and foremost a return to the shaping of the ground, to the creation of earthworks and landform. Water urbanism can result from a diachronic and synchronic reading of the territory and an understanding of the contested territories of the region's politics, economics and ecology through a new sectional interplay of landscape, infrastructure and urbanization. Expanding cities and their peripheral territories can be planned as juxtapositions of characters and scales, resulting from an orchestration of the infrastructural net, natural (green and blue) systems, topographical differences, related soil conditions and, finally, the programmatic destinations allocated to them.

The landscape architect can explicitly design sectional richness: the creation of micro-topographies is for water to naturally flow; for steps and ramps for people to gather and circulate; for dry season programmes that are yet floodable; and for architecture to have a topographical character and animate sites. Intelligent cut-and-fill operations can create vital ecosystems/migration corridors and 'safe' lands and elevations for vital infrastructure and settlement.

New energy landscapes and transport towards deep decarbonization

Renewable energy and new developments in transport—including electric self-driving cars, e-bikes and more efficient public transport—will radically transform the territory. As more hydro, solar, wind, biomass and other energy technologies come online, and as costs, particularly of solar and wind energy, are predicted to fall precipitously until 2030 due to increased developments that come out of the Paris Agreement pledges,[20] there is the opportunity for landscape architects and urbanists to work with engineers and other experts (including agronomists, ecologists, biologists, etc.) to hybridize programmes and economies. Rather than merely utilitarian civil infrastructure, energy landscapes can become civic, and can also be developed from visual/aesthetic perspectives. At the same time, a great deal of the globe's twentieth-century highway infrastructure can be replaced by more resourceful networks, releasing vast tracks of space for conversion to other uses.

The requalification of decommissioned highway sites can significantly contribute to a reinvigorated public realm, productive landscapes, and space needed for requalifying habitat.

Simultaneously, the creation of new energy and transport infrastructure should no longer simply be considered as the accumulation of technical objects in isolation from their surroundings. Landscape and infrastructure can merge, become incubators of collective life and enhance the quality of the landscape.[21] Hence, conceiving infrastructure blends with generating architecture, building more productive landscapes and producing more qualitative urban and rural settings and living environments.

The landscape architect can engage with the formulation of new energy landscapes and infrastructure in order to include the social and imaginative dimensions as much as engineering. The conception of contemporary landscapes of infrastructure can, from the onset, include state-of-the-art technological aspects which interweave with generative ecological systems.

Accentuate agro-ecological regions to address food and water security

For the past thirty years, the International Institute for Applied Systems Analysis (IIASA) and the Food and Agriculture Organization of the United Nations (FAO) have been developing the agro-ecological zone (AEZ) methodology in order to assess agricultural resources and potential. Due to technological innovation, since 2000 they have been able to carry out these assessments globally. AEZs are geographical areas exhibiting similar climatic conditions that determine their ability to support rainfed agriculture. At a regional scale, AEZs are influenced by latitude, elevation and temperature as well as seasonality, and by rainfall amounts and distribution during the growing season.[22] Since IIASA and FAO have now identified the globe's AEZs, context-embedded development can be anchored quite precisely to very specific environmental characteristics, not only allowing an extremely logical and pragmatic land use, but also accentuating the identity of various regions by underscoring their 'natives'. The AEZs serve as a particular type of landscape assessment, since they highlight suitability of crops and plants based on soils, etc.

The landscape architect would have the know-how to include the AEZ methodology in the regular analysis of sites in order to address the 'wicked problems' of food and water security, and also to re-establish core identities of landscapes and territories. The recognition of AEZs can counterbalance the relative homogeneity of regional urbanism and architecture—which has tended to 'flatten out' due to processes of neoliberal economics—and rearticulate landscapes with specific ecological qualities and processes. Natural locational assets can redefine levels of differentiation and hierarchy in the landscape. Landscape, urban and rural morphologies can also be designed to work with soil types, water flows, sedimentation, erosion, subsidence, etc., and with future predictions of the consequences of climate change (Figure 16.3).

Forest urbanism to create micro-climates and carbon sinks

Forests operate as self-regenerating ecosystems. For centuries, forests were planned, systematically exploited and maintained; often this management was more extensive and sophisticated than the town planning of the same period.[23] Throughout the history of urbanism, there has been an interweaving of structures of forests and trees with urban armatures and tissues. Forests have traditionally formed the counterfigure of the city, been embedded in the city or complemented the city. At the same time, lines of trees planted in public spaces—including along promenades, *alleés* and boulevards, on walls and ramparts, beside moats and canals, in city squares, and parallel to city streets and railways— have for millennia been systematically planned and constructed at national scales, and directly organized by central and municipal governments. Historically, in Europe for example, urban forests had strong ties to the Church and state, as well as with private landowners. In medieval times, monastic orders, monarchs, nobility and municipal authorities

Figure 16.3 Accentuating, Differentiating and Intensifying Six Agro-Ecological Regions, Mekong Delta, Vietnam, 2014–17. For the Revision of the Mekong Delta Region Plan 2030, Vision to 2050, infrastructure and development are realigned to take advantage of climate change and the opportunities of the landscape. The delta is optimized as a productive landscape. New state-of the-art typologies are conceived for living with the forces of nature. © RUA: Shannon, Kelly and Bruno De Meulder with Claudia Lucia Rojas Bernal and Tracy Collier, Eric Heikkila, Chris Keseteloot, Michael Waibel, Donielle Kaufman and Christina Hood as collaborators (as consultants for the Ministry of Construction, Hanoi and SISP (Southern Institute of Spatial Planning, HCMC)).

owned forests; other forests were part of the commons for communities.[24] Urban forestry draws from deep traditions of classical forestry management, and also from environmental science, but it is constantly inventing new techniques for various urban contexts.

In the 'United States Mid-Century Strategy for Deep Decarbonization' there is the stated goal to upscale forest restoration and afforestation on federal and private lands in order to sequester carbon.[25] Increased street-tree planting and trees in the city are also increasingly important to counter pollution, work as a counterbalance to heat islands and provide necessary shade, but also for generating spatial form, species occupation and new ecologies, and even supporting informal economies.

The landscape architect needs to reinvest in the design of the urban forest, which ultimately must be understood as a critical component of contemporary urban infrastructure. Contemporary research in land management, forestry and urban ecology can inform design by landscape architects of urban forestry. The radical redesign of urban profiles might consider asymmetrical vegetation to respond to prevailing winds, shading needs and water-harvesting potentials. Clearly, cities need trees. However, the urban forest must also be understood as a larger system that extends as a continuous, interconnected regional system, in the manner that E.O. Wilson envisions.

Climate changes, landscapes evolve: Landscape architects must act

As Jedediah Purdy explains in *After Nature: A Politics for the Anthropocene*, humanity has outstripped geology.[26] Humans, as the force shaping the planet, have a responsibility for its change. The contemporary landscape architect needs to rise to the challenge and act. Now. Politics, economics and ecology are all in near-perpetual crisis. Purdy's book underscores the fact that landscapes contain and are shaped by clashing visions of nature, their political uses and legal embodiments. The trap of a 'neoliberal Anthropocene' is an increasingly constant reality. It is in this context of climate change that landscape architecture can make its greatest and most distinctive intervention.

Landscape architecture, particularly at the territorial scale, is an ever more urgent necessity. It must reclaim its position as the key discipline uniquely able to synthesize ecological systems, scientific data, engineering methods, social practices and cultural values, and to integrate them all into the design of the built environment. Bold visions must combine the tangible and the imaginary to provoke conversations that promote social equity and environmental justice as well as manifesting landscape's transformative power. It is time once again to make a strong call to arms for the discipline to more actively engage with the contemporary challenges of the world and prove that landscape architecture is a necessity. Its political agency must be forcibly reactivated, and the power of landscape architecture engaged, to engage and create interventions at the scale of territories and ecological systems in order to qualitatively transform forms of living.

Notes

1 Thomas Piketty, 'We Must Rethink Globalization, or Trumpism Will Prevail', *The Guardian*, 16 November 2016; 'Donald Trump's Victory Challenges the Global Liberal Order', *Financial Times*, 10 November 2016.
2 Andrea Thompson, '99 Percent Chance 2016 Will Be the Hottest Year on Record', *Scientific American,* 18 May 2016.
3 UNISDR and CRED, *The Human Cost of Weather-Related Disasters 1995–2015* (Geneva: UNISDR, 2016), 5.
4 Ibid., 23.
5 Elizabeth Kolbert, *The Sixth Extinction: An Unnatural History* (New York: Henry Holt and Co., 2014).
6 *The Guardian* (2015) 'The day in climate change rallies', 30 November 2015, https://www.theguardian.com/environment/live/2015/nov/29/global-peoples-climate-change-march-2015-day-of-action-live (accessed 20 December 2016).
7 By 2 February 2018, the United States Climate Alliance included California, New York, Washington, Connecticut, Rhode Island, Massachusetts, Vermont, Oregon, Hawaii, Virginia, Minnesota, Delaware, Puerto Rico, Colorado, North Carolina and Maryland (see https://www.usclimatealliance.org/).
8 Pope Francis, *Encyclical on Climate Change and Inequality: On Care for Our Common Home* (Brooklyn: Melville House Publishing, 2015).
9 Goldenberg and Kirchgaessner, 'Pope Francis Demands.'
10 See https://leapmanifesto.org/en/the-leap-manifesto/.
11 'Facts About Water: Statistics of the Water Crisis', The Water Project, , last modified 31 August 2016, https://thewaterproject.org/water-scarcity/water_stats (accessed January 2017).
12 Edward O. Wilson, *Half-Earth: Our Planet's Fight for Life* (New York: W.W. Norton and Co., 2016).

13 Andrea Wulf, *The Invention of Nature: Alexander von Humboldt's New World* (New York: Alfred A. Knopf, 2015).
14 Joel Snyder, 'Territorial Photography', in *Landscape and Power*, ed. W.J.T. Mitchell (Chicago: University of Chicago Press, 1994); James Corner, 'Aerial Representation: Irony and Contradiction in an Age of Precision', in *The Landscape Imagination: Collected Essays of James Corner 1990–2010*, eds. James Corner and Alison Bick Hirsch (New York: Princeton Architectural Press, 2014); Denis Cosgrove, 'Liminal Geometry and Elemental Landscape: Construction and Representation', in *Recovering Landscape: Essays in Contemporary Landscape Architecture*, ed. James Corner (New York: Princeton Architectural Press, 1999).
15 Chordata are birds, mammals and amphibians; Arthropoda are insects, spiders, mites, millipedes, centipedes and crustaceans; Mollusca are snails and slugs; Annelida are earthworms; Nematoda are roundworms; Tardigrada are bear animalcules; Rotifera are rotifers. Wilson, *Half-Earth*, 114–15.
16 Bruno De Meulder and Kelly Shannon, 'Mangroving Ca Mau, Vietnam: Water and Forest as Development Frames', in *Water Urbanisms East*, eds. Bruno De Meulder and Kelly Shannon (Zurich: Park Books, 2013).
17 Brahma Chellaney, *Water: Asia's New Battleground* (Washington, DC: Georgetown University Press, 2011), 268.
18 Bruno De Meulder and Kelly Shannon, 'Water and the City: The "Great Stink" and Clean Urbanism', in *Water Urbanisms*, eds. Kelly Shannon et al. (Amsterdam: Sun, 2008); Kelly Shannon, 'Eco-Engineering for Water: From Soft to Hard and Back,' in *Resilience in Ecology and Urban Design: Linking Theory and Practice for Sustainable Cities*, eds. Steward Pickett, M.L. Cadenasso and Brian McGrath (London: Springer, 2013).
19 De Meulder and Shannon, 'Water and the City'; Shannon, 'Eco-Engineering'.
20 The White House, 'United States Mid-Century Strategy for Deep Decarbonization', Washington, DC: The White House, November 2016.
21 Kelly Shannon and Marcel Smets, *Landscape of Contemporary Infrastructure* (Rotterdam: NaI Publishers, 2010).
22 IIASA/FAO, *Global Agro-Ecological Zones: GAEZ v3.0* (Laxenburg and Rome: IIASA and FAO, 2012).
23 J.B. Bridel, *Manuel pratique du forestier: Ouvrage dans lequel on traite de l'estimation, exploitation, conservation, aménagement, repeuplement, des semis and plantations des forêts, avec les moyens de prévenir la disette des bois de construction and de chauffage* (Paris: Baudelot and Eberhart, 1798).
24 Bruno De Meulder and Kelly Shannon, 'Forests and Trees in the City: Southwest Flanders and the Mekong Delta', in *Revising Green Infrastructure: Concepts Between Nature and Design*, eds. Daniel Czechowski, Thomas Hauck and Georg Hausladen (London: CRC Press, 2014).
25 White House, 'Deep Decarbonization'.
26 Jedediah Purdy, *After Nature: A Politics for the Anthropocene* (Cambridge, MA: Harvard University Press, 2015).

Bibliography

Bridel, J.B. *Manuel pratique du forestier: Ouvrage dans lequel on traite de l'estimation, exploitation, conservation, aménagement, repeuplement, des semis and plantations des forêts, avec les moyens de prévenir la disette des bois de construction and de chauffage*. Paris: Baudelot and Eberhart, 1798.

Chellaney, Brahma. *Water: Asia's New Battleground*. Washington, DC: Georgetown University Press, 2011.

Corner, James. 'Aerial Representation: Irony and Contradiction in an Age of Precision'. In *The Landscape Imagination: Collected Essays of James Corner 1990–2010*, edited by James Corner and Alison Bick Hirsch, 133–60. New York: Princeton Architectural Press, 2014.

Cosgrove, Denis. 'Liminal Geometry and Elemental Landscape: Construction and Representation'. In *Recovering Landscape: Essays in Contemporary Landscape Architecture*, edited by James Corner, 103–19. New York: Princeton Architectural Press, 1999.

De Meulder, Bruno, and Kelly Shannon. 'Water and the City: The "Great Stink" and Clean Urbanism'. In *Water Urbanisms*, edited by Kelly Shannon, Bruno De Meulder, Viviana D'Auria and Janina Gosseye, 5–9. Amsterdam: Sun, 2008.

De Meulder, Bruno, and Kelly Shannon. 'Mangroving Ca Mau, Vietnam: Water and Forest as Development Frames'. In *Water Urbanisms East*, edited by Bruno De Meulder and Kelly Shannon, 118–37. Zurich: Park Books, 2013.

De Meulder, Bruno, and Kelly Shannon. 'Forests and Trees in the City: Southwest Flanders and the Mekong Delta'. In *Revising Green Infrastructure: Concepts Between Nature and Design*, edited by Daniel Czechowski, Thomas Hauck and Georg Hausladen, 427–49. London: CRC Press, 2014.

Duarte, Carlos M., Iñigo J. Losada, Iris E. Hendriks, Inés Mazarrasa and Núria Marbà. 'The Role of Coastal Plant Communities for Climate Change Mitigation and Adaptation'. *Nature Climate Change* 3 (2013): 961–8.

Financial Times. 'Donald Trump's Victory Challenges the Global Liberal Order'. 10 November 2016. https://www.ft.com/content/a4669844-a643-11e6-8b69-02899e8bd9d1 (accessed January 2017).

Francis (pope). *Encyclical on Climate Change and Inequality: On Care for Our Common Home*. Brooklyn: Melville House Publishing, 2015.

Goldenberg, Suzanne, and Stephanie Kirchgaessner. 'Pope Francis Demands UN Respect Rights of Environment over "Thirst for Power"'. *The Guardian*, 25 September 2015. https://www.theguardian.com/world/2015/sep/25/pope-francis-asserts-right-environment-un (accessed January 2017).

IIASA/FAO. *Global Agro-Ecological Zones: GAEZ v3.0*. Laxenburg and Rome: IIASA and FAO, 2012.

Klein, Naomi. *This Changes Everything: Capitalism vs. the Climate*. New York: Simon and Schuster, 2014.

Kolbert, Elizabeth. *The Sixth Extinction: An Unnatural History*. New York: Henry Holt and Co., 2014.

McKibben, Bill. *The End of Nature*. New York: Random House, 1989.

McKibben, Bill. *Oil and Honey: The Education of an Unlikely Activist*. New York: St Martin's Press, 2013.

Oreskes, Naomi, and Erik Conway. *The Collapse of Western Civilization: A View from the Future*. New York: Columbia University Press, 2014.

Piketty, Thomas. 'We Must Rethink Globalization, or Trumpism Will Prevail'. *The Guardian*, 16 November 2016. https://www.theguardian.com/commentisfree/2016/nov/16/globalization-trump-inequality-thomas-piketty (accessed January 2017).

Purdy, Jedediah. *After Nature: A Politics for the Anthropocene*. Cambridge, MA: Harvard University Press, 2015.

Shannon, Kelly. 'Eco-Engineering for Water: From Soft to Hard and Back'. In *Resilience in Ecology and Urban Design: Linking Theory and Practice for Sustainable Cities*, edited by Steward Pickett, M.L. Cadenasso and Brian McGrath, 163–82. Springer: London, 2013.

Shannon, Kelly, and Marcel Smets. *Landscape of Contemporary Infrastructure*. Rotterdam: NaI Publishers, 2010.

Snyder, Joel. 'Territorial Photography'. In Landscape and Power, edited by W.J.T. Mitchell, 175–201. Chicago: University of Chicago Press, 1994.

The Guardian (2015) 'The day in climate change rallies', 30 November 2015. https://www.theguardian.com/environment/live/2015/nov/29/global-peoples-climate-change-march-2015-day-of-action-live (accessed 20 December 2016).

The Water Project. 'Facts About Water: Statistics of the Water Crisis'. Last modified 31 August 2016. https://thewaterproject.org/water-scarcity/water_stats. (accessed January 2017).

Thompson, Andrea. '99 Percent Chance 2016 Will Be the Hottest Year on Record'. *Scientific American*, 18 May 2016. https://www.scientificamerican.com/article/99-percent-chance-2016-will-be-the-hottest-year-on-record/. (accessed January 2017).

UNISDR and CRED. *The Human Cost of Weather-Related Disasters 1995–2015*. Geneva: UNISDR, 2016. http://www.unisdr.org/files/46796_cop21weatherdisastersreport2015.pdf (accessed January 2017).

White House. 'United States Mid-Century Strategy for Deep Decarbonization'. Washington, DC: The White House, November 2016.

Wilson, Edward O. *Half-Earth: Our Planet's Fight for Life*. New York: W.W. Norton and Co., 2016.

Wulf, Andrea. *The Invention of Nature: Alexander von Humboldt's New World*. New York: Alfred A. Knopf, 2015.

17

LANDSCAPE ARCHITECTURE AND SOCIAL SUSTAINABILITY IN AN AGE OF UNCERTAINTY

The need for an ethical debate

Shelley Egoz

Landscape has become an embracing entity, a source of interest and stimulation for scholarly enquiry across several disciplines. While landscape is a framework for investigation in fields as diverse as geography, history, ecology, archaeology, anthropology, art history, sociology, political sciences, psychology, cultural heritage, and more, only the discipline of landscape architecture explicitly contains the term 'landscape' in its name, indicating that it is the applied science—one that engages with interventions in and management of landscapes, converting visions into physical spaces.[1] Landscape architecture hence bears responsibility for materializing the aspirations for individual and social well-being embedded in the European Landscape Convention (ELC)[2] described as '[a] unique setting and meeting place for populations, landscape is a key factor in the physical, mental and spiritual wellbeing of individuals and societies. A source of inspiration, it takes us on a journey, both individual and collective, through time, space and imagination.'[3]

Every landscape is dynamic: each landscape has a history, a present condition, and a future, indicating that change is inherent to landscape and constitutes the essence of both the concept and the physical entity. Uncertainties have always been an innate challenge of landscape architecture as an agent of change. Dealing with landscape transformations necessitates addressing the intricate nature of ecological systems. It requires considering the complex unforeseen consequences of local and global social, economic, and political drivers, and their impact on landscape.[4]

In the twenty-first century, we are witnessing an increased sense of geopolitical uncertainty and unprecedented economic disparity. These trends have been heightened by the election of President Trump and his unpredictable policies in the US, as well as the Brexit vote in the UK.[5] At the same time, and related to those political agendas, insecurity is growing due to climate change, displacing populations and causing conflict over resources.

Signs of this unprecedented state of the world were first identified by scholars around two decades ago. Sociologist Zygmunt Bauman coined the term 'liquid modernity', employing the metaphor of fluidity to reflect the condition of our era, 'where the absence of political control makes the newly emancipated powers [of capital] into a source of profound and in principle untameable uncertainty'.[6] Bauman and several other critics of neoliberal capitalism, most notably historian Tony Judt, economic sociologist Wolfgang Streeck, and geographer David Harvey,[7] have highlighted the impact of the last four decades of capitalist ideology on environmental and human well-being. Landscape, I maintain, is our life-place infrastructure: it is the physical environment that supports livelihood and contains intangible qualities and meanings. Landscape is moulded partly by ideological drivers—any outcome of spatial planning exists within a political context. To that end, the events and processes of the past decades have also shaped the discourses, scope, and type of landscape architecture practised. Against this backdrop of social and political uncertainty coupled with environmental crisis, this chapter reflects on some of the prevailing discourses in landscape architecture. The main focus is on the term 'social sustainability', arguing that the ethical axiom of the universal right to landscape[8] is a constant value that needs to be taken on board, particularly in times of uncertainty.

Sustainability: The ethos of landscape architecture in the twenty-first century

Sustainability, a concept borrowed from the science of ecology, has been defined as 'the property of biological systems to remain diverse and productive indefinitely'.[9] Sustainability is hence an ideal for an infinite and balanced process that supports life. This ideal, it is believed, can be achieved through particular practices and behaviours.[10]

A recognition that natural resources are limited can be traced back to ancient philosophers, but only surfaced in a significant manner in the twentieth century. In response to accelerated development propelled by the Industrial Revolution in the nineteenth and early twentieth centuries, the idea of the need for conservation emerged. In the late nineteenth century, awareness of the dangers of overuse of natural resources was notably promulgated by George Perkins Marsh.[11] Discourses on ecology and threats posed to the natural environment by accelerated human development sprang from 1960s countercultural movements and 1970s alternative lifestyles, and led to the rise of the environmental movement and mainstreaming of green politics. By the late 1990s a notion of sustainable development had evolved to counter the modern idea of progress.[12] The scientific community's increasing warnings that use of fossil fuels would cause global warming (later renamed climate change) gained momentum. Academic disciplines from the hard sciences to the humanities were reframed in novel ways (e.g. environmental science, landscape ecology, environmental ecology, political ecology, and environmental law). Avant-garde environmental art that evolved in the 1970s adopted landscape as its linchpin. It was hence logical that landscape architecture, as the discipline that engages with and shapes the natural environment, embraced the emerging culture of environmental sustainability. It has become a disciplinary ethos—not a style or applied technology, but an underpinning value explicitly stated in charters of international and national professional organizations, such as the International Federation of Landscape Architects (IFLA) and the American Society of Landscape Architects (ASLA).

Landscape frameworks such as 'green infrastructure' and 'landscape urbanism', which I elaborate on later in this chapter, are examples of the widespread adoption of environmental sustainability values in landscape architecture discourse and practice.

The precise meaning of sustainability is open to interpretation, contestation, and manipulation. Whether any design is really true to its claims of sustainability is debatable. These days the term 'sustainable landscape design' should be seen as a tautology—it is unlikely that any landscape architect will declare their work 'unsustainable landscape design'. In the late twentieth century, Ian Thompson identified sustainability as one of several discourses in landscape architecture:

> For landscape architects, the concept of sustainability meshes easily with a pre-existing idea that good design will be enduring [... T]he discipline [...] has a temporal dimension.

> Plants take time to grow. Landscapes take time to mature. It has been easy, therefore, to assimilate the idea that they are designing not just for a present generation of users but for their descendants.[13]

In landscape planning and design, the work of formative figures such as Olmsted[14] and McHarg[15] predated the late twentieth-century sustainability paradigm. It thus makes sense that the environmental sustainability paradigm shift has been embraced in landscape architecture. Planners and urban designers have adopted schemes in line with sustainability principles: compact city planning, pedestrian-first cities, biodiversity in cities, and so on. Landscape architects have developed 'technocratic accommodation':[16] technical engineering solutions for healthy environments through the design of rain gardens, green roofs, surface storm water design and management, wetland construction, native plantings, and more.

Sustainability is generally described as consisting of three main pillars, environmental, economic, and social, often portrayed in a Venn diagram.[17] Incorporating environmental sustainability into landscape architecture is clear and straightforward.[18] The meanings and interpretations assigned to social and economic aspects of sustainability are more nebulous. The focus of this chapter is on the most vaguely defined aspect: social sustainability. Despite the fog around the concept, social dimensions have been recognized as instrumental for achieving the goal of sustainability.[19]

Social sustainability

> 'If you wish peace, care for justice.'
>
> —Averred ancient wisdom; and unlike knowledge, wisdom does not age[20]

Sustainability is one of the catchphrases of the twenty-first century. Nonetheless, how can we claim social sustainability is an attainable goal when the wealth gap today is increasing to unprecedented levels? Several scholars[21] have addressed this disturbing economic gap and explained its origins in globalized free-market capitalism. The crux is that sustainable social existence and a stable society require an ethical commitment to values of equity, democracy, and social justice. In order for social sustainability to persist, this ethos must reach beyond empty rhetoric, individual beliefs, and personal ethical choices. Social sustainability will have to be instituted through a just political and economic system, and preserved according to a social contract between citizens and their governing bodies, one where justice is uncontested.

Such a commitment indeed appears in IFLA's International Landscape Convention: 'The aim is [...] to stimulate a more integrated, democratic approach that establishes the landscape as a holistic tool for planning, managing and creating sustainable development.'[22] This is not surprising, since landscape architects are instrumental in landscape change that impacts on the public arena. Therefore, landscape architects play a significant political role, whether they are aware of that role or not.[23] Brown and Jennings critique the collective denial of landscape architecture's political role, particularly regarding issues of social justice, and present a conceptual model for educating landscape architects in social consciousness.[24] Discourses on landscape justice and landscape democracy are still emerging and are yet to be defined.[25] The ELC, The European Landscape Convention, a document that endorses the Council of Europe's humanist and democratic values, sets a high bar for landscape architects who practise in an environment of social and economic uncertainty that is increasingly inflicting economic disparity. I will return to the challenges facing landscape architecture in these times of liquid modernity later in this chapter.

At this point, in order to address social sustainability, it is essential that we claim landscape as a vital component for well-being, by introducing the concept of the right to landscape.

The right to landscape

The idea of the right to landscape relies on the premise that landscape is a common good and an infrastructure for well-being, a life setting for everyday human practices.[26] Such an understanding of landscape practices is in line with theory developed in the last three decades by cultural geographers, notably Kenneth Olwig.[27] This stance also underlines the ELC and is relevant to landscape architecture.

Yi-Fu Tuan suggests that the notion of the Right to Landscape 'promises to transform "landscape" from a concept in cultural geography and landscape architecture to a concept indispensable to the probing of human nature and human wellbeing'.[28] Tuan's words assert the claim for landscape as a human right. The doctrine of human rights is associated with the basic, universal rights of individual human beings for subsistence and dignity, but at the same time it recognizes that the individual is also part of society,[29] and it addresses communal social rights.[30]

Landscape is our infrastructure for survival. This corroborates the imperative that landscape is a universal realm, in line with the universal nature of human rights. The articles of the Universal Declaration of Human Rights (UDHR) represent a moral standard that transcends local or national law. These are rights to basic material requirements for survival, coupled with rights to the emotional, cultural, and social dynamics that support human dignity. Landscape, like human rights, encompasses both tangible and intangible dimensions (Figure 17.1). Its overall quality is essential for securing the human well-being referred to in the UDHR.

Landscape architecture is the practice of engaging with and creating a life-supporting material system. Landscape is site-specific, and landscape architects have developed methods to study and analyse the landscape in order to offer relevant solutions. At the same time, just as every individual is different but entitled to universal rights, so does landscape have a universal dimension, since every human being lives, uses, and depends on a landscape for survival and well-being.

The principle of the right to landscape is the cornerstone of landscape justice.[31] It is at once an abstract concept, an ethical commitment, and a political statement. I have argued elsewhere that landscapes are political, and that landscape architects are sometimes complicit in injustices by serving undemocratic political agendas.[32] Gailing and Leibenath also claim that 'political aspects can be discovered in virtually all landscapes as they are inevitably imbued with politics, antagonistic dimensions and power', and that 'some scholars from fields such as landscape ecology or landscape architecture neglect the political dimensions of landscapes and employ "landscape" as an entirely apolitical concept.'[33]

Although these topics are currently at the margins of landscape architecture's professional discourse, there are a few burgeoning examples of a counter-movement. Most notable in North America are Anne Whiston Spirn[34] and Randolph Hester.[35] Hester argued for the pertinence of taking a political stance in landscape architecture. In November 2014, a presentation entitled 'Social Justice: The New Green Infrastructure', on methods of inclusive and socially just design, was included in the annual meeting of ASLA.[36] Since then, several blogs and ASLA publications have addressed social justice and landscape architecture.[37] Another example of landscape justice applied through ecological design is the work of Beirut-based landscape architect Jala Makhzoumi. Underpinned by principles of social justice, Makhzoumi applied ecological design strategies and actions to the sustainable post-war recovery of marginalized communities in southern Lebanon.[38]

Despite the above examples and IFLA's international declarations, recognition, understanding, and engagement with political aspects is currently at the fringes of landscape architecture's

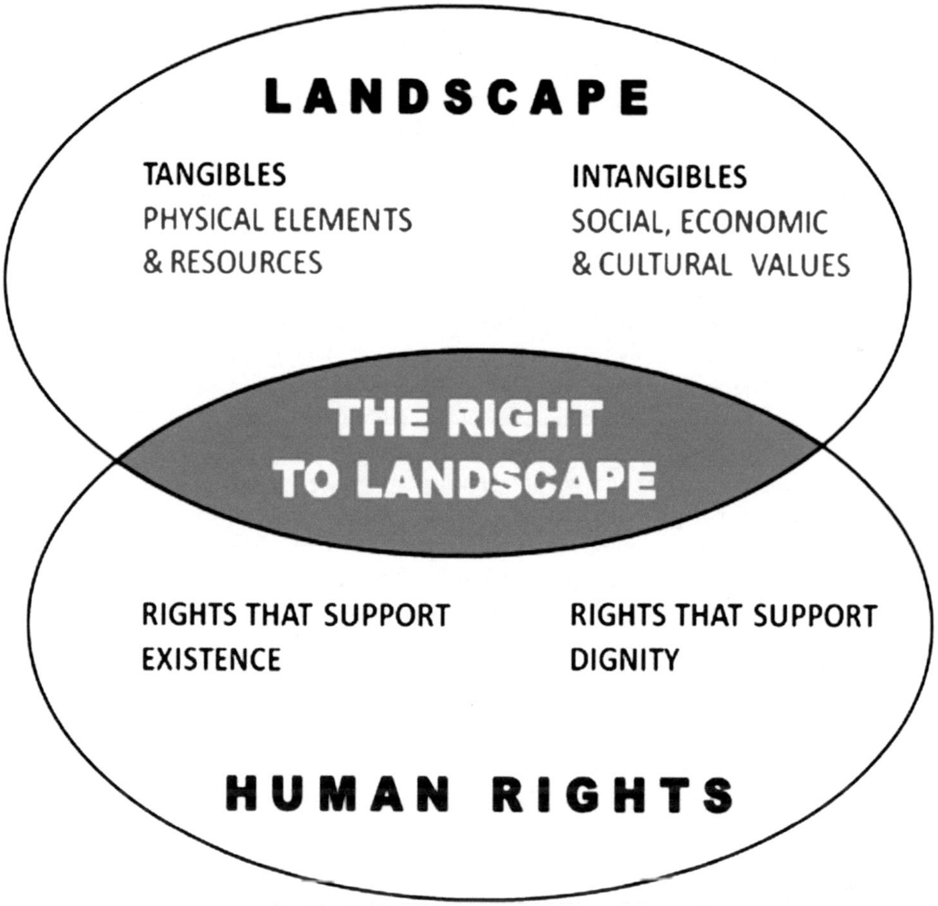

Figure 17.1 Representation of the conceptual overlap between landscape and human rights, 2011. © Egoz et al. *The Right to Landscape: Contesting Landscape and Human Rights.* Surry and Vermont: Ashgate, 2011.

professional discourse. The political argument for the right to landscape conceives of landscape as a common good, and as an infrastructure that supports life. Two other contemporary approaches that represent physical dimensions of landscape as infrastructure, green infrastructure and landscape urbanism, are now brought into the discussion to explore their political facets and meanings in these times of uncertainty.

Green infrastructure

Green infrastructure, 'an interconnected network of waterways, wetlands, woodlands, wildlife habitats, and other natural areas, as well as greenways, parks and other conservation areas',[39] has been gaining much attention in sustainability discourse and practice.[40] In spatial planning, the model of an interconnected network of green spaces derives its inspiration from landscape architecture, most notably Olmsted's Boston Emerald Necklace in the late nineteenth century and McHarg's pioneering *Design with Nature* in 1969, as well as the contemporary incorporation of ecology into landscape design. This model capitalizes on the multifunctionality embedded in such structures. Experts have largely argued for the ecological biodiversity value—as well as the potential social and economic advantages—of sustainable engineering solutions such as storm water management or improving air quality through planting.[41] Across scientific disciplines, the hard sciences are also addressing nature conservation, biodiversity in cities (also called urban ecology), and various ecosystem services that represent a technocratic instrumentalization of nature and thus of 'landscape'.

According to Benedict and McMahon, values of social sustainability are embedded in the concept of green infrastructure:

> The functions, values and benefits of green infrastructure are available for everyone. Creating interconnected green space systems benefits communities by providing land for resource protection and restoration, recreation and other public values. More important, strategic placement of green infrastructure reduces the need for some gray infrastructure, freeing up public funds for other community needs.[42]

The same authors also argue that green infrastructure is an inclusive project—it requires diverse groups to participate in creating a shared vision. This assertion situates the green infrastructure approach as neutral and apolitical, a pitfall in terms of landscape justice. Evidence shows that providing qualities marketed as 'liveability'—which contemporary cities take pride in and promote in order to attract residents—through green recreational amenities often leads to social injustices through 'green gentrification' (sometimes called environmental gentrification).[43]

Environmental psychologist Melissa Checker highlights the 'paradoxical politics of urban sustainability'.[44] Checker adopted a multidisciplinary approach to ethnographic research in the New York City neighbourhood of Harlem from 2007 to 2011. She describes the phenomenon of environmental gentrification when urban redevelopment endorses ecology and environmental activism in an era of advanced capitalism. Her main claim is that such a process 'subordinates equity to profit-minded development'.[45] These dynamics correlate with what social philosopher Slavoj Žižek calls the 'post-political era', where politics are practised by professional experts: a process of negotiating interests to arrive at a 'consensus' disguises power structures and silences disadvantaged citizens.[46] In this scenario, landscape architects join the 'enlightened technocrats' observed by Žižek. Bauman explained the context for this in his theory of 'liquid modernity' as a state of constant flux and fragmentation in structures.[47] Bauman identified a steady process of abandoning institutions that were set up to provide a safety net for individuals. The particular depoliticized circumstances we are experiencing these days are relevant to landscape architecture and its embrace of sustainability.

Inherent to both sustainability and landscape is an implicit acceptance of the intrinsic value of nature. This 'goodness' is an ideology[48] that may mask, or even legitimize, discriminatory dimensions: as Checker emphasizes, 'sustainability planning becomes part of a post-political project based on technocratic deliberation and consensus, which sidelines questions of real political inclusion and justice.'[49] The technical dimensions of sustainability, including green infrastructure planning, are the bread and butter of landscape architecture's professional practice. They are a critical means of materializing environmental sustainability, but en route to the ideal of sustainability they often compromise the social component, delinking sustainability from justice. As argued earlier in this chapter, landscape is the infrastructure for social life, and decoupling landscape and politics is, at best, naive.

Landscape as infrastructure is also the underlying theme of a controversial suggestion for a novel, ecologically driven method for organizing contemporary cities: landscape urbanism.

Landscape urbanism

The term 'landscape urbanism' was popularized by Charles Waldheim in the late 1990s, and in the twenty-first century is promulgated as 'a disciplinary realignment [...] in which landscape replaces architecture as the basic block of contemporary urbanism'.[50] A plethora of writing has been produced on landscape urbanism.[51] Some have explained its emergence as a reaction to the 1980s North American new urbanism movement. New urbanism principles included recommendations for neighbourhood planning, and policies that supported and enhanced sustainable lifestyles (such as mixed uses, diverse populations, pedestrian-first planning, universal access to public space, local architecture, and responses to local climates and ecologies). New urbanism arose in reaction to what was seen as rigid modernist planning, believed to be detrimental to the well-being of the environment and social life. Yet new urbanism has been criticized as an unimaginative, stale, structured way of thinking about contemporary challenges that accommodates 'reactionary cultural politics and nostalgic sentiment'.[52] Instead, some critics of new urbanism propose a transdisciplinary, visionary, bold way of flexible thinking: landscape urbanism. Proponents present the new theory as cutting-edge, embracing the unpredictable and emphasizing process over form, positioning landscape urbanism as an innovative organizing method associated with nature and landscape.[53] This began a debate in academic literature:[54] landscape urbanism has been accused of being a provocation, vague, pretentious, evasive, and a shallow stylistic embellishment, rather than a genuine and rigorous contribution to addressing the complexity of the challenges that contemporary cities face.[55] Critics argue that landscape urbanism embraces urban sprawl rather than the reshaping of the city, in an almost 'Daoist notion of going with the flow',[56] says Thompson also questions whether the trend hides a political facet. Indian planner Leon Morenas went further and boldly stated that landscape urbanism caters to an elite at the expense of weaker groups, and fails to address 'social justice, political emancipation, or ecologically saner designs, as its proponents have argued'.[57] Grahame Shane[58] too, in his review of *Stalking Detroit*, doubted the claim that landscape urbanism would achieve social justice and equity.[59]

US-based landscape architect James Corner's influential essay 'Terra Fluxus' has inspired many landscape architects[60] to explore the viability of landscape urbanism in a global context.[61] Morenas nonetheless examines Corner's acclaimed design of the High Line in Manhattan, a project regarded as a good example of landscape urbanism, and exposes 'the lesser-known narratives on its "impact" on NY's urban context'.[62] One concern is that prestigious projects such as the High Line deplete public resources that could have supported bigger parks in underprivileged neighbourhoods. In addition, Morenas points to disturbing fiscal arrangements that benefit 'philanthropically inclined individuals'.[63] Despite Corner's stated opposition to 'uncontrolled capital accumulation, backed by class privilege and gross inequalities of political economic power',[64] Morenas claims that in effect Corner has contributed to such a process, as the influence of the High Line has accelerated new development that leads to exclusive gentrification (Figure 17.2).

Morenas also questions whether Waldheim's argument that 'infrastructural systems and the landscapes they engender "shape and modify" the organization of urban settlement and its inevitably indeterminate economic, political, and social futures'[65] is indeed credible.

Nevertheless, Waldheim and Corner argue that models of flow, flexibility, and unpredictability that are at the core of landscape urbanism will accommodate social justice. Corner even suggests that Marxist geographer David Harvey's call for city planners and designers to embrace 'more socially just, politically emancipatory and ecologically sane mix(es) of spatio-temporal production processes'[66] could materialize in the adoption of landscape urbanism's principles of

Figure 17.2 New luxury apartment building in Highline Park, Manhattan, October 2017. © Williams, Tim.

'process over form'. Waldheim also highlighted this point when claiming that 'too much of the main body of mainstream urban design practice has been concerned with the crafting of "look and feel" of environments for destination consumption by the wealthy.'[67]

Against the above claims, more critiques emerged. Architect Eva Castro[68] made a pertinent observation about the permeability of the design industry to neoliberalism, and the ways in which planners, architects, urban designers, and landscape architects have embraced neoliberalism by serving the interests of private developers. In effect, the flexibility that landscape urbanism promulgates as its main contribution goes hand in hand with the philosophy of the free market, argues Castro.

Architectural theorist Douglas Spencer[69] proclaims that champions of landscape urbanism fail to critically reflect on ongoing accelerated transformations of the global landscape. Like Castro, he sees landscape urbanism's celebration of unpredictability as an invitation to 'the full force of neoliberal mechanisms of creative destruction'.[70] Rapid landscape transformations in India and China have caused mass displacement, devastated environments, and boosted speculation, investments, and development that profits elites at the expense of the underprivileged majority. Spencer claims that landscape urbanism has failed to address these issues.

The elephant in the room is landscape urbanism's avoidance of political probing into its compliance with neoliberalism. It sounds at best naive, if not completely cynical, for landscape urbanism to make a claim for social justice when neoliberal economics have been detrimental to social justice.[71] The core metaphors of landscape urbanism, its principles of 'flow', 'flexibility', and 'liquidity', reinforce what Bauman describes as the problem of the current post-political stage: '"society" is increasingly viewed and treated as a "network" rather than a structure [... I]t is perceived and treated as a matrix of random connections and disconnections and of an essentially infinite volume of possible permutations.'[72] He sees the condition as 'the collapse of long-term thinking, planning and acting', arguing that 'the virtue proclaimed to serve the individual's interests best is not conformity to rules but flexibility: a readiness to change tactics and style at short notice.'[73]

Conclusion

The above three examples of discourses on infrastructure have demonstrated that social sustainability is not a goal that can be achieved through osmosis. Landscape and society do not flow onto utopian ground—in every society, we face competition, power struggles, and conflicts. Apolitical landscape architecture is a trap. For landscape architects to rely on rhetorics and metaphors of ecology, nature, and landscape alone, while glossing over the political consequences of their work, raises deep ethical challenges.

We cannot deny the value of 'greening' our cities, or landscape architects' major contributions to environmental sustainability. Yet in some cases, the pursuit of urban sustainability can provide a platform for what Gould and Lewis term 'the green growth machine'. They argue that when 'the coalition of real-estate developers and political elites [...] transform[ed] brownfields [...] through real-estate development associated with the greening' the result was that 'the green growth machine turned a profit and demographically transformed neighborhoods through this process of green gentrification'.[74]

Unfortunately, such critical thinking emerges mainly in the work of sociologists and anthropologists, while it is landscape architects who are responsible for materializing the aspirations for individual and social well-being that are embedded in the ELC.

The role of the landscape architect may be to create quality physical settings for public engagement and provide opportunities for everyone to enjoy a commonly shared landscape design—indeed, the design of public spaces is instrumental in enabling social interaction. Bauman suggests that designers can contribute to inclusiveness in our cities: 'architects and urban planners could do quite a lot to assist the growth of mixophilia and minimize the occasions for mixophobic responses to the challenges of life.'[75] Harvey argues: 'The right to the city [...] is [...] a common rather than an individual right. [...] The freedom to make and remake our cities and ourselves is [...] one of the most precious yet most neglected of our human rights.'[76]

The human right to landscape is at the core of building a credible professional ethic. Landscape architects might argue that the 'good' design of public space is their contribution to social sustainability. There is intrinsic value in a well-informed, inclusive process that results in places in the city that contribute to environmental well-being and in which people find pleasure. However, an apolitical approach is disturbing, because there are political consequences to every design. Visionary thinking and creative alternative proposals are a professional skill, but without incorporating ethical values, they will not achieve sustainability. Social sustainability merits critical ethical discussion before landscape architects make claims for it.

Notes

1 One can argue that landscape ecology holds similar status, and indeed it does contribute towards the same goals. However, its disciplinary field is ecology, of which landscape ecology is a subfield, whereas landscape architecture is the main field (which on many occasions relies on expert advice from landscape ecologists).

2 Treaty no. 176, 'European Landscape Convention', Council of Europe, signed 20 October 2000, https://rm.coe.int/1680080621.

3 Maguelonne Déjeant-Pons, 'The European Landscape Convention', *Landscape Research* 31, no. 4 (2006): 363.

4 Thomas Germundsson, Peter Howard and Kenneth R. Olwig, 'Reassessing Landscape Drivers and the Globalist Environmental Agenda', *Landscape Research* 36, no. 4 (2011): 395.
5 See for example 'Political Risk Map 2017 Update: Adapting to an Unpredictable World', Marsh, accessed 29 November 2017, https://www.marsh.com/us/campaigns/political-risk-map-2017.html.
6 Zygmunt Bauman, *Liquid Modernity* (Cambridge: Polity, 2000), 2.
7 Tony Judt, *Ill Fares the Land* (London: Penguin Press, 2010); Wolfgang Streeck, 'How Will Capitalism End?', *New Left Review* 87 (2014), https://newleftreview.org/II/87/wolfgang-streeck-how-will-capitalism-end; David Harvey, 'Neoliberalism is a Political Project', *Jacobin*, 23 July 2016, https://www.jacobinmag.com/2016/07/david-harvey-neoliberalism-capitalism-labor-crisis-resistance.
8 Maguelonne Déjeant-Pons, 'The European Landscape Convention: From Concepts to Rights', in *The Right to Landscape: Contesting Landscape and Human Rights*, eds. Shelley Egoz, Jala Makhzoumi and Gloria Pungetti (Farnham: Ashgate, 2011), 51–6.
9 'Sustainability', Wikipedia, last modified 18 January 2018, https://en.wikipedia.org/wiki/Sustainability.
10 The overused term 'sustainability' has been criticized as lacking content and open to political manipulation: see Kate Carter and Stuart Moir, 'Diagrammatic Representations of Sustainability: A Review and Synthesis', in *Proceedings 28th Annual ARCOM Conference 3–5 September 2012*, ed. Simon D. Smith (Edinburgh: ARCOM, 2012), 1479–89. The argument that the mainstream has 'hijacked' sustainability without actually changing, including in the prevalent 'greenwashing' by corporations, is also part of the discourse on sustainability (see Adrian Parr, *Hijacking Sustainability* (Cambridge, MA: MIT Press, 2009)).
11 George Perkins Marsh, *Man and Nature* (Charleston, SC: Nabu Press, 2011).
12 Jacobus A. Du Pisani, 'Sustainable Development: Historical Roots of the Concept', *Environmental Sciences* 3, no. 2 (2006): 83.
13 Ian Thompson, 'Aesthetic, Social and Ecological Values in Landscape Architecture: A Discourse Analysis', *Ethics, Place and Environment*, Vol. 3, No.3 (2000): 280.
14 Anne W. Spirn, 'Constructing Nature: The Legacy of Frederick Law Olmsted', in *Uncommon Ground: Reinventing Nature*, ed. William Cronon (New York: W.W. Norton & Company, 1995), 91–113.
15 Susan Herrington, 'An Ontology of Landscape Design', in *The Routledge Companion to Landscape Studies*, eds. Peter Howard, Ian Thompson and Emma Waterton (London: Routledge, 2013), 355–65; Sue Kidd, 'Landscape Planning: Reflections on the Past, Directions for the Future', in Howard, Thompson and Waterton, *Routledge Companion*, 366–82.
16 Ian Thompson, 'Aesthetic, Social and Ecological Values in Landscape Architecture: A Discourse Analysis', *Ethics Place and Environment* 3, no. 3 (2000): 269.
17 Carter and Moir, 'Diagrammatic Representations', 1479.
18 However, practitioners might not always adhere to sustainable practices with full integrity. Examples include the use of imported materials such as marble from China for a design in Scandinavia, and similar practices where decisions are driven by finances, rather than by consideration of environmental impact.
19 Efrat Eizenberg and Yosef Jabareen, 'Social Sustainability: A New Conceptual Framework', *Sustainability* 9, no. 68 (2017).
20 Zygmunt Bauman, *Liquid Times: Living in an Age of Uncertainty* (Cambridge: Polity, 2007): 5.
21 These include Bauman, *Liquid Modernity*; David Harvey, 'The Right to the City', *New Left Review* 53 (2008); Judt, *Ill Fares the Land*; Streeck, 'How Will Capitalism End?' Most prominent is Thomas Piketty, *Capital in the 21st Century* (Cambridge, MA: Belknap Press and Harvard University Press, 2014).
22 'International Landscape Convention', IFLA, accessed 4 April 2017, HYPERLINK "http://iflaonline.org/projects/ilc" http://iflaonline.org/projects/ilc.
23 Kyle D. Brown and Todd J. Jennings, 'Social Consciousness in Landscape Architecture Education: Toward a Conceptual Framework', *Landscape Journal* 22, no. 2 (2003): 99; Randolph Hester, 'Whose Politics Do You Style? If All Designers Serve Some Political Will, Whose Will Do You Serve?' *Landscape Architecture* 95, no. 12 (2005): 72–9; Shelley Egoz, 'Deconstructing the Hegemony of Nationalist Narratives Through Landscape Architecture', *Landscape Research* 33, no. 10 (2008): 29.
24 Brown and Jennings, 'Social Consciousness', 99.
25 See Shelley Egoz, ed., *Hva betyr landskapsdemokrati? Defining Landscape Democracy*, (Ås: Centre for Landscape Democracy, 2015); Anna Jorgensen, 'Editorial 2016: Landscape Justice in an Anniversary Year', *Landscape Research* 41, no. 1 (2016): 1; Shelley Egoz and Alessia De Nardi, 'Defining Landscape Justice: The Role of Landscape in Supporting Wellbeing of Migrants', *Landscape Research* 42, sup. 1 (2017): s74; Shelley Egoz, Karsten Jørgensen and Deni Ruggeri, eds., *Defining Landscape Democracy: A Path to Spatial Justice* (Cheltenham: Edward Elgar, 2018).

26 Egoz, Makhzoumi and Pungetti, *Right to Landscape*.
27 Kenneth R. Olwig, 'Recovering the Substantive Nature of Landscape', *Annals of the Association of American Geographers* 86, no. 4 (1996): 630; Olwig, 'The Practice of Landscape "Conventions" and the Just Landscape: The Case of the European Landscape Convention', *Landscape Research* 32, no. 5 (2007): 579; Olwig, 'Epilogue to Landscape as Mediator: The Non-Modern Commons Landscape and Modernism's Enclosed Landscape of Property', in *Landscape as Mediator, Landscape as Commons: International Perspectives on Landscape Research,* eds. Benedetta Castiglioni, Fabio Parascandolo and Marcello Tanca (Padova: Cleup, 2015), 197–214.
28 Quoted on the back cover of Egoz, Makhzoumi and Pungetti, *Right to Landscape*.
29 Noam Chomsky, David Barsamian and Arthur Naiman, *The Common Good* (Monroe, ME: Odonian Press, 1998).
30 One example is the cultural rights or land rights of indigenous people.
31 Egoz and De Nardi, 'Defining Landscape Justice', s72.
32 Egoz, 'Deconstructing', 29; Shelley Egoz and Racheli Merhav, 'Ruins, Archaeology and the Other in the Landscape: The Case of Zippori National Park, Israel', *JoLA* 8 (2009): 56.
33 Ludger Gailing and Markus Leibenath, 'Political Landscapes Between Manifestations and Democracy, Identities and Power', *Landscape Research* 42, no. 4 (2017): 337.
34 Anne W. Spirn, 'Restoring Mill Creek: Landscape Literacy, Environmental Justice and City Planning and Design', *Landscape Research* 30, no. 3 (2005): 395.
35 Hester, 'Whose Politics?'
36 'Social Justice: The New Green Infrastructure', ASLA, modified 27 January 2014, https://asla.org/uploadedFiles/CMS/Meetings_and_Events/2014_Annual_Meeting_Handouts/SAT-A10_Social%20Justice%20The%20New%20Green%20Infrastructure.pdf.
37 For example, 'Environmental Justice', ASLA, accessed 23 January 2018, https://asla.org/contentdetail.aspx?id=46027.
38 Jala Makhzoumi, 'Marginal Landscapes, Marginalized Rural Communities: Sustainable Postwar Recovery in Southern Lebanon', in *Lessons in Postwar Reconstruction: Case Studies from Lebanon in the Aftermath of the 2006 War*, ed. Howayda Al Harithy (London: Routledge, 2010), 127–57.
39 Mina Di Marino and Kimmo Lapintie, 'Exploring the Concept of Green Infrastructure in Urban Landscape: Experiences from Italy, Canada and Finland', *Landscape Research* 43, no. 1 (2018).
40 Maria Ignatieva, Shelley Egoz, Diane Menzies and Irinia Melnichuk, eds., *Green Infrastructure: From Global to Local* (St Petersburg: St Petersburg Publishing House of Polytechnic University, 2012); Ian C. Mell, 'Green Infrastructure: Reflections on Past, Present and Future Praxis', *Landscape Research* 42, no. 2 (2017): 135.
41 See 'The Multifunctionality of Green Infrastructure: Science for Environment Policy', European Commission, March 2012, http://ec.europa.eu/environment/nature/ecosystems/docs/Green_Infrastructure.pdf.
42 Mark A. Benedict and Edward T. McMahon, 'Green Infrastructure: Smart Conservation for the 21st Century', May 2001, http://www.sprawlwatch.org/greeninfrastructure.pdf.
43 Melissa Checker, 'Wiped Out by the "Greenwave": Environmental Gentrification and the Paradoxical Politics of Urban Sustainability', *City & Society* 23, no. 2 (2011): 210; Andrew Newman, *Landscape of Discontent* (Minneapolis: University of Minnesota Press, 2015); Kenneth A. Gould and Tammy L. Lewis, *Green Gentrification, Urban Sustainability and the Struggle for Environmental Justice* (New York: Routledge, 2017).
44 Checker, 'Wiped Out'.
45 Ibid., 212.
46 Quoted in ibid.
47 Bauman, *Liquid Times*.
48 See Neil Everden, *The Social Construction of Nature* (Baltimore: John Hopkins University Press, 1992).
49 Checker, 'Wiped Out', 213.
50 Charles Waldheim, ed., *The Landscape Urbanism Reader* (New York: Princeton Architectural Press, 2006), 11.
51 For example Charles Waldheim, 'On Landscape, Ecology, and Other Modifiers to Urbanism', *Scenario 01: Landscape Urbanism* (2012); Peggy Tully, 'On Landscape Urbanism,' in Howard, Thompson and Waterton, *Routledge Companion*, 438–49; Matthew Heins, 'Finding Common Ground Between New Urbanism and Landscape Urbanism', *Journal of Urban Design* 20, no. 3 (2015); David De la Peña, 'New Landscape Urbanism: Promising New Paths for Urban Design', *Journal of Urban Design* 20, no. 3 (2015):

314; Julian Bolleter, 'Charting the Potential of Landscape Urbanism in Dubai', *Landscape Research* 40, no. 5 (2015): 621.

52 Waldheim, 'On Landscape'.

53 Stan Allen, *Points + Lines: Diagrams and Projects for the City* (New York: Princeton Architectural Press, 1999); Alex Wall, 'Programming the Urban Surface', in *Recovering Landscape: Essays in Contemporary Landscape Architecture*, ed. James Corner (New York: Princeton Architectural Press, 1999), 233–50; James Corner, 'Terra Fluxus', in Waldheim, *Landscape Urbanism Reader*, 21–33; Richard Weller, 'Landscape (Sub)Urbanism in Theory and Practice', *Landscape Journal* 27, no. 2 (2008): 247.

54 Heins, 'Finding Common Ground'.

55 Douglas Kelbaugh, 'The Environmental Paradox of the City: Landscape Urbanism and New Urbanism', *Consilience: The Journal of Sustainable Development* 13, no. 1 (2014): 1; Thompson, 'Ten Tenets', 7; Leon A. Morenas, 'Critiquing Landscape Urbanism: A View on New York's High Line', *Economic & Political Weekly* 47, no. 7 (2012): 19–22; Douglas Spencer, 'The Obdurate Form of Landscape Urbanism: Neoliberalism, Design, and Critical Agency', Critical Grounds, 26 May 2011, https://terraincritical.wordpress.com/2011/05/26/the-obdurate-form-of-landscape-urbanism-neoliberalism-design-and-critical-agency.

56 Thompson, 'Ten Tenets', 16.

57 Morenas, 'Critiquing Landscape Urbanism', 19.

58 Grahame Shane, 'The Emergence of "Landscape Urbanism": Reflections on *Stalking Detroit*', *Harvard Design Magazine*, Fall – Winter 2003, http://crtl-i.com/PDF/GrahamShane_OnLandscape.pdf.

59 Georgia Daskalakis, Charles Waldheim and Jason Young, eds., *Stalking Detroit* (Barcelona: ACTAR, 2001).

60 'Terra Fluxus' received about 250 citations in the last decade, according to Google Scholar. This is significant considering the small scope of the landscape architecture scientific community.

61 For example, Julian Bolleter, 'Para-Scape: Landscape Architecture in Dubai', *JoLA* 4, no. 1 (2009): 28; Weller, 'Landscape (Sub)Urbanism'.

62 Morenas, 'Critiquing Landscape Urbanism', 21.

63 Ibid.

64 Corner, 'Terra Fluxus', 28.

65 Waldheim, 'Landscape as Urbanism', 39.

66 Quoted in Corner, 'Terra Fluxus', 28.

67 Waldheim, 'On Landscape'.

68 Eva Castro and Alfredo Ramirez, 'Multiplying the Ground', in *Design Innovation for the Built Environment: Research by Design and the Renovation of Practices*, ed. Michael U. Hensel (London: Routledge, 2012), 205–20.

69 Spencer, 'Obdurate Form'; Spencer, 'Agency and Artifice in the Environment of Neoliberalism', in *Landscape and Agency Critical Essays*, eds. Ed Wall and Tim Waterman (London: Routledge, 2018), 177–8.

70 Spencer, 'Obdurate Form'.

71 Bauman, *Liquid Times*; Judt, *Ill Fares the Land*; Streeck, 'How Will Capitalism End?'; Harvey, 'Neoliberalism'; Harvey, 'Right to the City'.

72 Bauman, *Liquid Times*, 3.

73 Ibid., 4.

74 Gould and Lewis, *Green Gentrification*, 5.

75 Bauman, *Liquid Times*, 90.

76 Harvey, 'Right to the City', 23.

Bibliography

Allen, Stan. *Points + Lines: Diagrams and Projects for the City.* New York: Princeton Architectural Press, 1999.

ASLA. 'Environmental Justice'. Accessed 23 January 2018. https://asla.org/contentdetail.aspx?id=46027.

ASLA. 'Social Justice: The New Green Infrastructure'. Modified 27 January 2014. https://asla.org/uploadedFiles/CMS/Meetings_and_Events/2014_Annual_Meeting_Handouts/SAT-A10_Social%20Justice%20The%20New%20Green%20Infrastructure.pdf.

Bauman, Zygmunt. *Liquid Modernity.* Cambridge: Polity, 2000.

Bauman, Zygmunt. *Liquid Times: Living in an Age of Uncertainty*. Cambridge: Polity, 2007.

Benedict, Mark A., and Edward T. McMahon. 'Green Infrastructure: Smart Conservation for the 21st Century'. May 2001. http://www.sprawlwatch.org/greeninfrastructure.pdf.

Bolleter, Julian. 'Para-Scape: Landscape Architecture in Dubai'. *JoLA* 4, no. 1 (2009): 28–41.
Bolleter, Julian. 'Charting the Potential of Landscape Urbanism in Dubai'. *Landscape Research* 40, no. 5 (2015): 621–42. doi:10.1080/01426397.2014.967189.
Brown, Kyle D., and Todd J. Jennings. 'Social Consciousness in Landscape Architecture Education: Toward a Conceptual Framework'. *Landscape Journal* 22, no. 2 (2003): 99–112. doi:10.3368/lj.22.2.99.
Carter, Kate, and Stuart Moir. 'Diagrammatic Representations of Sustainability: A Review and Synthesis'. In *Proceedings 28th Annual ARCOM Conference 3–5 September 2012*, edited by Simon D. Smith, 1479–89. Edinburgh: ARCOM, 2012. http://www.research.ed.ac.uk/portal/files/6320768/ARCOM_2012_paper.pdf.
Castro, Eva, and Alfredo Ramirez. 'Multiplying the Ground'. In *Design Innovation for the Built Environment: Research by Design and the Renovation of Practices*, edited by Michael U. Hensel, 205–20. London: Routledge, 2012.
Checker, Melissa. 'Wiped out by the "Greenwave": Environmental Gentrification and the Paradoxical Politics of Urban Sustainability'. *City & Society* 23, no. 2 (2011): 210–29. doi:10.1111/j.1548-744X.2011.01063.x.
Chomsky, Noam, David Barsamian and Arthur Naiman. *The Common Good*. Monroe, ME: Odonian Press, 1998.
Corner, James. 'Terra Fluxus'. In *The Landscape Urbanism Reader*, edited by Charles Waldheim, 21–33. New York: Princeton Architectural Press, 2006.
Council of Europe. Treaty no. 176, 'European Landscape Convention', 20 October 2000. https://rm.coe.int/1680080621.
Daskalakis, Georgia, Charles Waldheim and Jason Young, eds. *Stalking Detroit*. Barcelona: ACTAR, 2001.
Déjeant-Pons, Maguelonne. 'The European Landscape Convention'. *Landscape Research* 31, no. 4 (2006): 363–84. doi:10.1080/01426390601004343.
Déjeant-Pons, Maguelonne. 'The European Landscape Convention: From Concepts to Rights'. In *The Right to Landscape: Contesting Landscape and Human Rights*, edited by Shelley Egoz, Jala Makhzoumi and Gloria Pungetti, 51–6. Farnham: Ashgate, 2011.
De la Peña, David. 'New Landscape Urbanism: Promising New Paths for Urban Design'. *Journal of Urban Design* 20, no. 3 (2015): 314–17. doi:10.1080/13574809.2015.1030989.
Di Marino, Mina, and Kimmo Lapintie. 'Exploring the Concept of Green Infrastructure in Urban Landscape: Experiences from Italy, Canada and Finland'. *Landscape Research* 43, no. 1 (2018): 139–49. doi:10.1080/01426397.2017.1300640.
Du Pisani, Jacobus A. 'Sustainable Development: Historical Roots of the Concept'. *Environmental Sciences* 3, no. 2 (2006): 83–96. doi:10.1080/15693430600688831.
Egoz, Shelley. 'Deconstructing the Hegemony of Nationalist Narratives Through Landscape Architecture'. Landscape Research 33, no. 10 (2008): 29–50.
Egoz, Shelley, ed. *Hva betyr landskapsdemokrati? Defining Landscape Democracy*. Ås: Centre for Landscape Democracy, 2015.
Egoz, Shelley, and Alessia De Nardi. 'Defining Landscape Justice: The Role of Landscape in Supporting Wellbeing of Migrants'. *Landscape Research* 42, sup. 1 (2017): s72–s89. doi:10.1080/01426397.2017.1363880.
Egoz, Shelley, Karsten Jørgensen and Deni Ruggeri, eds. *Defining Landscape Democracy: A Path to Spatial Justice*. Cheltenham: Edward Elgar, 2018.
Egoz, Shelley, Jala Makhzoumi and Gloria Pungetti, eds. *The Right to Landscape: Contesting Landscape and Human Rights*. Farnham: Ashgate, 2011.
Egoz, Shelley, and Racheli Merhav. 'Ruins, Archaeology and the Other in the Landscape: The Case of Zippori National Park, Israel'. *JoLA* 8 (2009): 56–69.
Eizenberg, Efrat, and Yosef Jabareen. 'Social Sustainability: A New Conceptual Framework'. *Sustainability* 9, no. 68 (2017). doi:10.3390/su9010068.
European Commission. 'The Multifunctionality of Green Infrastructure: Science for Environment Policy'. March 2012. http://ec.europa.eu/environment/nature/ecosystems/docs/Green_Infrastructure.pdf
Everden, Neil. *The Social Construction of Nature*. Baltimore: John Hopkins University Press, 1992.
Gailing, Ludger, and Markus Leibenath. 'Political Landscapes Between Manifestations and Democracy, Identities and Power'. *Landscape Research* 42, no. 4 (2017): 337–48. doi:10.1080/01426397.2017.1290225.
Germundsson, Thomas, Peter Howard and Kenneth R. Olwig. 'Reassessing Landscape Drivers and the Globalist Environmental Agenda'. *Landscape Research* 36, no. 4 (2011): 395–99.
Gould, Kenneth A., and Tammy L. Lewis. *Green Gentrification, Urban Sustainability and the Struggle for Environmental Justice*. New York: Routledge, 2017.

Harvey, David. 'The Right to the City'. *New Left Review* 53 (2008).
Harvey, David. 'Neoliberalism is a Political Project'. *Jacobin*, 23 July 2016. https://www.jacobinmag.com/2016/07/david-harvey-neoliberalism-capitalism-labor-crisis-resistance.
Heins, Matthew. 'Finding Common Ground Between New Urbanism and Landscape Urbanism'. *Journal of Urban Design* 20, no. 3 (2015). doi:10.1080/13574809.
Herrington, Susan. 'An Ontology of Landscape Design'. In *Routledge Companion to Landscape Studies*, edited by Peter Howard, Ian Thompson and Emma Waterton, 355–65. London: Routledge, 2013.
Hester, Randolph. 'Whose Politics Do You Style? If All Designers Serve Some Political Will, Whose Will Do You Serve?' *Landscape Architecture* 95, no. 12 (2005): 72–9.
IFLA. 'International Landscape Convention'. Accessed 4 April 2017. http://iflaonline.org/projects/ilchttp://iflaonline.org/projects/ilc.
Ignatieva, Maria, Shelley Egoz, Diane Menzies and Irina Melnichuk, eds. *Green Infrastructure: From Global to Local*. St Petersburg: St Petersburg Publishing House of Polytechnic University, 2012.
Jorgensen, Anna. 'Editorial 2016: Landscape Justice in an Anniversary Year'. *Landscape Research* 41, no. 1 (2016): 1–6.
Judt, Tony. *Ill Fares the Land*. London: Penguin, 2010.
Kelbaugh, Douglas. 'The Environmental Paradox of the City: Landscape Urbanism and New Urbanism'. *Consilience: The Journal of Sustainable Development* 13, no. 1 (2014): 1–15.
Kidd, Sue. 'Landscape Planning: Reflections on the Past, Directions for the Future'. In *The Routledge Companion to Landscape Studies*, edited by Peter Howard, Ian Thompson and Emma Waterton, 366–82. London: Routledge, 2013.
Makhzoumi, Jala. 'Marginal Landscapes, Marginalized Rural Communities: Sustainable Postwar Recovery in Southern Lebanon'. In *Lessons in Postwar Reconstruction: Case Studies from Lebanon in the Aftermath of the 2006 War*, edited by Howayda Al Harithy, 127–57. London: Routledge, 2010.
Marsh. 'Political Risk Map 2017 Update: Adapting to an Unpredictable World'. Accessed 29 November 2017. https://www.marsh.com/us/campaigns/political-risk-map-2017.html.
Marsh, George Perkins. *Man and Nature*. Charleston, SC: Nabu Press, 2011.
McHarg, Ian. *Design with Nature*. New York: Doubleday, 1969.
Mell, Ian C. 'Green Infrastructure: Reflections on Past, Present and Future Praxis'. *Landscape Research* 42, no. 2 (2017): 135–45. doi:10.1080/01426397.2016.1250875.
Morenas, Leon A. 'Critiquing Landscape Urbanism: A View on New York's High Line'. *Economic & Political Weekly* 47, no. 7 (2012): 19–22.
Newman, Andrew. *Landscape of Discontent*. Minneapolis: University of Minnesota Press, 2015.
Olwig, Kenneth R. 'Recovering the Substantive Nature of Landscape'. *Annals of the Association of American Geographers* 86, no. 4 (1996): 630–53. doi:10.1111/j.1467-8306.1996.tb01770.x.
Olwig, Kenneth R. 'The Practice of Landscape "Conventions" and the Just Landscape: The Case of the European Landscape Convention'. *Landscape Research* 32, no. 5 (2007): 579–94.
Olwig, Kenneth R. 'Epilogue to Landscape as Mediator: The Non-Modern Commons Landscape and Modernism's Enclosed Landscape of Property'. In *Landscape as Mediator, Landscape as Commons: International Perspectives on Landscape Research*, edited by Benedetta Castiglioni, Fabio Parascandolo and Marcello Tanca, 197–214. Padova: Cleup, 2015.
Parr, Adrian. *Hijacking Sustainability*. Cambridge, MA: MIT Press, 2009.
Piketty, Thomas. *Capital in the 21st Century*. Cambridge, MA: Belknap Press and Harvard University Press, 2014.
Shane, Grahame. 'The Emergence of "Landscape Urbanism": Reflections on *Stalking Detroit*'. *Harvard Design Magazine*, Fall – Winter 2003. http://crtl-i.com/PDF/GrahamShane_OnLandscape.pdf.
Spencer, Douglas. 'The Obdurate Form of Landscape Urbanism: Neoliberalism, Design, and Critical Agency'. Critical Grounds, 26 May 2011. https://terraincritical.wordpress.com/2011/05/26/the-obdurate-form-of-landscape-urbanism-neoliberalism-design-and-critical-agency.
Spencer, Douglas. 'Agency and Artifice in the Environment of Neoliberalism'. In *Landscape and Agency: Critical Essays*, edited by Ed Wall and Tim Waterman, 177–8. London: Routledge, 2018.
Spirn, Anne W. 'Constructing Nature: The Legacy of Frederick Law Olmsted'. In *Uncommon Ground: Reinventing Nature*, edited by William Cronon, 91–113. New York: W.W. Norton & Company, 1995.
Spirn, Anne W. 'Restoring Mill Creek: Landscape Literacy, Environmental Justice and City Planning and Design'. *Landscape Research* 30, no. 3 (2005): 395–413.
Streeck, Wolfgang. 'How Will Capitalism End?' *New Left Review* 87, 2014. https://newleftreview.org/II/87/wolfgang-streeck-how-will-capitalism-end.

Thompson, Ian. 'Aesthetic, Social and Ecological Values in Landscape Architecture: A Discourse Analysis'. *Ethics Place and Environment* 3, no. 3 (2000): 269–87. doi:10.1080/713665903.

Thompson, Ian. 'Ten Tenets and Six Questions for Landscape Urbanism'. *Landscape Research* 37, no. 1 (2012): 7–26. doi:10.1080/01426397.2011.632081.

Tully, Peggy. 2013 'On Landscape Urbanism'. In *The Routledge Companion to Landscape Studies*, edited by Peter Howard, Ian Thompson and Emma Waterton, 438–49. London: Routledge, 2013.

Waldheim, Charles. 'Landscape as Urbanism'. In *The Landscape Urbanism Reader*, edited by Charles Waldheim (New York: Princeton Architectural Press, 2006), 35–55.

Waldheim, Charles. 'On Landscape, Ecology, and Other Modifiers to Urbanism'. Scenario 01: Landscape Urbanism. Accessed 5 February 2017. http://scenariojournal.com/article/topos-landscape-urbanism.

Wall, Alex. 'Programming the Urban Surface'. In *Recovering Landscape: Essays in Contemporary Landscape Architecture*, edited by James Corner, 233–50. New York: Princeton Architectural Press, 1999.

Weller, Richard. 'Landscape (Sub)Urbanism in Theory and Practice'. *Landscape Journal* 27, no. 2 (2008): 247–67.

Wikipedia. 'Sustainability'. Last modified 18 January 2018. https://en.wikipedia.org/wiki/Sustainability.

18

COUPLING ENVIRONMENTAL AND SOCIOCULTURAL SUSTAINABILITY FOR BETTER DESIGN

A case study of Emirati neighbourhoods and landscape

Sneha Mandhan and Alan M. Berger

Sustainable development has been defined by various scholars and practitioners since its initial formal conception in 1987. Authors have written about its necessity, its principles, and the indices that aim to measure its ecological, social, and economic parameters (also known as the three pillars of sustainability or the 'triple bottom line').[1] There have also been theoretical and practice-based translations of sustainable development principles into the physical planning and design of urban and rural landscapes. However, a majority of these focus on individual pillars of sustainability, with environmental aspects taking precedence. Popular initiatives such as Leadership in Energy and Environmental Design (LEED) use complex environmental models to calculate levels of sustainability based on metabolic parameters of energy, water, transport, and waste. While these are imperative in leading development towards a more environmentally sustainable future, they often result in the creation of parametrically modelled global designs that are socially and culturally inappropriate when applied to local cases.

Emphasizing better liveability of urban environments, how can designers combine aspects of sustainability to create solutions that are holistic and comprehensive? This chapter uses this primary question as a premise to present ideas for operationalizing sustainable development by coupling sociocultural and environmental parameters in urban settings. We challenge the alleged difficulty of incorporating social parameters into sustainable design,[2] and ask: how, as designers, can we provide sociocultural and environmental sustainability simultaneously? How can we minimize the negative externalities on social and cultural aspects of designs, especially for residential neighbourhoods, while optimizing environmental performance? How can global sustainability knowledge be transferred and adapted to the specific sociocultural needs of local communities?

To explore these questions, we use the example of Emirati residential neighbourhoods in Abu Dhabi, United Arab Emirates, in the hyper-arid desert region of the Arabian Peninsula.

Home to native communities that have substantial social and cultural needs associated with the influence of Islamic Shari'a law, this case presents a unique opportunity to couple sociocultural and environmental features to develop a comprehensively sustainable urban form. To do this, we propose a starting point: deriving inspiration from traditional city forms, while understanding underlying values for spatial typologies, and the influence of evolving technologies and increasing globalization.

The concept of sustainability

The idea of sustainable development was popularized in 1987 by *Our Common Future* (the Brundtland Report). This report, formulated by the World Commission on Environment and Development, defined sustainable development as that which 'meets the needs of the present without compromising the ability of future generations to meet their own needs'.[3] This broad definition aims to reconcile the needs of communities with the biophysical goals of environmental management. It encompasses the tangible and less tangible necessities for sustaining life. The Agenda 21 report, adopted by the United Nations in 1992 at the UN Conference on Environment and Development, outlines the three spheres of sustainability: ecological, economic, and social. For any development to be sustainable, it must integrate qualities associated with the overlap of these three pillars: it must be equitable, liveable, and viable.[4] In addition, the concept of cultural sustainability became a focus of planning, policy, and design efforts following its definition by the World Commission on Culture and Development in 1995 as 'inter and intra-generational access to cultural resources'.[5]

Scholars and various global and local agencies have developed sets of indicators and indices to measure sustainability. Alberti outlines their necessity by emphasizing that 'conventional measures of economic performance and urban quality of life are inadequate to capture the interdependence between urban society, economic development, and the environment.'[6] While these indicators go a step further than the definitions in operationalizing the concept of sustainable development, there is still a considerable disconnect between the principles of sustainability and the actionable steps a designer can take towards creating genuinely sustainable environments.

The case study: Abu Dhabi and the Gulf Cooperation Council states on the Arabian Peninsula

The countries in the Gulf Cooperation Council (GCC), located on the southern edge of the Arabian Gulf, are eighty per cent urbanized, and will be home to fifty-three million people by the year 2020.[7] Hyper-aridity, sandy soil, and extremely warm temperatures characterize the region.[8] Due to their high per-capita resource usage rate and annual greenhouse gas (GHG) emissions, many of the GCC countries have recognized the need to develop sustainably, and have initiated efforts to make sustainability central to their urban agendas.

The region's urban history is brief. Spurred by the discovery of oil in the region in the mid-twentieth century, previously nomadic settlements grew into sprawling metropolises at an accelerated pace. Due to the small size of local communities, foreign labour was invited to drive growth and construction.[9] With this influx came Western planning and design influences—urban grids, superblocks, and villa typologies.[10] Villa neighbourhoods, initially developed by oil companies to house their employees, soon became the 'ideal' and 'modern' model of residential development that local communities and governments aspired to.[11] These typologies have now become ingrained in the urban diction of the region, enabled and promoted through building standards and regulations. One example is the requirement of setbacks on all sides of the residential plot, which effectively results in the creation of the detached house (villa) typology. Scholars attribute the introduction of such requirements into local regulations to Western influence.[12]

The rapidity of change has resulted in the creation of dichotomies between traditional and modern, new and old, and global and local, presenting an ongoing challenge for urban development in the region.[13] It has left little scope for the natural evolution of climatically responsive and culturally appropriate vernacular typologies. As a result, urbanization here is continually tackling two main socio-environmental challenges—high carbon footprints caused by low resource capacities and high consumption rates, and the influence of globalization and large expatriate populations on the native cultural identity.

Abu Dhabi is the capital, and the largest emirate of the seven that form the United Arab Emirates. It is spread over 67,340 square kilometres and has a population of 2.78 million.[14] It comprises the cities of Abu Dhabi and Al Ain, and the Western Gharbia region. Abu Dhabi's urban development started as recently as the 1960s, when the erstwhile ruler Sheikh Zayed established policies and programmes for urban growth. It included the demolition of all existing structures and the installation of a new layout of arterial streets and superblocks on the island.[15] Since then, the city has grown rapidly, with an average annual growth rate of 9.6 per cent since 1960. Historical data from the emirate indicates a ninety-nine-fold population increase between 1960 and 2010.[16] Currently, native Emiratis form only nineteen per cent of the population in Abu Dhabi.[17] However, the urban landscape of the city is dominated by the typical low-rise and low-density villa neighbourhoods that house this Emirati population, as shown in Figure 18.1 (approximately fifty-five per cent of the land area consists of villa neighbourhoods, according to initial calculations of geographic information system spatial data).

The impacts of climate change, exacerbated by the emission of GHG into the atmosphere, threaten to increase temperatures in the region beyond human physiological tolerance, threatening

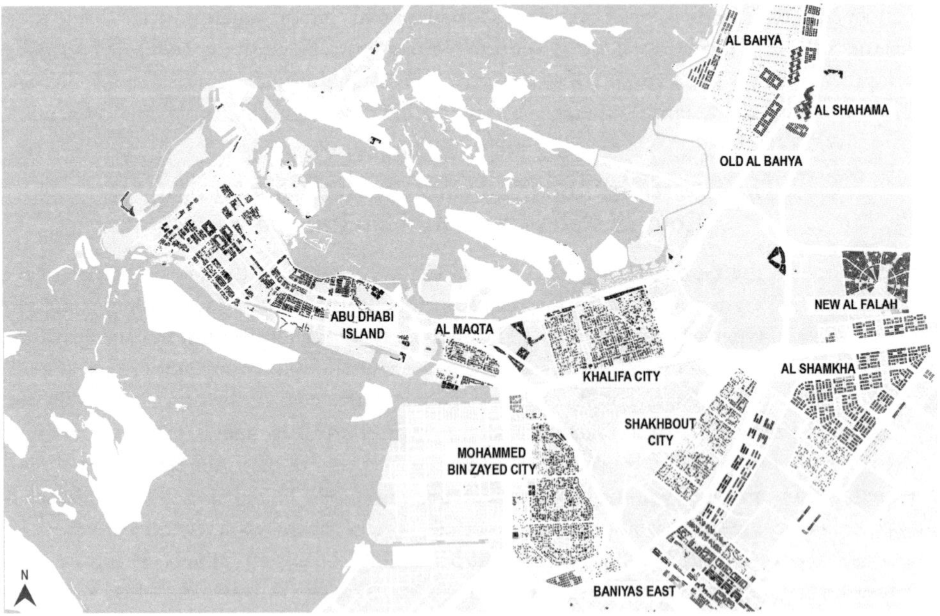

Figure 18.1 Map showing the low-rise neighborhoods spread across the Abu Dhabi city island and mainland. *These neighborhoods occupy over fifty percent of the urban landscape in the city.* Data Source: Abu Dhabi Municipality.

the very existence of these settlements.[18] Awareness of these factors has led to several recent government initiatives to educate consumers and promote environmentally sustainable building practices. In Abu Dhabi, the Estidama ('sustainability' in Arabic) programme, which includes the Pearl Rating System, tackles sustainability through the definition of best practice and minimum requirements for buildings, villas, and communities.[19] This programme, implemented in 2010, aims to create a new urban sustainability framework in the form of a tangible rating system (ranging from one to five Pearls) based on quantitative metrics. Research has shown that by achieving a one Pearl rating, a villa will reduce energy usage by twenty per cent and water usage by thirty per cent below a baseline of good practice.[20] One of the biggest strengths of this programme is its contextual specificity for the extreme heat and humidity of a coastal desert climate. However, in many ways the villa typology is itself ill-suited to the sociocultural needs of Emirati communities, as described later in this chapter, limiting these strategies' impact on addressing sustainability concerns.

Emirati villa neighbourhoods

The relationship between citizens and government in Abu Dhabi is established on strong principles of welfare. The government is responsible for providing housing, education, and healthcare to all Emiratis. The strength of the welfare system in these states is attributed to their origins as tribal communities, in which tribal leaders were responsible for providing for their communities.[21] The Abu Dhabi Housing Authority (ADHA) provides Emiratis in Abu Dhabi with plots and low-interest loans for construction, or with ready-built homes. Real-estate developers work with the ADHA to design and construct neighbourhoods in the city. Sites for these are selected based on the citywide 'Plan Abu Dhabi 2030', developed by the Abu Dhabi Urban Planning Council.[22]

A typical neighbourhood, as shown in Figure 18.2, consists of arterial and sub-arterial streets that define neighbourhood blocks, with a central space allotted for a community mosque, small

Figure 18.2 Satellite imagery (2015) and photographs of Khalifa City (2016), Abu Dhabi, showing the typical villa and block typologies, landscape characteristics, and open space configurations of typical, contemporary Emirati neighborhoods. © Google Earth, 2015; © Mandhan, Sneha and Alan M. Berger, 2016.

retail, and public open space. In some neighbourhoods, narrow alleys (*sikkas*) create a void between adjacent plots. Streets are often planted with date palms. Blocks consist of arrays of equally dimensioned plots. They are individually allotted to Emirati families, with the provision of a low-interest loan for construction. Each parcel typically contains a detached, single-family villa unit, surrounded by boundary walls along the edges of the plot. In some cases, if adjoining plots are allotted to members of the same extended family, they are combined to form a larger complex of similar villas, with a single boundary wall along the edge of the combined plot. Alternatively, for ready-built housing neighbourhoods, a similar layout is adopted, but the design vocabulary of individual villas is kept comparably uniform.

Generally, villas are extroverted, with all walls exposed to sunlight. They are also relatively large. The open land between the walls of the villa and the outer boundary walls is often vegetated—containing lush and water-intensive plants such as turf and date palms. The floor plans of the villas show clear segregation of semi-private and private spaces. The *majlis*, a gathering space for men, is located near the entrance to enable a separation between visitors and the private realm of the family.[23]

Sociocultural aspects of Emirati neighbourhoods

Traditionally, Emirati communities, which are primarily Muslim, have lived according to customs that derive from Shari'a law.[24] These principles, and the resulting cultural needs, have guided socio-spatial relationships, and shaped neighbourhood and urban form.[25] Abu-Lughod summarizes the spatial principles that constitute the 'deep grammar' of the Islamic city: (1) residential superblocks enclosing semi-public space, (2) the secondary circulation system of the city, with the exception of major routes to city gates, (3) neighbourly cooperation with regard to property rights and community policing, and (4) a highly fragmented system of property rights.[26] She and other scholars have emphasized the need for a trifold division of space in the Arab-Muslim city, from private to semi-private to public. This relates to the gendering of space that is prevalent in these communities, where the private home is considered the domain of the female members, while the males occupy semi-private and public areas.

Contemporary neighbourhoods have often been criticized for not attributing the same level of importance to these needs.[27] The extroverted nature of the villa typology provides fenestrations on external walls, creating opportunities for outsiders to peek into the private spaces of the house. As a result, most houses have high boundary walls and keep windows screened through the use of blinds or curtains. Most villas range from one to three storeys tall. Differences in the heights of adjacent villas allow visibility into neighbouring plots, thus hindering privacy. This often results in the creation of ad hoc roof-screening mechanisms.[28] The open space between the villa and the surrounding boundary wall, being outside the private area of the home, becomes an underutilized space, only intermittently used by the children for play. Women, who are often confined to enclosed spaces inside the house, lack access to private, open areas for recreation.[29]

When describing reasons for the increasing size of villas in Saudi Arabia, Bahammam mentions the duplication of rooms for male and female use. He attributes this to the fast pace of construction and the lack of user participation in the planning and design process, resulting in families overestimating their spatial needs. Traditionally, the process of home building was incremental, with rooms being added as families grew. The standardized procedure of housing/plot allocation and construction financing has led to families building houses as large as they can afford at the time. Bahammam discusses the influence of neighbours and other social connections, and the potential impact of the idealization of various global contexts. He also comments on the equation of the largeness of the dwelling with the public display of social status and economic prosperity.[30] Although he is writing about this in the context of Saudi Arabian cities, several of these factors apply to most cities in the region, including Abu Dhabi. Conversations with local real-estate developers confirm these observations.

The need for the coupling of sociocultural and environmental sustainability

Local communities in this region present a unique opportunity for the development of innovative housing design that caters to environmental specificities such as low biocapacities and high temperatures, as well as to sociocultural specificities such as the need for privacy and the gendering of spaces. To enhance liveability, these factors must be considered on an equal footing. This will allow the origination of designs that lower the carbon footprints of the houses and provide for all of the cultural customs that pertain to daily familial life.

As described above, efforts are being made, through the implementation of various environmental programmes and initiatives, to reduce the resource consumption of these low-rise and low-density villa neighbourhoods. However, the large-scale provision of housing, the cultural embeddedness of typologies imported from Western contexts in the 1960s and 1970s, and the extreme privacy of local Emirati communities create hurdles for the generation of innovative housing design.[31] A large percentage of the urban planning, design, and architecture professionals responsible for designing these neighbourhoods are expatriates. In such cases, it is essential to question not only the alignment of the built form and landscape with principles of sustainability, but also the transfer of knowledge and sensitivity to local traditions, customs, and lifestyles. This makes designing for sociocultural sustainability an even more challenging prospect, thus making academic inquiries increasingly imperative.

Deriving inspiration from traditional urban form and landscape

Traditional Arab-Muslim cities reflected strong socio-spatial relationships based on the cultural needs of local communities. In his book *Urban Form in the Arab World*, Bianca describes the influence of Islam on urban form as one that 'has given birth to a comprehensive and integrated cultural system by totally embedding the religious practice in the daily life of the individual and the society [... and] providing a matrix of behavioral archetypes which, by necessity, generated correlated physical patterns'.[32] Islam emphasizes the family as the essential unit in the community, and neighbourliness as the backbone of the community.[33] Thus, traditional cities exhibited a strong mode of bottom-up community organization.[34] This was reflected in the cellular structure of the Arab-Muslim city, where neighbourhoods were created by the aggregation of residential units around narrow, gated, semi-public paths.[35] These settlements also showed high levels of sensitivity to the environmental context, including the limited access to natural resources and the extreme climatic conditions.

The physical translations of these, as shown in Figure 18.3, included: (1) courtyard houses with strictly controlled access, (2) semi-private spaces (*fina'* and *majlis*), (3) a network of semi-private and semi-public pathways, and (4) a central community space containing a mosque and market (*souq*). Introverted houses accommodated privacy while allowing adequate light and ventilation from an interior courtyard. The courtyard, being a cool, shaded, open space within the private realm of the house, provided a safe recreational space for women and children. Roof heights of adjacent houses were controlled, in the spirit of neighbourly cooperation, to

Figure 18.3 Satellite imagery (2015) and photographs of the historic Al-Bastakiya neighborhood in Dubai (2016) showing courtyard houses with native gaf trees, hierarchies of streets, dense neighborhood form, minimal fenestrations on external walls, and vernacular features such as the wind tower. © Google Earth, 2015; © Mandhan, Sneha and Alan M. Berger, 2016.

minimize sightlines into neighbours' houses. External entrances were staggered, and opened onto blank walls and secondary entryways, to protect interior spaces from the gaze of passers-by. Fenestrations on exterior walls were kept to a minimum, which provided privacy and reduced solar heat gain. Any exterior windows were screened with wooden screens (*mashrabiyya*) for privacy. These also helped to filter out sand particles from incoming breezes. Housing units clustered and developed incrementally to define pathways which were often gated for protection of the neighbourhood. The widths of paths signalled their level of publicness—narrower streets meant higher levels of privacy—and semi-private streets ended in dead ends. These narrow streets also facilitated wind circulation and shade. Houses also featured several architectural elements for passive cooling, such as wind towers and thick stone or adobe walls.[36]

Concerning landscapes, Shari'a law establishes the role of humans as stewards of the earth. It emphasizes principles of sustainability and resource management, to ensure that current and future generations have access to natural resources and are respectful of them.[37] The common historical narrative of landscape design in Arab-Muslim cities has comprised the geographically sensitive design of oases and Islamic gardens.[38] Makhzoumi, writing about contemporary landscape practice in the Middle East, emphasizes the problematic translation of the Western concept of 'landscape' into the Arab-Muslim context. Along with Nassauer and Naveh, Makhzoumi highlights the importance of the local culture in the interpretation of landscape design.[39]

These principles outline the traditional customs that are associated with historical Arab-Muslim neighbourhoods. Their vernacular spatial translations exhibit a unique combination of context-specificity and sustainability for environmental and sociocultural needs. They had several years to develop and evolve according to the needs of local communities and their climatic and geographic contexts. However, cities such as Abu Dhabi have had much shorter development trajectories, which have also in many cases been synthetic and manufactured.[40] Some of the social and cultural conditions that were prevalent in these historic communities remain, while some have evolved with changes in lifestyle, an increased individualism that dominates community life, and technological developments such as private transport and air conditioning.[41]

There is significant potential in deriving inspiration from the coupled approach to sociocultural and environmental sensitivity observed in traditional Arab-Muslim settlements. This realization has driven efforts by the Urban Planning Council in Abu Dhabi to incorporate features of the traditional neighbourhood (*fareej*), with narrow, shaded pedestrian streets (*sikkas*) and shaded private and semi-private courtyards.[42] Engaging the end users, the Emirati communities, especially women, will also provide insights into cultural needs that are often invisible to designers and planners. Understanding, decoding, and inventorying the spatial vocabulary of these settlements, and the underlying environmental and sociocultural forces that shaped them, will yield an essential starting point for contemporary designers to develop housing schemes that are suited to current conditions in the region.

Conclusion

Cities in the GCC states, often referred to as 'manufactured cities', are developing at a dizzying pace. As a result of globalization, local lifestyles are evolving, creating a tension between the preservation of tradition and the adoption of global ways of living. The rapidity of development, with no long-standing urban typological history to draw from, has resulted in these cities borrowing development ideals from various global contexts, such as Singapore and Vancouver.[43] In many cases, this has resulted in the sociocultural needs of local communities getting lost in translation, in the planning and design not only of the built form but also of the landscape. The specific environmental context, when coupled with the sociocultural complexities of the population in this region, calls for extreme sensitivity in the design process and the designed architecture. For example, the unique affordances of privacy and the gendering of space exemplify the need to rethink conventional landscape practices, expanding the dichotomy of public and private landscapes to accommodate a spectrum between the two.

While this chapter explores the combination of environmental and sociocultural parameters for sustainable design by using the case of Emirati neighbourhoods in Abu Dhabi, the concept applies at a global scale. Academic research and design have focused prominently on the environmental aspects of sustainability. Modelling tools such as Envi-Met, Urban Modeling Interface (UMI), Energy Plus, and BEopt are frequently used by designers and engineers to develop environmentally sustainable building and landscape solutions, and to run their cost-benefit analyses. However, due to the relatively qualitative and intangible nature of social and cultural sustainability, local community needs often get incorporated to a much lesser degree. Masdar City in Abu Dhabi is one such example. It has been applauded for its innovative advancement of technologies to decrease resource consumption, capture energy from renewable sources, and subsequently reduce the carbon footprint of the campus. However, it has also been sharply criticized for not catering to the social needs of residents and visitors, being over-engineered, being located far away from the urban core and amenities of Abu Dhabi city, and not providing well-designed interior spaces.[44]

The primary concept outlined in this chapter is simple—that sustainable development can no longer overlook the social and cultural contexts within which it exists. This is not to discount the emphasis on the reduction of carbon footprints and GHG emissions, but to argue for an increased focus on improving liveability by designing for the nuances of daily customs that drive public and private life in cities. The ideas proposed provide mere starting points for designers to think about social and cultural factors comprehensively and relate them to environmental parameters to create resilient cities that are holistically sustainable.

Acknowledgements

We would like to thank David Birge for providing constructive, critical feedback which contributed to the evolution of the ideas presented in the chapter. This work has been made possible, in part, by the Cooperative Agreement between the Masdar Institute of Science and Technology, Abu Dhabi, UAE and the Massachusetts Institute of Technology, Cambridge, MA, USA.

Notes

1 Marina Alberti, 'Measuring Urban Sustainability', *Environmental Impact Assessment Review*, 16, no.s 4–6 (1996): 381–424; Herman E. Daly, 'Toward Some Operational Principles of Sustainable Development', *Ecological Economics* 2, no. 1 (1990): 1–6; Bill Hopwood, Mary Mellor, and Geoff O'Brien, 'Sustainable Development: Mapping Different Approaches', *Sustainable Development* 13, no. 1 (2005): 38–52; Lu Huang, Jianguo Wu, and Lijiao Yan, 'Defining and Measuring Urban Sustainability: A Review of Indicators', *Landscape Ecology* 30, no. 7 (2015): 1175–93; Georges A. Tanguay et al., 'Measuring the Sustainability of Cities: An Analysis of the Use of Local Indicators', *Ecological Indicators* 10, no. 2 (2010): 407–18.

2 Stephen McKenzie, 'Social Sustainability: Towards Some Definitions', Hawke Research Institute Working Paper Series No. 27, Magill, University of South Australia, 2004, http://w3.unisa.edu.au/hawkeinstitute/publications/downloads/wp27.pdf.

3 Gru Brundtland et al., 'Our Common Future ("Brundtland Report")' (Oxford, UK: Oxford University Press, USA, May 21, 1987), 41. http://www.bne-portal.de/fileadmin/unesco/de/Downloads/Hintergrundmaterial_international/Brundtlandbericht.File.pdf?linklisted=2812.

4 Tanguay et al., 'Sustainability of Cities'.

5 Robert Axelsson et al., 'Social and Cultural Sustainability: Criteria, Indicators, Verifier Variables for Measurement and Maps for Visualization to Support Planning', *Ambio* 42, no. 2 (2013): 215–28.

6 Alberti, 'Measuring Urban Sustainability', 381

7 Abbas El-Zein et al., 'Health and Ecological Sustainability in the Arab World: A Matter of Survival', *The Lancet* 383, no. 9915 (2014): 458–76.

8 Abu Dhabi Urban Planning Council, 'Abu Dhabi Vision 2030', n.d.

9 Mustapha Ben Hamouche, 'The Changing Morphology of the Gulf Cities in the Age of Globalisation: The Case of Bahrain', *Habitat International* 28, no. 4 (2004): 521–40.

10 Mohammed Abdullah Eben Saleh, 'Privacy and Communal Socialization: The Role of Space in the Security of Traditional and Contemporary Neighborhoods in Saudi Arabia', *Habitat International* 21, no. 2 (1997): 167–84.

11 Ali Bahammam, 'Factors Which Influence the Size of the Contemporary Dwelling: Riyadh, Saudi Arabia', *Habitat International* 22, no. 4 (1998): 557–70.

12 Saleh Ali Al-Hathloul, *The Arab-Muslim City: Tradition, Continuity and Change in the Physical Environment* (Riyadh: Dar Al Sahan, 1996).

13 Sareh Moosavi, Jala Makhzoumi, and Margaret Grose, 'Landscape Practice in the Middle East Between Local and Global Aspirations', *Landscape Research* 41, no. 3 (2016): 265–78.

14 'Abu Dhabi Emirate: Facts and Figures', Abu Dhabi Digital Government, accessed 11 July 2016, https://www.abudhabi.ae/portal/public/en/abu-dhabi-emirate/abu-dhabi-emirate-facts-and-figures.

15 Yasser Elshestawy, 'Urban Dualities in the Arab World: From a Narrative of Loss to Neo-Liberal Urbanism', in *The Urban Design Reader*, eds. Michael Larice and Elizabeth Macdonald, The Routledge Urban Reader Series (Abingdon, Oxon ; New York : Routledge, 2011), 475–96.

16 Statistics Center of Abu Dhabi, accessed 4 October 2017, https://www.scad.ae.

17 'Abu Dhabi Emirate: Facts and Figures'.

18 Jeremy S. Pal and Elfatih A. B. Eltahir, 'Future Temperature in Southwest Asia Projected to Exceed a Threshold for Human Adaptability', *Nature Climate Change* 6, no. 2 (2016): 197–200.

19 Abu Dhabi Urban Planning Council and Municipality of Abu Dhabi City, 'Estidama: A Comprehensive Guide of Procedures for Implementing Estidama in the Municipality of Abu Dhabi City', 2010.

20 Ibid.

21 R. A. McDonnell, 'Circulations and Transformations of Energy and Water in Abu Dhabi's Hydrosocial Cycle', *Geoforum* 57 (2014): 225–33.

22 Abu Dhabi Urban Planning Council, 'Plan Abu Dhabi 2030', accessed 4 October 2017, https://www.ecouncil.ae/PublicationsEn/plan-abu-dhabi-full-version-EN.pdf.

23 Adel M. Remali et al., 'A Chronological Exploration of the Evolution of Housing Typologies in Gulf Cities', *City, Territory and Architecture* 3 (2016): 14.
24 Janet L. Abu-Lughod, 'Contemporary Relevance of Islamic Urban Principles (Extract)', *Ekistics* 280 (1980): 6.
25 Al-Hathloul, *Arab-Muslim City*; Stefano Bianca, *Urban Form in the Arab World: Past and Present* (London; New York: Thames & Hudson, 2000); Eben Saleh, 'Privacy and Communal Socialization'.
26 Abu-Lughod, 'Contemporary Relevance'.
27 Kheir Al-Kodmany, 'Residential Visual Privacy: Traditional and Modern Architecture and Urban Design', *Journal of Urban Design* 4, no.3 (1999): 283–311; Bahammam, 'Factors'.
28 Al-Hathloul, *Arab-Muslim City*.
29 Ibid.; Eben Saleh, 'Privacy and Communal Socialization'; Remali et al., 'Chronological Exploration'.
30 Bahammam, 'Factors'.
31 Pieter W. Germeraad, 'Islamic Traditions and Contemporary Open Space Design in Arab-Muslim Settlements in the Middle East', *Landscape and Urban Planning* 23, no. 2 (1993): 97–106; Gamal Taha Mohammed and Kevin Thwaites, 'An Exploratory and Reflective Process of Urban Spatial Morphology within Social Sustainability: Lessons from Middle Eastern Islamic Tradition', *Digest of Middle East Studies* 19, no. 2 (2010): 249–67; Ahmed Farid Moustapha and Frank J. Costa, 'Al Jarudiyah: A Model for Low Rise/High Density Development in Saudi Arabia', *Ekistics* 48, no. 287 (1981): 100–8.
32 Bianca, *Urban Form*.
33 Mohammed Eben Saleh, 'The Impact of Islamic and Customary Laws on Urban Form Development in Southwestern Saudi Arabia', *Habitat International* 22, no. 4 (1998): 537–56.
34 Bianca, *Urban Form*.
35 Abu-Lughod, 'Contemporary Relevance'.
36 Janet L. Abu-Lughod, 'The Islamic City: Historic Myth, Islamic Essence, and Contemporary Relevance', *International Journal of Middle East Studies* 19, no. 2 (1987): 155–76; Al-Hathloul, *Arab-Muslim City*; Al-Kodmany, 'Residential Visual Privacy'; Mustapha Ben Hamouche, 'Climate, Cities and Sustainability in the Arabian Region: Compactness as a New Paradigm in Urban Design and Planning', *International Journal of Architectural Research: ArchNet-IJAR* 2, no. 2 (2014): 196–208; Bahammam, 'Factors'; Bianca, *Urban Form*; Eben Saleh, 'Impact.'
37 Germeraad, 'Islamic Traditions'; Othman Llewellyn, 'Shari'ah Values Pertaining to Landscape Planning and Design', in *Islamic Architecture and Urbanism: Selected Papers from a Symposium Organized by the College of Architecture and Planning*, ed. Aydin Germen (Dammam, Saudi Arabia: King Faisal University, 1983).
38 Llewellyn, 'Shari'ah Values'; Jala M. Makhzoumi, 'Landscape in the Middle East: An Inquiry', *Landscape Research* 27, no. 3 (2002): 213–28.
39 Makhzoumi, 'Landscape in the Middle East'; Joan Iverson Nassauer, 'Culture and Changing Landscape Structure', *Landscape Ecology* 10, no. 4 (1995): 229–37; Zev Naveh, 'Interactions of Landscapes and Cultures', *Landscape and Urban Planning* 32, no. 1 (1995): 43–54.
40 Yasser Elsheshtawy, *The Evolving Arab City: Tradition, Modernity and Urban Development* (London; New York: Routledge, 2008).
41 Ben Hamouche, 'Climate, Cities and Sustainability'.
42 Abu Dhabi Urban Planning Council, 'Neighborhood Planning, Abu Dhabi Vision 2030', n.d.
43 Michael Cameron Dempsey, *Castles in the Sand: A City Planner in Abu Dhabi* (Jefferson, North Carolina: McFarland, 2014).
44 Patrick Kingsley, 'Masdar: The Shifting Goalposts of Abu Dhabi's Ambitious Eco-City', Wired UK, December 2013, http://www.wired.co.uk/article/reality-hits-masdar; Nicolai Ouroussoff, 'In Arabian Desert, a Sustainable City Rises', *The New York Times*, 25 September 2010, https://www.nytimes.com/2010/09/26/arts/design/26masdar.html.

Bibliography

Abu Dhabi Digital Government. 'Abu Dhabi Emirate: Facts and Figures'. Accessed 11 July 2016. https://www.abudhabi.ae/portal/public/en/abu-dhabi-emirate/abu-dhabi-emirate-facts-and-figures.

Abu Dhabi Urban Planning Council. 'Abu Dhabi Vision 2030', n.d.

Abu Dhabi Urban Planning Council. 'Neighborhood Planning, Abu Dhabi Vision 2030', n.d.

Abu Dhabi Urban Planning Council. 'Plan Abu Dhabi 2030'. Accessed 4 October 2017. https://www.ecouncil.ae/PublicationsEn/plan-abu-dhabi-full-version-EN.pdf.

Abu Dhabi Urban Planning Council and Municipality of Abu Dhabi City. 'Estidama: A Comprehensive Guide of Procedures for Implementing Estidama in the Municipality of Abu Dhabi City', 2010.

Abu-Lughod, Janet L. 'Contemporary Relevance of Islamic Urban Principles (Extract)'. *Ekistics* 280 (1980): 6.

Abu-Lughod, Janet L. 'The Islamic City: Historic Myth, Islamic Essence, and Contemporary Relevance'. *International Journal of Middle East Studies* 19, no. 2 (1987): 155–76.

Alberti, Marina. 'Measuring Urban Sustainability'. *Environmental Impact Assessment Review* 16, no.s 4–6 (1996): 381–424. doi:10.1016/S0195-9255(96)00083-2.

Al-Hathloul, Saleh Ali. *The Arab-Muslim City: Tradition, Continuity and Change in the Physical Environment.* Riyadh: Dar Al Sahan, 1996.

Al-Kodmany, Kheir. 'Residential Visual Privacy: Traditional and Modern Architecture and Urban Design'. *Journal of Urban Design* 4, no. 3 (1999): 283–311. doi:10.1080/13574809908724452.

Axelsson, Robert, Per Angelstam, Erik Degerman, Sara Teitelbaum, Kjell Andersson, Marine Elbakidze, and Marcus K. Drotz. 'Social and Cultural Sustainability: Criteria, Indicators, Verifier Variables for Measurement and Maps for Visualization to Support Planning'. *Ambio* 42, no. 2 (2013): 215–28. doi:10.1007/s13280-012-0376-0.

Bahammam, Ali. 'Factors Which Influence the Size of the Contemporary Dwelling: Riyadh, Saudi Arabia'. *Habitat International* 22, no. 4 (1998): 557–70. doi:10.1016/S0197-3975(98)00018-6.

Bianca, Stefano. *Urban Form in the Arab World: Past and Present.* London; New York: Thames & Hudson, 2000.

Brundtland, Gru, Mansour Khalid, Susanna Agnelli, Sali Al-Athel, Bernard Chidzero, Lamina Fadika, Volker Hauff, et al. 'Our Common Future ("Brundtland Report")'. Oxford, UK: Oxford University Press, USA, May 21, 1987. http://www.bne-portal.de/fileadmin/unesco/de/Downloads/Hintergrundmaterial_international/Brundtlandbericht.File.pdf?linklisted=2812.

Daly, Herman E. 'Toward Some Operational Principles of Sustainable Development'. *Ecological Economics* 2, no. 1 (April 1990): 1–6. https://doi.org/10.1016/0921-8009(90)90010-R.

Dempsey, Michael Cameron. *Castles in the Sand: A City Planner in Abu Dhabi.* Jefferson, North Carolina: McFarland, 2014.

Eben Saleh, Mohammed Abdullah. 'Privacy and Communal Socialization: The Role of Space in the Security of Traditional and Contemporary Neighborhoods in Saudi Arabia'. *Habitat International* 21, no. 2 (1997): 167–84. doi:10.1016/S0197-3975(96)00055-0.

Eben Saleh, Mohammed. 'The Impact of Islamic and Customary Laws on Urban Form Development in Southwestern Saudi Arabia'. *Habitat International* 22, no. 4 (1998): 537–56. doi:10.1016/S0197-3975(98)00015-0.

Elsheshtawy, Yasser. *The Evolving Arab City: Tradition, Modernity and Urban Development.* London; New York: Routledge, 2008.

Elshestawy, Yasser. 'Urban Dualities in the Arab World: From a Narrative of Loss to Neo-Liberal Urbanism'. In *The Urban Design Reader*, edited by Michael Larice and Elizabeth Macdonald, 475–96. The Routledge Urban Reader Series. Abingdon, Oxon; New York: Routledge, 2011.

El-Zein, Abbas, Samer Jabbour, Belgin Tekce, Huda Zurayk, Iman Nuwayhid, Marwan Khawaja, Tariq Tell, et al. 'Health and Ecological Sustainability in the Arab World: A Matter of Survival'. *The Lancet* 383, no. 9915 (2014): 458–76. doi:10.1016/S0140-6736(13)62338-7.

Germeraad, Pieter W. 'Islamic Traditions and Contemporary Open Space Design in Arab-Muslim Settlements in the Middle East'. *Landscape and Urban Planning* 23, no. 2 (1993): 97–106. doi:10.1016/0169-2046(93)90110-Y.

Hamouche, Mustapha Ben. 'The Changing Morphology of the Gulf Cities in the Age of Globalisation: The Case of Bahrain'. *Habitat International* 28, no. 4 (2004): 521–40. doi:10.1016/j.habitatint.2003.10.006.

Hamouche, Mustapha Ben. 'Climate, Cities and Sustainability in the Arabian Region: Compactness as a New Paradigm in Urban Design and Planning'. *International Journal of Architectural Research: ArchNet-IJAR* 2, no. 2 (2014): 196–208.

Hopwood, Bill, Mary Mellor, and Geoff O'Brien. 'Sustainable Development: Mapping Different Approaches'. *Sustainable Development* 13, no. 1 (2005): 38–52. doi:10.1002/sd.244.

Huang, Lu, Jianguo Wu, and Lijiao Yan. 'Defining and Measuring Urban Sustainability: A Review of Indicators'. *Landscape Ecology* 30, no. 7 (2015): 1175–93. doi:10.1007/s10980-015-0208-2.

Kingsley, Patrick. 'Masdar: The Shifting Goalposts of Abu Dhabi's Ambitious Eco-City'. Wired UK, December 2013. http://www.wired.co.uk/article/reality-hits-masdar.

Llewellyn, Othman. 'Shari'ah Values Pertaining to Landscape Planning and Design'. In *Islamic Architecture and Urbanism: Selected Papers from a Symposium Organized by the College of Architecture and Planning*, edited by Aydin Germen. Dammam, Saudi Arabia: King Faisal University, 1983.

Makhzoumi, Jala M. 'Landscape in the Middle East: An Inquiry'. *Landscape Research* 27, no. 3 (2002): 213–28. doi:10.1080/01426390220149494.

McDonnell, R. A. 'Circulations and Transformations of Energy and Water in Abu Dhabi's Hydrosocial Cycle'. *Geoforum* 57 (2014): 225–33. doi:10.1016/j.geoforum.2013.11.009.

McKenzie, Stephen. 'Social Sustainability: Towards Some Definitions'. Hawke Research Institute Working Paper Series No. 27, Magill, University of South Australia, 2004. http://w3.unisa.edu.au/hawkeinstitute/publications/downloads/wp27.pdf.

Mohammed, Gamal Taha, and Kevin Thwaites. 'An Exploratory and Reflective Process of Urban Spatial Morphology within Social Sustainability: Lessons from Middle Eastern Islamic Tradition'. *Digest of Middle East Studies* 19, no. 2 (2010): 249–67. doi:10.1111/j.1949-3606.2010.00033.x.

Moosavi, Sareh, Jala Makhzoumi, and Margaret Grose. 'Landscape Practice in the Middle East Between Local and Global Aspirations'. *Landscape Research* 41, no. 3 (2016): 265–78. doi:10.1080/01426397.2015.1078888.

Moustapha, Ahmed Farid, and Frank J. Costa. 'Al Jarudiyah: A Model for Low Rise/High Density Development in Saudi Arabia'. *Ekistics* 48, no. 287 (1981): 100–8.

Nassauer, Joan Iverson. 'Culture and Changing Landscape Structure'. *Landscape Ecology* 10, no. 4 (1995): 229–37. doi:10.1007/BF00129257.

Naveh, Zev. 'Interactions of Landscapes and Cultures'. *Landscape and Urban Planning* 32, no. 1 (1995): 43–54. doi:10.1016/0169-2046(94)00183-4.

Ouroussoff, Nicolai. 'In Arabian Desert, a Sustainable City Rises'. *The New York Times*, 25 September 2010. https://www.nytimes.com/2010/09/26/arts/design/26masdar.html.

Pal, Jeremy S., and Elfatih A. B. Eltahir. 'Future Temperature in Southwest Asia Projected to Exceed a Threshold for Human Adaptability'. *Nature Climate Change* 6, no. 2 (2016): 197–200. doi:10.1038/nclimate2833.

Remali, Adel M., Ashraf M. Salama, Florian Wiedmann, and Hatem G. Ibrahim. 'A Chronological Exploration of the Evolution of Housing Typologies in Gulf Cities'. *City, Territory and Architecture* 3 (2016): 14. doi:10.1186/s40410-016-0043-z.

Tanguay, Georges A., Juste Rajaonson, Jean-François Lefebvre, and Paul Lanoie. 'Measuring the Sustainability of Cities: An Analysis of the Use of Local Indicators'. *Ecological Indicators* 10, no. 2 (2010): 407–18. doi:10.1016/j.ecolind.2009.07.013.

19

PLANNING WITH HERITAGE

A critical debate across landscape architecture practice and heritage theory

Svava Riesto and Anne Tietjen

Today, landscape architects engage more profoundly than ever with the transformation of existing urban landscapes. Since the 1980s, cities and regions in Europe and North America have shifted their development focus from urban expansion to adaptation, revitalization and redevelopment, and landscape architects have become vital players in urban and regional transformation processes. In these processes, not only built structures and landscapes but also stories and cultural traditions inherited from the past are increasingly used as a resource for future development—to promote sustainability and enhance local identity, to make places attractive for residents and tourists, and to design and brand cities and entire regions.[1] In short, heritage has become a central issue in the intersection of landscape architecture and spatial planning. While there is a growing body of research on how landscape architects design with heritage,[2] the role and agency of landscape architecture in heritage-led spatial development is still little theorized. Yet the relationship between heritage, spatial planning and landscape architecture is anything but straightforward. We believe that working with heritage in a spatial planning context raises new questions about heritage values and the political role of landscape architecture. By addressing these questions, this article aims to contribute to a critical debate across landscape architecture practice and critical heritage theory.

Our starting point is that heritage is not one, definable stock of objects for preservation, but a malleable concept that can serve many different purposes. Specific ideas about what heritage is, what aims it should serve and for whom have evolved and diversified over time, together with changing societal challenges and values.[3] This article first outlines the historical trajectory along which heritage, spatial planning and landscape architecture became increasingly intertwined, to shed light on different contemporary approaches to heritage. On this basis, we examine three landscape projects from Denmark, Switzerland and Germany, ranging from a heritage-led strategic urban development to an unconventional river restoration and the design of a temporary urban space. These projects show how landscape architects actively engage with heritage, and how that can contribute to spatial development in general while also provoking critical discussion and productive conflict over spatial development. Guided by critical heritage theory, we will discuss the processes and products of heritage-making in these projects. Our aim is twofold. We want to enable practitioners to work reflectively with heritage by stimulating critical thinking and inspiring landscape architects to consider a broad range of ways of working with heritage. We also wish to add to heritage theory by questioning landscape architecture practice in heritage-led spatial development.

How heritage moved to centre stage in landscape architecture, and changed in the process

Engaging with the past in the built environment, narrating history, and reusing old buildings or parts of them has a long tradition. Yet the idea of preserving historical monuments was first established in Europe in the late eighteenth century, in the context of modern nation-building and as a reaction to the massive societal and spatial changes that occurred with industrialization and rapid urban growth.[4]

In parallel with the formation of urban planning and landscape architecture as professional disciplines, a body of specialized laws, organizations, museums and educations was created to safeguard selected historical objects, monuments and buildings from unwanted change. These heritage artefacts—cathedrals, castles and monuments—were believed to have a special ability to represent historical development, a canon and societal norms.[5] Authorized experts—e.g. connoisseurs, and later academically trained learned societies—determined which objects contained such heritage value.[6] The focus was on protecting these objects from loss or change, in order to preserve the values believed to be inherent in them for posterity, i.e. ideally forever. Only very few buildings and monuments—preferably 'the grand, the old, and the beautiful'[7]—were seen as heritage worthy of preservation; the largest part of the built environment was considered *non-heritage*, and thus could be freely disposed of. This created a dichotomy between city-building, which was essentially about creating the new, and preservation, which was essentially about safeguarding what were believed to be the most valuable parts of the old city. In line with this, heritage management and spatial planning developed as two separate and often conflicting sectors.[8]

It was also characteristic of this heritage approach that it focused exclusively on monuments and individual buildings, and paid little attention to larger urban environments, open spaces or landscapes. In parallel, another, similar body of laws, institutions and practices emerged around the notion of nature conservation, which focused on preserving natural value, for example in heathlands and forests. Ultimately two systems emerged for cultural and natural heritage, promoting a clear separation between culture and nature. In Denmark, for instance, independent laws for listed buildings and monuments and for nature conservation were established in 1917 and 1918.[9] Landscape architecture was hardly ever valued by any of these systems, and thus fell between two stools. From a cultural heritage perspective, a designed landscape could only be valued and preserved as an appendix to a building, while some designed landscapes could be appreciated and protected on the basis of their natural value (such as specific flora, fauna and geological features), but not for their design. Landscape architects criticized this separation between nature conservation and cultural heritage right from the start,[10] but only in recent years has it become a central theme in heritage theory and practice.[11]

In the aftermath of World War II, new approaches to heritage developed. Severe physical damage to historic European cities and an accelerating pace of urbanization led to an increasing 'anxiety over losing valuable objects, buildings, landscapes and traditions'.[12] Heritage became a matter of concern for ordinary people, as stated for instance in the Venice Charter on heritage.[13] From urbanist Jane Jacob's[14] early civil resistance against clean-slate urban renewal in New York to the squatter movement in European cities in the 1970s and 1980s, citizens increasingly opposed the demolition of mundane historical buildings and neighbourhoods, while countless local preservation societies worked for their recognition as heritage. In parallel with this growing public concern with heritage, architects and theorists who critiqued modernist urbanism bemoaned the loss of distinct local identities thanks to universalist approaches,[15] calling for a new 'regionalism'.[16]

From being a separate object for protection, cultural heritage from the 1980s onwards increasingly became an integrated factor in local spatial development processes.[17] Ideas about what was worthy of preservation were gradually broadened to include younger and ordinary buildings of local significance, such as farms and residential and industrial buildings. Furthermore, the scope of heritage conservation expanded from monumental objects to a more holistic idea of heritage landscape in the revitalization of historic city centres, towns and villages, including immaterial aspects such as stories and cultural traditions.[18] The idea increasingly became to layer the old with the new. Making historical development legible in the urban fabric was seen as a way to enhance local identity, which was and still is seen as a prerequisite for attractive, liveable urban landscapes.[19] With these new objectives for heritage management in an increasingly public domain, landscape-based approaches gained ground. In Denmark, for example, rivers and terrain forms became recognized as constitutive of townscapes more broadly, which led to more inclusive ways of Surveying Architectural (heritage) Values in the Environment, the 'SAVE' method.[20] Heritage preservation and future development began to go hand in hand, to the extent that heritage today is used as a resource and driver for urban and even regional development. Specifically, the redevelopment of post-industrial areas, most famously the IBA Emscher Park in the German Ruhr district (1989–99), fuelled a new landscape-based approach to designing with heritage, where the remains of industrial production are being reused, reprogrammed and transformed as a component of spatial quality embedded in new large-scale regeneration schemes.[21]

Over the last three to four decades, the role of heritage has changed so fundamentally that some scholars speak of a 'new heritage' to distinguish contemporary practices from historical approaches.[22] Rather than safeguarding selected buildings and monuments from change, 'new heritage' is concerned with managing change in the whole built environment and *using* heritage for the purposes of local development. Therefore, 'new heritage' potentially embraces everything, be it young or old, tangible objects or intangible stories and traditions, as long as it can serve present-day needs and development objectives, from place-making and city-branding to sustainability. In line with this, the ways of dealing with heritage have expanded, from preservation and reconstruction to reuse, transformation and even demolition, as well as immaterial practices such as storytelling or heritage events.[23] Finally, heritage is no longer exclusively defined and managed by heritage experts. Rather, it is made in inclusive, collaborative processes involving many actors—planners, politicians, citizens, investors, special-interest organizations—with different and often conflicting ideas about heritage and the development objectives at hand.[24] In these complex heritage-making processes, landscape architects and landscape-based approaches have come to play a vital role.

In order to make sense of the role of landscape architecture in heritage-led spatial development, we propose a conceptual model developed by heritage scholars John Tunbridge and Gregory Ashworth.[25] They see heritage-making as a process that involves two main activities: heritage *selection* and heritage *targeting*. Heritage selection starts from a 'quarry' of possible heritage resources—such as 'past events, personalities, folk memories, mythologies, literary associations, surviving physical relics'[26]—from which the heritage makers select the most useful in relation to the desired development objectives and users. Heritage targeting *assembles* these resources, that is, heritage makers interpret and package the selected heritage with other resources into a tailored 'heritage product'[27] to target these objectives and users.

As in any other spatial development process, there are conflicting views about heritage-making and what heritage to select for whose objectives. Tunbridge and Ashworth have coined the notion 'dissonant heritage' to emphasize the contested nature of heritage, stressing that 'all heritage is someone's heritage and therefore logically not someone else's.'[28] Such dissonances are not always laid open when heritage is part of place-making and spatial development processes that aim for collective narratives, yet they are an inescapable aspect of the reality that landscape architects work with.

In the following, we examine how three different landscape projects select, target and assemble heritage resources into heritage products in relation to contemporary spatial development tasks involving different ideas of heritage and a broad range of actors with different interests and values.

Heritage as development driver: 'Steely Town' Frederiksværk, Denmark

'Steely Town' Frederiksværk is an example of the heritage-led strategic development of a whole town, in active collaboration with a broad range of local actors. Frederiksværk has about 12,000 inhabitants and is located between Lake Arre and Roskilde Fjord, about one hour's drive from the Danish capital, Copenhagen. Established as a royal cannon foundry in the mid-1700s, Frederiksværk is associated with the metal industry like no other town in Denmark. While there is still one active steelworks, most of the production disappeared in the 1970s. Significant industrial buildings have since been demolished, and together with new road infrastructure and buildings, this has created a somewhat fragmented appearance and weakened the link between the town and the landscape.

In 2007, the Danish Cultural Heritage Agency declared Frederiksværk an Industrial Heritage Site of National Significance. Thus encouraged, Halsnæs municipality became a so-called cultural heritage municipality in 2008, and made heritage—namely Frederiksværk's industrial heritage—pivotal to municipal development in general.[29]

The architecture competition 'Steely Urban Spaces Frederiksværk' was announced to develop a strategic spatial development plan, to identify key sites for design interventions, to develop design proposals for selected sites, to draw up a comprehensive heritage communication strategy to make Frederiksværk's industrial history legible for residents and visitors, and to create a framework for future spatial development and citizen involvement. By using 250 years of industrial history as a development narrative, the municipality aimed 'to strengthen the town's profile and create world-class experiential qualities' for residents and visitors.[30]

Five multidisciplinary teams, including landscape architects, urban planners, cultural historians, engineers and communication experts, participated in the two-phase competition. These teams were provided with abundant knowledge about the relevant heritage quarry—inventories of listed and potentially valuable historical buildings, landscape elements and historical narratives—as well as knowledge about contemporary challenges and municipal planning goals, e.g. better pedestrian and cycle connections, and more active urban spaces.

With 'Landscape, Works and Town', the winning team of SLETH, Brandt Dam et al. proposed a landscape-based approach to revitalize historical relationships between the town, the industrial sites and the landscape as a basis for future urban development (Figure 19.1). The human-made canal from Lake Arre to the fjord, which used to be the backbone of the steel industry but now has lost its function, will be revitalized and staged as the constituting element of the steelworks and a large-scale link between the town and the landscape. Tree plantings will trace and communicate Frederiksværk's historical structure and create spatial transitions and variation. Finally, new recreational paths will create new opportunities for movement, making the town and landscape more accessible. Within this new spatial framework, selected key sites will communicate Frederiksværk's industrial history and be transformed into everyday urban spaces. These strategic physical interventions will be accompanied by a multidimensional communication and participation strategy using social media, websites and apps; planning and active citizen involvement; and education, activities, events and temporary projects.[31]

Figure 19.1 Illustration from the design proposal 'Landscape, Works and Town', for Frederiksværk, 2014. © SLETH/Erik Brandt Dam et al.

'Landscape, Works and Town' builds on a thorough reading of historical traces, existing spatial and aesthetic qualities, and not least, existing social and cultural activities related to the town's industrial history. History is here understood as a continuous narrative including past, present and future. In consequence, heritage comprises old and new tangible structures, as well as intangible cultural activities, traditions and stories. From this quarry of heritage resources, the design team chose those resources which fitted the guiding narrative (industrial heritage) and present-day urban development objectives and reassembled those resources into a focused heritage product. Even future constructions can be made part of this heritage narrative. The designers proposed, for example, to build a new recycling station, and to stage it as a contemporary version of the historical industrial production, an architectural landmark and an inviting meeting place for local inhabitants.

After the competition, the municipality took over work on the heritage product. Since 2015 the municipality has been implementing the strategic plan and communication strategy in collaboration with public institutions, citizens, associations and local businesses. Children have planted trees; together with teenagers, the municipality established a temporary urban space on the renamed Valsetorvet ('Rolling Square'); and numerous 'steely' events are taking place. In 2017, the first urban space project achieved financing.[32]

Integrating natural and cultural heritage: Aire River Garden, Switzerland

Similarly to Frederiksværk, heritage played a key role in the renaturation of the River Aire in Switzerland. Here too a multidisciplinary team, Superpositions—this time of landscape architects, architects, biologists and civil and hydraulic engineers—worked together with local actors for the reorganization of a large territory. Yet while the work in Frederiksværk is guided by one heritage narrative of the town's industrial history, the Aire River Garden orchestrates different and potentially conflicting historical narratives as a basis for future landscape development.

For more than 100 years, the small River Aire was rectified and ran in a human-made canal to control the course of the water and allow intensive agricultural production in the Geneva region. Yet by the end of the twentieth century, flooding had become a regular problem in a now heavily urbanized landscape, the water quality had deteriorated, ecological and landscape quality was impoverished, and access to a potentially attractive recreational landscape was limited. In 2000 the Canton of Geneva therefore initiated a project for the restoration of the Aire, with three main objectives: flood safety for persons and goods, nature and landscape protection, and better access to the water.[33]

The authorities expected the meandering riverbed to be restored to its condition before the establishment of the canal. However, Superpositions refused to 'restore' the river to a supposed 'original form'.[34] Instead, they proposed to preserve the canal as a clearly human-made artefact, and to superimpose on it a new meandering riverbed, thus transgressing ideas of natural and cultural heritage. Parts of the retained canal now serve as a trail and parkland. A new meandering *lit majeur* with erodible banks has been established immediately south of the canal to give room for the new river channel, which is expected to wander by up to fifty metres in just one year. In the new riverbed, diamond-shaped depressions were excavated to form a 'lozenge' pattern, which provided some starting riverbed complexity and presented a forceful design for the initial landscape condition. Through the conversion of agricultural land into riparian habitat, an eighty-metre-wide wildlife corridor was created along the channel.

This design allows the Aire to recreate its own complex morphology, while providing access for the large nearby urban population, as well as flood retention[35] (Figure 19.2). Beyond functional solutions to technical problems, the Aire River Garden reassembles past, present and future moments in the life of the Aire into a new dynamic whole, telling the story of ongoing transformations: from a pre-industrial riparian landscape, to a rectified canal in an agricultural production landscape, and then to a restored ecological corridor and leisure landscape in a peri-urban region. In this new evolving landscape, the industrial artefact plays its part together with restored historical landscaping devices—ditches, hedges, groves, marshes—and contemporary

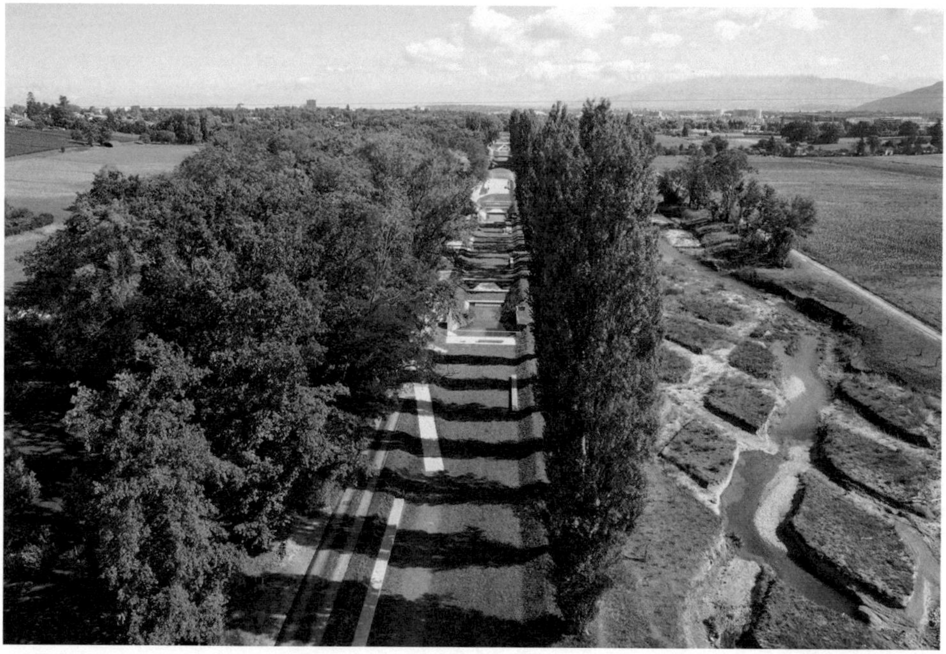

Figure 19.2 Aerial photo of the Aire River Garden under construction, 2015. © Chironi, Fabio. Superpositions, Planning with Heritage.

hydrological engineering infrastructure. By selecting natural and cultural heritage resources and assembling them into one tangible heritage product targeted at contemporary objectives, the potentially conflicting heritage values of local farmers, nature conservationists, landscape historians and outdoor recreationists are allowed to coexist.[36]

Dissonant heritage under construction: Castle Garden, Berlin

The ongoing reconstruction of Berlin Castle on the site of the Palace of the Republic, the former German Democratic Republic (GDR) Parliament building, was the subject of heated public heritage debate after the fall of the Berlin Wall.[37] The bone of contention was: should we reconstruct the castle, which had been blown up by the GDR government in 1950, to 'heal' the urban fabric as a symbol of the reunited Germany? Or should we retain the Palace of the Republic, which had lost its function and was asbestos-contaminated, but which also held cultural-historical values and many personal memories for former GDR citizens? Whose history should be told and how?

Unprecedentedly, the German parliament made the final heritage selection, in 2003. With reference to comprehensive expert reports, it decided to reconstruct Berlin Castle as a new building with historicizing facades and open spaces, and to tear down the Palace of the Republic. To mediate between the demolition and the reconstruction, the parliament recommended the establishment of a 'transitional garden' on the (de)construction site.[38]

In 2006 the City of Berlin, together with the Federal Ministry of Transport, Building and Urban Development, conducted a competition for a temporary public space on the castle terrain.[39] The winning proposal, by Relais landscape architects, made the shifting sites of the demolition and reconstruction pivotal to the design. The Castle Garden encompassed the entire terrain of the castle and its open spaces, of which the Palace of the Republic covered the eastern part towards the River Spree. In 2007 the park was first established as a green lawn, with wooden boardwalks on the remaining western terrain from which people could experience the progressive demolition of the palace and archaeological excavations of the remains of the medieval castle (Figure 19.3). Subsequently, the park was expanded until it incorporated the entire terrain. As the reconstruction of the castle began in 2012, the park was successively unbuilt until it had completely disappeared.

The landscape architects deliberately chose a simple grid design and rough materiality for the boardwalks to reflect the character of a construction site. Further, they decided to 'keep all physical remains of the past',[40] traces of both the castle and the GDR palace, and to incorporate them into their design. The remaining base of a monument to the German emperor, for example, became the base of a new wooden staircase for people to sit on. Parts of the foundations of the Palace of the Republic which could not be demolished were partially excavated and emphasized by a lawn sloping down towards the riverbank, forming a confined space between two walls.

The temporary design not only staged the physical demolition, excavation and new construction, but also enabled people to enter and literally experience heritage-making as an event in space and time. Although the heritage selection had already been made, the public heritage controversy continued—in on-site protests against the demolition of the palace, but also in many cultural events and art projects to reflect on and commemorate the Palace of the Republic. At the same time, people reappropriated this symbolically and geographically central location as an everyday urban space.

The Castle Garden was well used by residents and visitors. Yet its public reception remained controversial. The design received critical acclaim by the 2011 German Landscape Architecture Award jury for 'turning one of Berlin's most difficult historical sites into a tangible experience

Figure 19.3 The temporary Castle Garden, Berlin, 2008. © Müller, S.

and a useable public space'.[41] But critical voices dismissed the design as 'a nice-looking lawn, yet completely inappropriate to the location and the historical context'.[42]

New roles for landscape architects and landscape architecture

What, then, is the role of landscape architects in heritage-led spatial development, and what heritage products do they contribute?

In Frederiksværk the landscape architects entered the heritage process as part of a multidisciplinary team, with the task of proposing what to select from an abundant heritage quarry, and how to assemble and target those resources with other resources for predefined objectives. To do so, they collaborated with municipal planners, local heritage experts, citizens and stakeholders. The resulting heritage product was a strategic development plan, design proposals for selected urban spaces, and most importantly a strategy to engage a broad range of actors in place-making, guided by a heritage narrative.

In the River Aire project, the desired heritage selection was implicitly defined by the competition brief, which requested the 'restoration' of the meandering river channel. The design team contested this selection by proposing an alternative heritage product, which considerably enlarged the perceived heritage quarry to include not only natural heritage values, but also the cultural heritage of the human-made canal. The heritage product of the Aire River Garden is a tangible, dynamic landscape successively created by landscape preservation, reuse, demolition, transformation and construction, in dialogue with public authorities and local actors over a fifteen-year period.

In Berlin, the heritage product had already been decided upon before the landscape architects entered the process. The design articulated the process of dissonant heritage construction, i.e. deselecting and disassembling one heritage, the palace, and selecting and constructing another, the castle, as a spatial and temporal heritage event. The ephemeral heritage product made it possible to experience this controversial heritage process in a tangible way, and eventually for visitors to make up their own minds. The Castle Garden worked as an interlude which gave room for people to bid farewell to the Palace of the Republic and to follow the construction of the castle, while it also allowed new everyday uses on a contested heritage site.

The three cases exemplify that landscape architects can enter heritage-making processes with many different tasks and contribute to them in different ways. They can make proposals as to how to select and assemble heritage products according to predefined objectives to guide long-term spatial development. They can contest the heritage selection implied in a given task by suggesting alternative heritage products. And they can turn conflicted heritage-making into a tangible, spatial and temporal experience. While these few examples merely outline the broad range of possible tasks, they show that is it crucial to be aware of the inherent complexity and contested nature of heritage-making in order to be able to act in a reflective manner.

Working with heritage-led spatial development broadens the practice of landscape architecture, from designing open green spaces to formulating large-scale and long-term spatial development strategies, facilitating participatory processes, and narrating histories through design. This requires new types of transdisciplinary cooperation with a broad range of experts and laypeople.[43] As heritage scholars have pointed out, when a 'new past' is constructed in local communities, there is a need to understand what values are being articulated, how and why.[44] This requires landscape architects to listen carefully, but also to express their own values and argue for their choices in heritage-making. Yet it is important to keep in mind that reaching consensus on heritage values, and creating one shared heritage narrative and product, is not always possible. The Berlin case demonstrates that heritage-making can be a highly contested, political affair, which cannot be 'solved' by design.

Heritage-led spatial development clearly requires new ways of working in landscape architecture practice. But what can landscape architecture practice possibly add to the understanding of heritage and its uses in spatial planning?

Landscape-based approaches to new heritage

Heritage scholars emphasize a strong solidarity between new ways of conceptualizing heritage and landscape, specifically in the Faro Convention and the European Landscape Convention.[45] These conventions present 'new heritage' and landscape as holistic concepts 'which bring together previously separate aspects of the world into a stronger whole'.[46] Graham Fairclough observes that new heritage practices often draw on the idea of landscape, 'focusing on context rather than only the object itself, and recognizing other ways to achieve sustainable management of heritage than only the conventional approach of careful, conservative preservation or restoration'.[47] Moreover, both the Faro and European Landscape conventions promote a democratization of heritage and landscape in spatial planning.

Of our three cases, the Frederiksværk project most clearly demonstrates a 'new heritage' approach. Here landscape is used not only as a holistic concept, but as a physical frame for making historical development legible, while at the same time being open for adaptation. By juxtaposing past, present and future elements in a larger landscape connected by walking paths, the designers seek to enable tangible experiences of moments in the town's industrial history. Further, the design proposals for key urban spaces provide a pivot for engaging a broad range of actors in the democratic development of a heritage-based place narrative. Experiences from Frederiksværk demonstrate how well suited a combined landscape and new heritage approach is for collective place-making, yet it also exposes some of its pitfalls. Precisely because the process was so focused on the utility of heritage for place-making and branding, it resulted in a singular, authoritative narrative, leaving little room for other histories and prospects. Although this narrative is open-ended, it is entirely based on the industrial period, which the town has largely left behind. This could become a 'straitjacket of identity',[48] standing in the way of innovation in Frederiksværk's spatial, cultural and economic development. Further, the project constructs a purified image of a frictionless industrial past which leaves little room for the dirt, smell, danger and social conflict that were also part of the town's industrial and military past. In short, we see a risk of editing out negative and conflicted heritage. Finally, the only steelworks that is still productive—which today has a Russian owner and mainly Eastern European employees—plays no significant role in the narrative and strategic development of Frederiksværk as 'Steely Town'. This raises the question: who is part of the collective heritage-making process, and who is not?

These issues are not restricted to Frederiksværk. Rather, producing reductive historical narratives—which bracket out dissonant viewpoints and conflicted aspects of the past, as well as development possibilities that do not fit the narrative—are inherent risks of a new heritage approach.[49] With landscape architects increasingly working with such projects, a critical debate on these issues is needed across theory and practice.

Beyond natural and cultural heritage values

Today, no place on earth is untouched by humans, and there is an increasing realization that all human culture is entangled with natural processes. In line with this, heritage scholars strive to transgress the nature/culture dichotomy,[50] and established separate practices around natural and cultural heritage are being challenged and rethought.[51]

The Aire River Garden shows how natural and cultural heritage values can be combined in one designed landscape. Designing landscape with heritage here led to a dynamic heritage product, which not only juxtaposed different moments in the history of the River Aire, but also staged the dynamics of cultural and natural processes as a continuous transformation which could not be completely controlled. The project negotiated natural and cultural values in a tangible way, and thereby raised awareness and enabled people to engage in the development of this large-scale landscape. In this way, landscape architecture can help to problematize the established nature/culture dichotomy in a particular landscape.

Exposing dissonances

A starting point for this article was that heritage-making is always dissonant. While the conditions of the Berlin project are clearly exceptional, it articulates issues about dissonance that are at stake in every heritage-making process. Its public reception shows that articulating conflicting heritage values in a temporary landscape design can be interpreted as a way of acknowledging dissonances by providing space and time for remembrance, lament and farewell for those who will be disinherited. Yet it can also be interpreted as a way of pacifying and overruling the 'losing party'. At the same time, working with heritage in a temporary manner fundamentally challenges still-prevailing ideas about heritage as objects from the past that should be preserved for the future. Precisely by being ephemeral, the Castle Garden emphasizes that heritage is made for the present and in the present.

All three projects clearly show that heritage values are made for contemporary aims and planning objectives, ranging from place-making to flood safety and even the unification of a formerly divided city and nation. Landscape projects are clearly highly suitable to assemble heritage resources with other resources in future-oriented spatial development processes. Sometimes this can support the development of one collective narrative, such as 'Steely Town' Frederiksværk. Landscape architecture can also integrate natural and cultural heritage values, as in the Aire River Garden. Finally, landscape architecture can be used to lay open conflicting heritage interests. While the potential of landscape architecture to synthesize different viewpoints has been relatively well described, its capacity to expose dissonances is still little explored. We believe the potential to make dissonant heritage values tangible should be explored further, not only in projects where the decisions have already been made by other actors, but also where landscape architects raise critical questions and give voice to different positions in heritage-making processes.

Notes

1 Cornelius Holtorf and Graham Fairclough, 'The New Heritage and Re-Shapings of the Past', in *Reclaiming Archaeology: Beyond the Tropes of Modernity*, ed. Alfredo González Ruibal (London: Routledge, 2013), 197–210; Koos Bosma, 'Heritage Policy in Spatial Planning', in *The Cultural Landscape and Heritage Paradox: Protection and Development of the Dutch Archaeological-Historical Landscape and its European Dimension*, eds. J.H.F. Bloemers et al. (Amsterdam: Amsterdam University Press, 2010), 641–52.

2 Sébastien Marot, 'The Reclaiming of Sites', in *Recovering Landscape: Essays in Contemporary Landscape Architecture,* ed. James Corner (New York: Princeton Architectural Press, 1999), 45–66; Ellen Braae, *Beauty Redeemed: Recycling Post-Industrial Landscapes* (Aarhus: Ikaros Press, 2015); John Dixon Hunt, *Historical Ground: The Role of History in Contemporary Landscape Architecture* (London: Routledge, 2014); Marc Treib, ed., *Spatial Recall: Memory in Architecture and Landscape* (New York: Routledge, 2009).

3 Janssen et al., 'Heritage as Sector, Factor and Vector: Conceptualizing the Shifting Relationship Between Heritage Management and Spatial Planning', *European Planning Studies* 25, no. 9 (2017).

4 David Löwental, *The Heritage Crusade and the Spoils of History* (Cambridge: Cambridge University Press, 1996); Francoise Choay, *The Invention of the Historic Monument* (Cambridge: Cambridge University Press, 2001).

5 Jan Kolen and Hans Renes, 'Landscape Biographies: Key Issues', in *Landscape Biographies: Geographical, Historical and Archaeological Perspectives on the Production and Transmission of Landscapes,* eds. Jan Kolen, Hans Renes and Rita Hermans (Amsterdam: Amsterdam University Press, 2015), 13.

6 Jan Kolen, 'Rejuvenation of the Heritage', *Scape*, no. 2 (2006): 50.

7 Laurajane Smith, *Uses of Heritage* (London: Taylor and Francis, 2006), 11.

8 Janssen et al., 'Heritage as Sector'.

9 The first Danish Building Preservation Act came into effect on 12 March 1918 ('Bygningsfredningens historie', Kulturministeriet, accessed 21 February 2018, https://slks.dk/bygningsfredning/fredet-eller-bevaringsvaerdigt/bygningsfredningens-historie/). The first Danish Nature Conservation Act was adopted in 1917 ('Fredninger', Miljøstyrelsen, accessed 21 February 2018, http://mst.dk/natur-vand/natur/national-naturbeskyttelse/fredninger/).

10 Gudmund Nyeland Brandt, 'Christiansø', *Architekten*, no. 26 (1924): 29–48.

11 See for example Rodney Harrison, 'Beyond "Natural" and "Cultural" Heritage: Towards an Ontological Politics of Heritage in the Age of Anthropocene', *Heritage and Society* 8, no. 1 (2015). The many heritage conferences that have addressed the nature-culture divide in recent years include, e.g. the University of Amherst's Heritage and Society conference themed 'Nature and Culture' in Prague 2016, and L'Atelier Technique des Espaces Naturels in partnership with ICOMOS France on the theme 'Patrimoines naturels et culturels, enjeux et synergie'.

12 Kolen and Renes, 'Landscape Biographies', 41.

13 'International Charter for the Conservation and Restoration of Monuments and Sites (The Venice Charter 1964)', International Council on Monuments and Sites, 1964, https://www.icomos.org/charters/venice_e.pdf.

14 Jane Jacobs, *The Death and Life of Great American Cities* (New York: Random House, 1961).

15 See for instance Christian Norberg-Schulz, *Genius Loci: Towards a Phenomenology of Architecture* (London: Academy Editions, 1980); Gordon Cullen, *The Concise Townscape* (New York: Reinhold, 1961).

16 Kenneth Frampton, 'Towards a Critical Regionalism: Towards an Architecture of Resistance', in *The Anti-Aesthetic: Essays on Postmodern Culture*, ed. Hal Foster (Port Townsen: Bay Press, 1983), 16–31.

17 Janssen et al., 'Heritage as Sector', 662.

18 Marilena Vecco, 'A Definition of Cultural Heritage: From the Tangible to the Intangible', *Journal of Cultural Heritage* 11, no. 3 (2010).
19 See for instance Council of Europe, 'European Landscape Convention, Treaty no. 176', 20 October 2000; Council of Europe, 'Framework Convention on the Value of Cultural Heritage for Society, Treaty no. 199', 27 October 2005.
20 The SAVE method was developed by the National Planning Department to map buildings and built environments in one Danish municipality at a time. Danish Ministry of the Environment, *SAVE: Kortlægning og registrering af byers og bygningers bevaringsværdi og udarbejdelse af kommuneatlas: Vejledning* (Copenhagen: Miljø- og Energiministeriet, Skov- og Naturstyrelsen, 1992).
21 Jansen et al., 'Heritage as Sector', 10.
22 Holtorf and Fairclough, 'New Heritage'.
23 Council of Europe, 'Value of Cultural Heritage'; Graham Fairclough, 'New Heritage Frontiers', in *Heritage and Beyond*, ed. Robert Palmer (Strasbourg: Council of Europe, 2009), 29–84.
24 Svava Riesto and Anne Tietjen, 'Doing Heritage Together: New Heritage Frontiers in Participatory Planning', in *Heritage, Democracy and the Public: Nordic Approaches*, eds. Torgrim Sneve Guttormsen and Grete Swensen (London: Routledge, 2016), 159–75.
25 John E. Tunbridge and Gregory Ashworth, *Dissonant Heritage: The Management of the Past as a Resource in Conflict* (Chichester: Wiley, 1996).
26 Ibid., 7.
27 Ibid., 8.
28 Ibid., 21.
29 Kulturarvsstyrelsen and Realdania, *Kommune—Kend din Kulturarv!* (Copenhagen: Kulturarvsstyrelsen and Realdania, 2010), 14–23.
30 Halsnæs Kommune and Realdania, *Stålsatte byrum: En konkurrence om Frederiksværks industrielle kulturarv—juryens betænkning* (Halsnæs/Copenhagen: Halsnæs Kommune/Realdania, 2014), 3.
31 SLETH A/S, Erik Brandt Dam Arkitekter, Moe A/S, Martin Zerlang, Vera Noldus and Ole Puggård, *Stålsatte byrum: Landskab, Værk og By* (n.p.: Frederiksværks industrielle kulturarv, 2014), 13.
32 See Stålsat By website, https://www.staalsatby.dk/nyheder.
33 Klaus Holzhausen, 'Editorial', in *Schulthess Gartenpreis/Prix Schulthess des jardins*, ed. Groupement Superpositions (Zürich: Patrimoine Suisse/Schweizer Heimatschutz, 2012), 3.
34 Georges Descombes, 'Displacements: Canals, Rivers, and Flows', in Treib, *Spatial Recall*, 124–5.
35 G. Mathias Kondolf, 'Liberty and Human Access for a Peri-Urban River: Restoration of the Aire', in *Schulthess Gartenpreis*, ed. Groupement Superpositions, 20.
36 Thomas Clemmensen, 'The Management of Dissonance in Nature Restoration', *Journal of Landscape Architecture* 9, no. 2 (2014).
37 See, e.g. Alexander Barti, *Die Schlossplatzdebatte: Eine Deutsche Entscheidung* (Magdeburg: Südwestdeutscher Verlag für Hochschulschriften, 2012).
38 See Bundesministerium für Verkehr, Bau und Stadtentwicklung and Senatsverwaltung für Stadtentwicklung, *Temporäre Freiraumgestaltung Schlossareal: Berlin Mitte: Begrenzt offener freiraumplanerischer Realisierungswettbewreb: Auslobung* (Berlin: Bundesministerium für Verkehr, Bau und Stadtentwicklung and Senatsverwaltung für Stadtentwicklung, 2006), 49.
39 Ibid.
40 Gero Heck, Relais Landschaftsarchitekten partner, interview by Svava Riesto, 7 September 2017.
41 Our translation from the German: 'macht einen der geschichtsträchtigsten Orte Berlins nutz- und erlebbar und hat sich [...] als beliebter Aufenthaltsort etabliert'. See 'Übergangsnutzung Schlossareal', Landschaftsarkitektur heute, accessed 21 February 2018, http://www.landschaftsarchitektur-heute.de/themen/berliner-kulturlandschaften/landschaftsarchitektur-als-kulturelles-erbe/details/10.
42 Our translation from the German: 'eine nett anzusehende, dem Ort und dem historischen Kontext allerdings gänzlich unangemessene Rasenfläche'. Oliver Hamm, editor in chief of *Deutsches Architektenblatt*, quoted in Martin Prominski, 'Holzwege der Kritik: Übergangsnutzung Schlossareal', *Garten + Landschaft*, no. 12 (2006), 9.
43 Riesto and Tietjen, 'Doing Heritage Together'.
44 Graham Fairclough, Milena Dragićević - Šešić, Ljiljana Rogač - Mijatović, Elizabeth Auclair and Katriina Soini, 'The Faro Convention, a New Paradigm for Socially—and Culturally—Sustainable Heritage Action?', *Култура/Culture* 8 (2014): 12.
45 Council of Europe, 'European Landscape Convention'; Fairclough, 'New Heritage Frontiers'.
46 Fairclough, 'New Heritage Frontiers', 30.

47 Ibid.
48 Rem Koolhaas, 'The Generic City: Guide,' in Rem Koolhaas and Bruce Mau, *S, M, L, XL*, ed. Jennifer Sigler (New York: Monacelli Press, 1995), 1250.
49 See e.g. Kari Larsen and Sveinung Berg, 'The Making of a Cultural Heritage Site', *Nordic Journal of Architecture* 1, no. 1 (2011); Svava Riesto, *Biography of an Industrial Landscape: Carlsberg's Urban Spaces Retold* (Amsterdam: Amsterdam University Press, 2018), 178–80.
50 Harrison, 'Beyond "Natural"', 32.
51 See for example the project 'Connecting Practice', a novel collaboration in 2013–15 between the International Union for Conservation of Nature and the International Council on Monuments and Sites, which continues in follow-up projects and publications.

Bibliography

Barti, Alexander. *Die Schlossplatzdebatte: Eine Deutsche Entscheidung*. Magdeburg: Südwestdeutscher Verlag für Hochschulschriften, 2012.

Bosma, Koos. 'Heritage Policy in Spatial Planning'. In *The Cultural Landscape and Heritage Paradox: Protection and Development of the Dutch Archaeological-Historical Landscape and its European Dimension*, edited by Tom (J.H.F.) Bloemers, Henk Kars, Arnold van der Valk and Mies Wijnen, 641–52. Amsterdam: Amsterdam University Press, 2010.

Braae, Ellen. *Beauty Redeemed: Recycling Post-Industrial Landscapes*. Aarhus/Basel: Ikaros Press/Birkhäuser Verlag, 2015.

Brandt, Gudmund Nyeland. 'Christiansø'. *Architekten*, no. 26 (1924): 29–48.

Bundesministerium für Verkehr, Bau und Stadtentwicklung and Senatsverwaltung für Stadtentwicklung. *Temporäre Freiraumgestaltung Schlossareal: Berlin Mitte: Begrenzt offener freiraumplanerischer Realisierungswettbewreb: Auslobung*. Berlin: Bundesministerium für Verkehr, Bau und Stadtentwicklung and Senatsverwaltung für Stadtentwicklung, 2006.

Choay, Francoise. *The Invention of the Historic Monument*. Cambridge: Cambridge University Press, 2001.

Clemmensen, Thomas. 'The Management of Dissonance in Nature Restoration'. *Journal of Landscape Architecture* 9, no. 2 (2014): 54–63.

Cullen, Gordon. *The Concise Townscape*. New York: Reinhold, 1961.

Danish Ministry of the Environment. *SAVE: Kortlægning og registrering af byers og bygningers bevaringsværdi og udarbejdelse af kommuneatlas: Vejledning*. Copenhagen: Miljø- og Energiministeriet, Skov- og Naturstyrelsen, 1992.

Descombes, Georges. 'Displacements: Canals, Rivers, and Flows'. In *Spatial Recall: Memory in Architecture and Landscape*, edited by Marc Treib, 120–35. New York: Routledge, 2009.

Dixon Hunt, John. *Historical Ground: The Role of History in Contemporary Landscape Architecture*. London: Routledge, 2014.

Fairclough, Graham. 'New Heritage Frontiers'. In *Heritage and Beyond*, edited by Robert Palmer, 29–84. Strasbourg: Council of Europe, 2009.

Fairclough, Graham, Milena Dragićević - Šešić, Ljiljana Rogač - Mijatović, Elizabeth Auclair and Katriina Soini. 'The Faro Convention, a New Paradigm for Socially—and Culturally—Sustainable Heritage Action?'. *Култура/Culture* 8, 2014: 9–20.

Frampton, Kenneth. 'Towards a Critical Regionalism: Towards an Architecture of Resistance'. In *The Anti-Aesthetic: Essays on Postmodern Culture*, edited by Hal Foster, 16–31. Port Townsen: Bay Press, 1983.

Halsnæs Kommune and Realdania. *Stålsatte byrum: En konkurrence om Frederiksværks industrielle kulturarv—juryens betænkning*. Halsnæs/København: Halsnæs Kommune and Realdania, 2014. http://www.e-pages.dk/halsnaes/44/.

Harrison, Rodney. 'Beyond "Natural" and "Cultural" Heritage: Towards an Ontological Politics of Heritage in the Age of Anthropocene'. *Heritage and Society* 8, no. 1 (2015): 24–42.

Holtorf, Cornelius, and Graham Fairclough. 'The New Heritage and Re-shapings of the Past'. In *Reclaiming Archaeology: Beyond the Tropes of Modernity*, edited by Alfredo González Ruibal, 197–210. London: Routledge, 2013.

Holzhausen, Klaus. 'Editorial'. In *Schulthess Gartenpreis/Prix Schulthess des jardins*, edited by Groupement Superpositions, 3. Zürich: Patrimoine Suisse/Schweizer Heimatschutz, 2012. http://www.superpositions.ch/extras/prix-schulthess-des-jardins-2012.html.

Jacobs, Jane. *The Death and Life of Great American Cities*. New York: Random House, 1961.

Janssen, Joks, Eric Luiten, Hans Renes and Eva Stegmeijer. 'Heritage as Sector, Factor and Vector: Conceptualizing the Shifting Relationship Between Heritage Management and Spatial Planning'. *European Planning Studies* 25, no. 9 (2017): 654–72.

Kolen, Jan. 'Rejuvenation of the Heritage'. *Scape*, no. 2 (2006): 50.

Kolen, Jan, and Hans Renes. 'Landscape Biographies: Key Issues'. In *Landscape Biographies: Geographical, Historical and Archaeological Perspectives on the Production and Transmission of Landscapes*, edited by Jan Kolen, Hans Renes and Rita Hermans, 21–47. Amsterdam: Amsterdam University Press, 2015.

Kondolf, G. Mathias. 'Liberty and Human Access for a Peri-Urban River: Restoration of the Aire'. In *Schulthess Gartenpreis/Prix Schulthess des jardins*, edited by Groupement Superpositions, 20. Zürich: Patrimoine Suisse/Schweizer Heimatschutz, 2012. http://www.superpositions.ch/extras/prix-schulthess-des-jardins-2012.html.

Koolhaas, Rem. 'The Generic City: Guide'. In Rem Koolhaas and Bruce Mau, *S, M, L, XL*, edited by Jennifer Sigler, 1238–67. New York: Monacelli Press, 1995.

Kulturarvsstyrelsen and Realdania. *Kommune—Kend din Kulturarv!* Copenhagen: Kulturarvsstyrelsen and Realdania, 2010. http://bygningskultur2015.dk/fileadmin/user_upload/dokumenter/Kommune_-_kend_din_kulturarv.pdf.

Larsen, Kari, and Sveinung Berg. 'The Making of a Cultural Heritage Site'. *Nordic Journal of Architecture* 1, no. 1 (2011): 40–7.

Löwental, David. *The Heritage Crusade and the Spoils of History*. Cambridge: Cambridge University Press, 1996.

Marot, Sébastien. 'The Reclaiming of Sites'. In *Recovering Landscape: Essays in Contemporary Landscape Architecture*, edited by James Corner, 45–66. New York: Princeton Architectural Press, 1999.

Norberg-Schulz, Christian. *Genius Loci: Towards a Phenomenology of Architecture*. London: Academy Editions, 1980.

Prominski, Martin. 'Holzwege der Kritik: Übergangsnutzung Schlossareal'. *Garten + Landschaft*, no. 12 (2006): 9.

Riesto, Svava. *Biography of an Industrial Landscape: Carlsberg's Urban Spaces Retold*. Amsterdam: Amsterdam University Press, 2018.

Riesto, Svava, and Anne Tietjen. 'Doing Heritage Together: New Heritage Frontiers in Participatory Planning'. In *Heritage, Democracy and the Public: Nordic Approaches*, edited by Torgrim Sneve Guttormsen and Grete Swensen, 159–75. London: Routledge, 2016.

Sleth A/S, Erik Brandt Dam Arkitekter, Moe A/S, Martin Zerlang, Vera Noldus and Ole Puggård. *Stålsatte byrum: Landskab, Værk og By*. N.p.: Frederiksværks industrielle kulturarv, 2014.

Smith, Laurajane. *Uses of Heritage*. London: Taylor and Francis, 2006.

Treib, Marc, ed. *Spatial Recall: Memory in Architecture and Landscape*. New York: Routledge, 2009.

Tunbridge, John E., and Gregory Ashworth. *Dissonant Heritage: The Management of the Past as a Resource in Conflict*. Chichester: Wiley, 1996.

Vecco, Marilena. 'A Definition of Cultural Heritage: From the Tangible to the Intangible'. *Journal of Cultural Heritage* 11, no. 3 (2010): 321–32.

20

THE CASE TO SAVE SOCIALIST SPACE

Soviet residential landscapes under threat of extinction

Christina E. Crawford

After the fall of the Soviet Union in July 1991, privatization and disaggregation of state-owned property occurred at many scales throughout formerly socialist territories. National boundaries suddenly appeared on maps of the region; workers' clubs transformed into nightclubs. And yet, certain sites in the post-socialist sphere maintain spatial relationships established during the Soviet era. This essay explores the most salient of these site types: the typical mass housing complex constructed by the Soviet state in the 1920s and 1930s, and post-1950. These complexes, known in their time as housing combines (*zhilkombinaty*) or microregions (*mikroraiony*), were communities designed to meet residents' housing, educational, cultural, commercial, and recreational needs in all-inclusive precincts, as shown in Figure 20.1.

Their site plans are instantly recognizable due to the unusually capacious open space that flows between free-standing, standardized housing bars. Such porous planning flouts the most basic spatial expectation of capitalist real-estate development: site density. Building footprints in early *mikroraiony*, for instance, account for less than twenty per cent of overall site coverage. The remaining eighty per cent of the plan is a landscape that includes spaces and programmes of collective use: greenery and pathways, playgrounds, open-air theatres, fountains, benches, and even dedicated air-drying laundry zones, many of which remain in use today.

But now, more than twenty-five years after the fall of the Soviet Union, this exterior collective space that was a hallmark of Soviet socialist housing provision is under threat of extinction. In the spring of 2017, the Russian State Duma (parliament) took up a bill to demolish nearly 8,000 'decrepit' Moscow residential buildings, primarily the standardized five-storey concrete apartment houses from the Khrushchev era that constitute the architecture of first-generation *mikroraiony*.[1] There is little to recommend the 1950s buildings as architectural objects, and I do not strive to defend them here. Instead, I would like to make a case for saving the landscape that surrounds and supports these buildings, the remarkable eighty-per-cent-open site plan that would be nearly impossible to replicate under a fully capitalist land regime. While free-standing apartment buildings may be found in various contexts—Congrès Internationaux d'Architecture Moderne (CIAM), American urban renewal, and post-war European mass housing planning made certain of this—I will refer in this essay to the threatened *zhilkombinat* and *mikroraion* landscapes as 'socialist space', as this is how they are implicitly understood within their

Figure 20.1 Novye Cheremushki mikroraion, Moscow, 1959. © Rubanenko, B.R. Arkhitekturno-Planirovochnoe Upravlenie. 9-Ĭ Kvartal: Opytno-Pokazatel′noe Stroitel′stvo Zhilogo Kvartala V Moskve (Raĭon Novye Cheremushki). Moskva: Glavmosstroi, 1959, (used with permission under UCC / Russian copyright law (published pre-1973)).

post-Soviet context. My intention is not to dwell on socialist space as an ideological topic—although I will introduce its ideological underpinnings—but rather to consider it as a distinct physical condition. In this essay, I seek to provide a clear definition of what socialist space is and how it works, and to argue for its maintenance as an alternative bodily experience to the enclosed and exclusionary spaces of neoliberal capitalism.

The 1971 book *The Ideal Communist City* offers a glimpse into how socialist space differs conceptually and physically from its capitalist counterpart. A diagram within it shows systems of relationships in communism (the ultimate, unattained goal of Soviet socialism) in which Man, Social Units, Spatial Forms, and Settlement Forms are bundled together in a 'unified structure'.[2] The components that make up Spatial Form—our concern here—range from housing to education, production, and nature. Culture (the built) and nature (the unbuilt) are purposely conflated in this diagram of socialist spatial forms, which is to say that there is no strict delineation between 'architecture' and 'landscape', as is the case under capitalist development (the former being monetizable, the latter merely in support). Instead, in socialist space, all of the ingredients of Spatial Form flow together in a fluid continuum of common use. In terms of user experience, socialist space might also be characterized as Deleuzian 'smooth space', which is non-hierarchical, offers many entry and exit points, and allows infinite trajectories.[3] When the last of the *mikroraiony* are demolished and replaced by the traditional 'striated' and ordered space of Central European perimeter blocks, as promised by the Moscow government, the lingering haptic experience of socialist spatial fluidity will disappear simultaneously.

To argue for the preservation of socialist space in the post-Soviet sphere, I begin by introducing theoretical spatial precepts that emerged in the years immediately following the Bolshevik revolution. Early Soviets advocated a dispersed model of development that gathered culture and nature under the broad umbrella of socialism. I then offer two constructed examples that exemplify socialist spatial potential: a 1930 *zhilkombinat* built adjacent to a tractor factory in Kharkiv, Ukraine, and a 1960s *mikroraion* in Baku, Azerbaijan. In presenting the theory behind and exemplary models of socialist space, I strive to equip the actors in Moscow's current demolition drama with context for the condition that they may soon lose. The commonly traversable ground plane of Moscow's remaining *mikroraiony* is a spatial remnant of a bygone society, a rare material environment that permits real-time reoccupation of an alternative to capitalist land organization. Once parcellized, sold, and developed, the liberated landscapes that move through these housing complexes cannot be, and will not be, reassembled (barring another socialist revolution).

How might we maintain such a historically and politically situated physical construct in the post-socialist era? I return to this question at the conclusion of this essay.

Foundations of socialist space

In 1914, Vladimir Lenin, future Soviet premier, wrote that the socialization of labour would lead to 'redistribution of the human population (thus putting an end both to rural backwardness, isolation and barbarism, and to the unnatural concentration of vast masses of people in big cities)'.[4] On 26 October 1917, in their second official act, Lenin's triumphant Bolshevik Party adopted the 'Decree on the Land'. In the name of workers, soldiers, and peasants, it proclaimed that 'private ownership of land shall be abolished forever [...] All land, whether state, crown, monastery, church, factory, entailed, private, public, peasant, etc., shall be confiscated without compensation and become the property of the whole people.'[5] In one fell swoop, the patchwork of land holdings across the former Russian empire dissolved.

The main asset of the newly socialist state was its magnificently large landmass—one sixth of the world, as the West was frequently reminded—to be harnessed in the service of collective production goals.[6] Socialist space was boundless space (*neob"iatanyi prostor*), a commonly owned and roamed surface unimpeded by the boundaries of private land ownership.[7] Revolutionary poet Vladimir Mayakovsky envisioned newly Soviet citizens as 'conquerors of the space of the seas, the oceans, and the continents', free to move, colonize, and disperse over the horizontally extensive surface that stretched from Europe to the Pacific Ocean.[8] As Mayakovsky and other revolutionary intellectuals of his generation made clear, wild nature—seas, oceans, continents—could be gathered conceptually under the same category as industrial cities. All belonged to, and could be equally celebrated and exploited by, Soviet socialism. Early Soviet economists cognized this newly aggregated territory as a single building site that coincided with the continental scale, and spatial planners were tasked to organize this space to maximize productivity, equality, and collectivity.

Leonid Sabsovich, economist for the Supreme Soviet of the National Economy, was one of the first state actors to bring economic, social, and spatial concerns together in an actionable theoretical model. He proposed to create a socialist society through decentralization, effectively instantiating Lenin's prognosis of diffuse spatial organization under socialism.[9] Sabsovich envisaged that, starting with the first Five-Year Plan for industrial development from 1928 to 1932, new industrial-residential settlements would replace existing cities and villages altogether. Technology was the key to enacting this decentralized spatial model:

> The condition that will assist us in realizing the objectives [of the Plan], is above all the 'victory over distance' [*pobeda nad rasstoianiem*] [... With] the vast number of large power plants and the possibility to transmit energy over long distances, we can to a large extent free ourselves from the attachment between industry and the fuel base [...] We will build new factories, scattering them over a wide area, closer to nature.[10]

While Sabsovich conceded that general plans for electrification, transport, and communications were far from complete, these infrastructural systems would eventually make dispersed settlement possible. The rails, roads, and telephone/telegraph wires that criss-crossed the geographical expanse of the union would connect far-flung nodes.

Spatial diffusion of industry and population was the means to an end, namely to instil socialism among the proletariat and peasantry. Sabsovich stressed repeatedly that transformation of the everyday life (*byt*) of Soviet citizens had to proceed in tandem with industrialization: changing stubborn daily habits through carefully designed domestic environments was as important to the cause as constructing steel plants. 'Without the socialist transformation of the everyday *byt*, we will not be able to efficiently manage the millions of trained workers who are instrumental to our grand economic development, which is necessary to build socialism in our country,' he wrote in 1930.[11] The built environment had the capacity and the responsibility to change habitual behaviours. Stalin's first Five-Year Plan, the hyper-industrialization drive to Soviet economic self-sufficiency, tested Sabsovich's socialist spatial theory in practice. The map of the plan dispersed industrialized agriculture and machine-building factories widely across Soviet territories in a diffuse pattern that took advantage of the USSR's continental scale.

Sabsovich tackled the problem of the industrial-residential settlement—the dispersed node itself—in his 1930 book *Socialist Cities*. He argued that these nodes must be not only productive, but also designed to inculcate socialist relations. The elements of socialist space took shape in Sabsovich's text when he wrote:

> In the socialist city [*sotsgorod*], homes should be located among the green and must be sufficiently distant from each other (this mandates the destruction of existing city blocks, where buildings are side by side). There should be large parks, stadiums, places for engaging in water sports, etc. The residential quarters should be sufficiently separated from the industrial zones by a wide green area and connected to it by convenient means of transportation [...] We must take into account that *in a socialist city, public life and the collective private life of the population will be developed on an immeasurably larger scale than the space available in our cities*.[12]

Sabsovich's textual description, replete with italics, sets up certain spatial relationships to be installed in future *zhilkombinaty* and *mikroraiony*. First, all constructions are to be located among 'the green' (nature, broadly construed). Second, residential buildings are to be set sufficiently apart from one another (free-standing). Third, to encourage increased collective life, recreational and cultural programming is to be increased and freely distributed in the common landscape owned and used by all. Here is the recipe for socialist space.

Housing combine/*zhilkombinat*

It its earliest iteration from the early 1930s, the building block of the socialist city was a housing combine or *zhilkombinat*, a self-sufficient planned unit for 1,500 residents of all ages. It was composed of apartment buildings standing free in a shared landscape, plus expansive communal services such as public canteens and laundries, libraries and sport facilities, educational institutions and commune-run round-the-clock childcare to allow Soviet mothers to enter the workforce. The *zhilkombinat* block was to be replicated in a rational grid across the territory of the socialist city until the demographic target needed to support on-site industry was reached. Sabsovich provided ample programmatic recommendations but almost no visual documentation for his scheme. Experimentation with both socialist city and *zhilkombinat* spatial organization occurred on specific sites, such as the one designated for the Kharkiv Tractor Factory (KhTZ) in 1930.

KhTZ sat ten kilometres outside the then Ukrainian capital of Kharkiv, on an open, rural site ripe for experimental design configuration. The project team were given basic demographic benchmarks, programmatic parameters, and an extraordinarily short time frame in which to complete their design. The designers, led by Ukrainian architect-planner Pavel Aleshin, based the organization of their socialist city upon the linear city model proposed by Nikolai Miliutin,

A. Zelenko, and others around 1929. In his explanation of the model, Miliutin equated the productive city with the factory assembly line to arrive at an efficient 'flowing' plan concept. He wrote: 'the residential sector of the settlement must be set up parallel to the productive zone and must be separated from it by a green belt no less than 500 meters wide.'[13] The green belt was intended to act as the lungs of the project, to filter any stray industrial particulates that might drift from the factory toward the residential zone. The benefits of the parallel layering programme included the relative proximity between the factory and its settlement, so that each worker would have a short walk to work from his or her residential unit to the factory. In addition, the green zone structured rational linear growth of the sectors in either direction along its length while maintaining the optimal distance between them. The expansion possibilities of the scheme were virtually boundless: one could imagine sinuous lines of such development snaking across the map of the USSR.

The KhTZ site plan reveals this exact linear organization of programmatic sectors (Figure 20.2). It is divided into parallel zones: heavy rail swings to the north, with the tractor factory just below. A 500-metre-wide green band cuts through the middle of the plan, and the residential settlement comprising repeated *zhilkombinat* blocks—each a tall rectangle populated with residential and social infrastructure—marches south from the green strip and faintly but insistently eastward, in promise of further colonization of the countryside.[14] Residents of even the southernmost *zhilkombinat* block had no more than a twenty-minute walk to the factory.

Once the socialist city site plan was set, the designers exerted their efforts on the standardized *zhilkombinat* blocks. Long thin housing bars run along a north-south axis. Like their German brethren, the *zeilenbau*, these bars are heliotropic: aligned to offer their broad facades east and west to maximize solar exposure within the residential units.[15] At KhTZ, this orientation also ensures that the prevailing grain of each *zhilkombinat* block privileges the long north-south view to connect factory, green buffer, housing, and countryside. Clear lines of sight and passage charge the open field of interaction between programmatic zones. Communal service buildings such as the workers' clubs and canteens, the public laundries, and the school buildings are the only ones permitted to flout the north-south grain of the plan. Regardless of programme, each building stands free of its neighbours to allow circumambulation on any given *zhilkombinat* block and between blocks. Pedestrian connectivity is emphasized by the amount of open space provided, which tops eighty per cent in the typical *zhilkombinat* at KhTZ.

A hand-drawn aerial perspective demonstrates the lived interaction between housing, social services, and landscape in the first *zhilkombinat* block constructed. Narrow six-storey bars hold dormitory-style living cells for singles, while six four-storey bars hold multiroom family units. A workers' club, replete with communal dining hall, library, multipurpose recreational rooms, and a mechanized laundry, sits centrally at the edge of the block, and is distinguished volumetrically through idiosyncratic form. Four educational buildings—elementary schools, kindergartens, and nurseries—line the back of the block. The project brief stipulates that 'all rooms in the residential sector must be connected between themselves and the premises of the socialized sector by warm corridors.'[16] In design perspectives, those lines become second-floor glassed-in skyways sitting atop thin columns to permit the ground plane to remain freely traversable along the dominant north-south axis. While these skyways were never built, the project as constructed permitted even greater freedom of movement on the ground plane. The common landscape was designed with pedestrian pathways and tree-lined boulevards, and was stocked with benches and dedicated play zones for the neighbourhood's children. In this *zhilkombinat* example on the Ukrainian steppe, topographical variation on-site was limited, which provided pedestrians with long views along a flat, easily traversable ground plane. By all accounts, the fluid spatial relationships between realms of work, housing, and leisure worked as projected, with each family

(a)

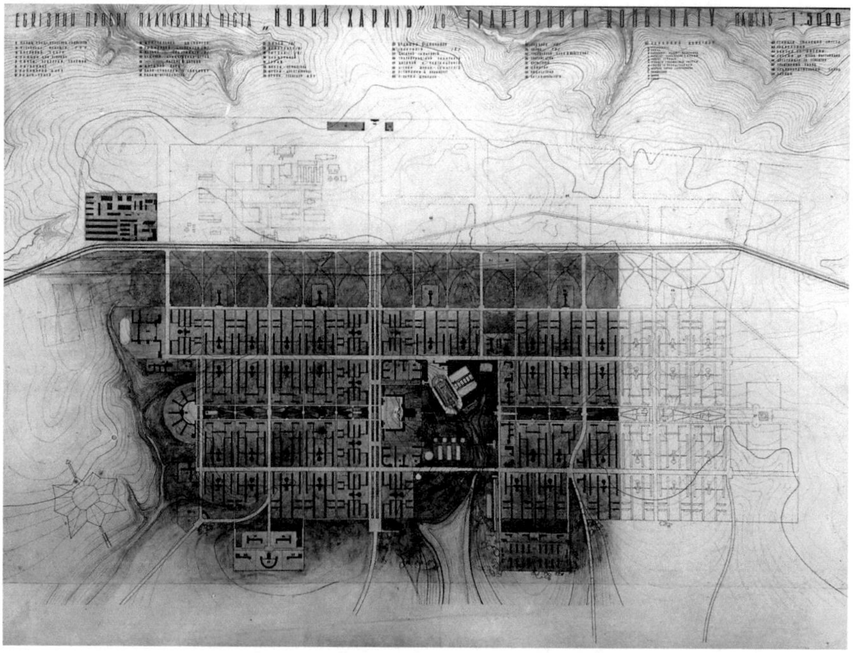

(b)

Figure 20.2a,b Kharkiv Tractor Factory *zhilkombinaty*. Kharkiv, Ukraine, 1930. © Central State Archives and Museum of Literature and Art of Ukraine, TsDAMLM.

member able to walk to his or her daytime occupation: the child to his or her school integrated into the *zhilkombinat* block, the parent through the green buffer to work at the tractor factory.

Experimentation with such novel spatial configurations to instil socialist values largely ceased upon the appointment of Stalin's right hand, Lazar Kaganovich, as First Secretary of the Moscow Committee. With Kaganovich at the helm of Moscow's planning efforts, the Central Committee issued a resolution denouncing 'utopian' urban theories and projects in favour of replanning efforts in existing cities.[17] In a mid-1931 speech, later published as *Socialist Reconstruction in Moscow and Other Cities in the USSR*, Kaganovich held up Moscow as the sole model for all future Soviet urbanism. To do so, he had to elide the issue of urban form, and assert that means of production alone made a context socialist: 'There are at present many who decline in every possible declension the formula, "we must build a socialist city." They forget one little trifle: that the cities of the USSR are already socialist cities. Our cities became socialist from the very moment of the October Revolution.'[18] This pronouncement signalled a shift in the state's priorities, from envisioning new spatial models to working within the traditions of urban planning from previous epochs. Moscow was a city of perimeter blocks, and this, Kaganovich implied, would be the model for Soviet housing going forward.

Microregion/*mikroraion*

The two decades between the end of the first Five-Year Plan and Josef Stalin's death (1932–53) saw the Soviet state abdicate responsibility for addressing the severe gap between housing need and capacity. Monumental state building projects, such as construction of the Moscow Metro in the 1930s, were given priority over the more mundane residential sector. Widespread destruction of Soviet urban fabric during World War II only exacerbated housing shortages, such that by 1951 each single-family apartment in Leningrad housed 3.3 families on average.[19]

In December 1954, newly installed Soviet premier Nikita Khrushchev spoke to the National Conference of Builders and Architects to acknowledge the housing shortage and to announce his campaign to address it. 'We have an obligation to significantly speed up, improve the quality of, and reduce the cost of construction,' Khrushchev proclaimed to applause. 'In order to do so, there is only one path—and that is the path of the most extensive industrialization of construction.'[20] In the span of eight years, from 1956 to 1964, fifty-four million Soviets—a quarter of the population—moved into the new five-storey prefabricated residential buildings that emerged from Khrushchev's industrialized construction campaign.[21] These standardized houses typically sat within a socialist superblock or *mikroraion*. Like the *zhilkombinat*, each *mikroraion* was designed as an all-inclusive pedestrian-centred superblock that provided the requisite commercial, cultural, educational, and recreational facilities within walkable distances. Unlike its 1930s predecessor, however, the *mikroraion* was almost always, and only, a residential enclave. Access to public transport was provided for these housing regions to connect to places of work, many of which were quite remote from the *mikroraion*.

The Novye Cheremushki district in south-western Moscow was the first experimental prefabricated housing precinct built in 1956, and it set certain standards for *mikroraion* site planning. Slim, rectangular, free-standing residential buildings were arranged in a staggered pattern that allowed diagonal view corridors across the site and freedom of pedestrian passage throughout the shared public zone. The buildings' entryways flipped and alternated to activate open spaces on all sides of the site plan, and provided easy access to services and amenities built into the block. The slang appellation for the standardized houses themselves—*khrushcheby*—was a play on the phonetic consonance of the name of the premier who had envisioned them with *trushchoby*, the Russian word for slums. Indeed, the prefabricated concrete residential buildings from the first phase of Khrushchev's mass housing campaign were poorly constructed, and the residential units extremely small.[22] Even promotional photos from the period invariably reveal misaligned concrete apartment units held together with sloppy lines of trowelled grout. The combination of unremarkable exterior architectural expression and cramped interiors put heavy stress on all other components of the residential precinct, especially the common ground plane between the buildings.

As intense architectural standardization took hold, the site plan of the *mikroraion* became virtually the only means for Soviet architects to engage in design as such, and the landscape surrounding the *khrushchevki* became imbued with creative expression. In plans for Moscow's Novye Cheremushki, the architects went to great lengths to qualify and quantify landscape elements for each programme within the neighbourhood unit. Particular attention was paid to the design of exterior spaces for children. The landscape of the typical kindergarten, for instance, covered 4,600 square metres, and included dedicated spaces for general, group, and sport play; a learning garden; a 'corner for young naturalists' (*ugolok iunykh naturalistov*); and an exterior space just for teachers in addition to a generous area of open, multifunctional green space.[23] The designers prepared similar charts to set landscape prescriptions for nurseries, schools for older children, and the *mikroraion* as a whole, which in many ways was designed *for* children as a respite from the chaos of urban life. Designers surrounded play areas with low hedges to mark out a safe precinct that also maintained visual connection to the wider common space. Near these play areas, integrated seating at each shared residential entryway hosted the grandmother-minders who socialized with each other—and marked comings and goings—as they looked over the hedges and low retaining walls toward their small charges. Photos from original publications about Novye Cheremushki show squealing children clambering over a rocky island in the middle of the neighbourhood pond, with no adults in sight. The inclusion of such common spaces that cater to the youngest residents demonstrates that in socialist residential landscape design children's needs were taken seriously. Contrast these images with the play spaces of neoliberalism, adjacent to shopping districts or viewable but locked to public use, which cater to parent-consumers more than to the children themselves.

To widen the geographical scope of this discussion, I introduce the first *mikroraion* built in Baku, Azerbaijan, from the early 1960s, designed after major kinks in Soviet architectural standardization were resolved. The designers of the first Baku *mikroraion* took the buildings as fixed types to be composed creatively within the residential precinct. The eighteen-hectare *mikroraion* for 6,500 residents was located in close proximity to the traditional urban core of Baku, and was bordered on its western side by a major vehicular thoroughfare.[24] The original site plan, first published in *Architecture of the USSR* (*Arkhitektura SSSR*) at the end of 1964, shows the majority of the residential bars oriented latitudinally along the east-west axis, so that the shortest ends of the buildings abut the busy street, and the long sides face the common open spaces of the *mikroraion*, as can be seen in Figure 20.3.

Like the KhTZ *zhilkombinat*, the first Baku *mikroraion* had a spatially fluid character created by the staggered, parallel siting of regular thin housing bars—here standardized five-storey apartment buildings known as the Series 460-AN. Two scales of open space resulted from a skipping site pattern. Pairs of bars placed close together created intimately scaled spaces, while between each set of pairs a larger space held a kindergarten and its dedicated play spaces. The Baku *mikroraion* navigated a significant grade change as it climbed from south to north. Each of the regular open spaces, captured between the long sides of the buildings, were graded flat. Short flights of steps and ramps connected these terraced zones. Each north-south strip of buildings and open spaces was repeated along the east-west axis, then shifted up, so that pedestrian movement was never impeded transversally. The designers for Baku set other types of planning rules to ensure the

(a)

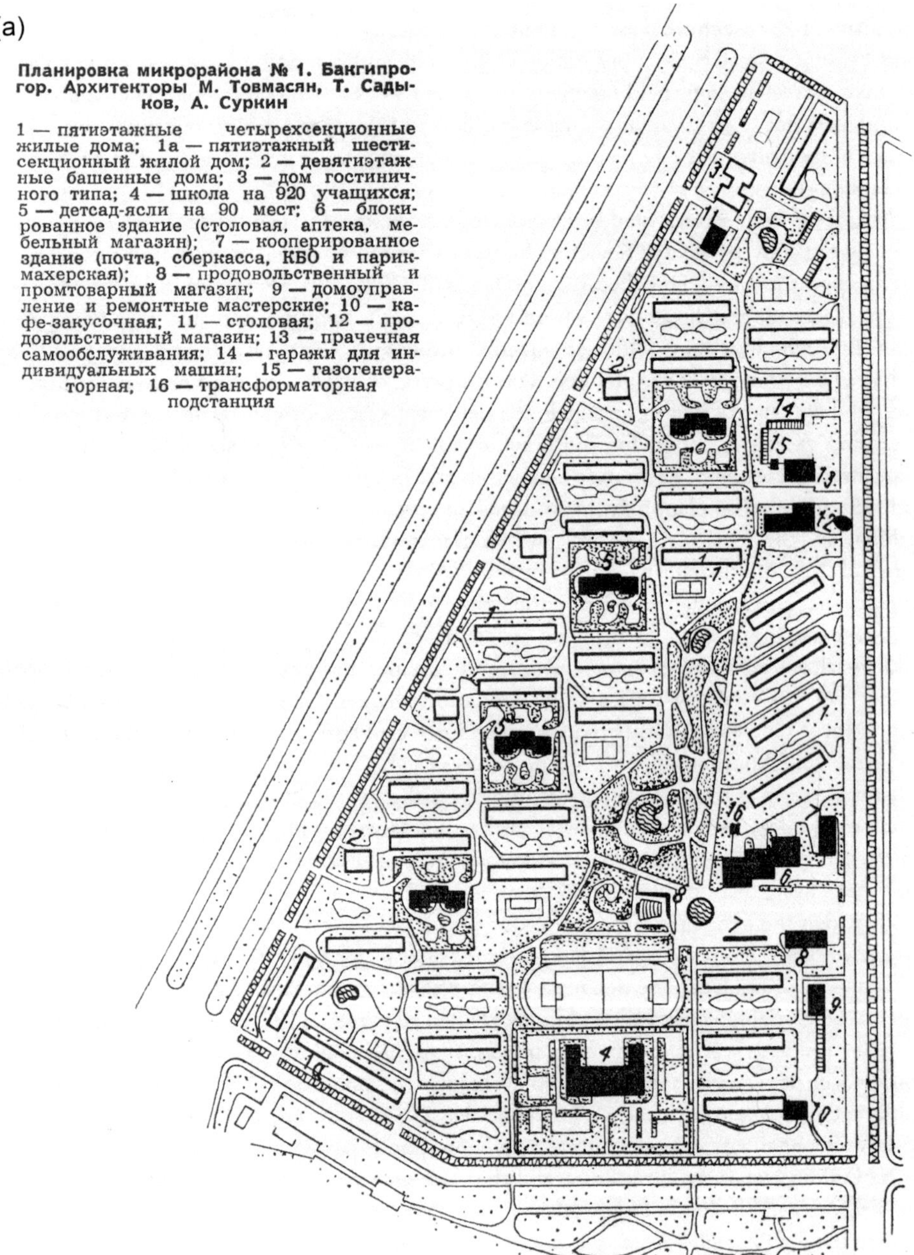
Планировка микрорайона № 1. Бакгипрогор. Архитекторы М. Товмасян, Т. Садыков, А. Суркин
1 — пятиэтажные четырехсекционные жилые дома; 1а — пятиэтажный шестисекционный жилой дом; 2 — девятиэтажные башенные дома; 3 — дом гостиничного типа; 4 — школа на 920 учащихся; 5 — детсад-ясли на 90 мест; 6 — блокированное здание (столовая, аптека, мебельный магазин); 7 — кооперированное здание (почта, сберкасса, КБО и парикмахерская); 8 — продовольственный и промтоварный магазин; 9 — домоуправление и ремонтные мастерские; 10 — кафе-закусочная; 11 — столовая; 12 — продовольственный магазин; 13 — прачечная самообслуживания; 14 — гаражи для индивидуальных машин; 15 — газогенераторная; 16 — трансформаторная подстанция

Figure 20.3a,b First Mikroraion. Baku, Azerbaijan, 1964. © Mkrtchian, R. "Osebennosti Zastroiki Pervogo Mikroraiona V Baku." Arkhitektura SSSR, no. 12 (1964): 31–36, (used with permission under UCC/Russian copyright law (published pre-1973)).

(b)

Figure 20.3a,b Continued.

'rhythmical spatio-volumetric composition' of the ensemble; for instance, the distance between a building's short ends could be no greater than its height, and its broad sides no closer than twice that height. In addition to the experiential variety that such a plan provided to residents, the designers asserted that the staggered massing pattern created favourable conditions for the ventilation and insulation of the buildings.[25]

These compositional and climatic justifications must also be contextualized within the logic of industrial construction techniques. Unlike the buildings of the *zhilkombinaty*, which were constructed of masonry or poured-in-place concrete, the standardized residential buildings constructed after 1954 were composed of factory-made panels trucked to and lifted into place on-site. The vacant space between the buildings bears the footprint of their means of production: the construction crane. Whether a standardized house was built by a stationary tower crane or a mobile crane on rails, the technological logic of assembly required space around the building.[26]

Like the Moscow example, the first Baku *mikroraion* was designed to accommodate the needs of its children. One upper school for 920 children anchors the bottom of the plan, and four kindergartens for 90 children each are planted diagonally across the precinct in the large regular gaps in the site plan's pattern. Each of these schools had an exterior precinct of its own comprising open play areas, shaded pavilions to shield the children during the hottest months of the year, learning gardens, and play structures. The fluid landscape of the *mikroraion* as a whole was initially solely pedestrian, to allow children to walk unaccompanied from home to school without crossing a vehicular street. In the original design, commercial amenities such as a grocery, canteen, laundry, and communal services building (including post office, telephone exchange, and hair salon) sat on a plaza on the eastern edge of the *mikroraion*, facing a fountain and exterior amphitheatre for community activities *en plein air*. Most importantly, the large green central park was 'free and open, like a natural landscape'. According to its designers, there were 'no main entrances or formal, geometrically planned gardens', but rather a 'picturesque' composition that was achieved through the use of polychromatic, loose plantings of various heights and volumes.[27] This bucolic urban landscape, populated by free-playing children who owned the paths, water features, and dedicated playscapes during the working day, is at the centre of all of the photographs in the inaugural 1964 article. The shared landscape holds the foreground in the images, and slides back and up into perspective to fill all of the gaps between the unremarkable standardized buildings, which are shown only in glancing views.

Socialist spatiality in the post-socialist period: What is to be done?

The designs of the KhTZ *zhilkombinat* and the Baku *mikroraion* were intended to instantiate communally minded behaviour by providing residents with unbounded, variegated landscapes open for common use.[28] The objectives of predictive planning, however, become fainter with time, and must jostle with the lived experience of the place and the aggregative memories of the past near-century. The Soviet state, with its desire for convenient access between working, living, and recreation sectors, was replaced in 1991 by free-market independent states. We cannot turn back the clock to reinstall the socialist condition in which the KhTZ *zhilkombinat* or the Baku *mikroraion* were built, but we can make an argument, as designers, for the unique qualities of these integrated residential landscapes. To do so, it is necessary to be clear about what so-called socialist space is, how it operates, and finally, what makes it worth preserving, as I have attempted to do here. As originally designed, the KhTZ *zhilkombinat* and the Baku *mikroraion* were spatial experiments that sought to erase the distinction between culture and nature. Their designers conceived of architecture and landscape, the built and the unbuilt, as belonging to a common category of shared space that had the capacity to offer a fluid haptic and social experience. This was, of course, a desire that could only be fulfilled under a wholly public property-ownership regime. So, what is to be done with these spaces now?

As the promised demolition of the Khrushchev-era housing blocks in Moscow has revealed, privatization is an unfinished project in many former Soviet states. Although the living units are by now largely privately owned in such residential communities, the land underneath the buildings remains controlled by local government. The legal implications of this ownership structure should be immediately obvious: if this land becomes desirable, there is no need for the government owner to resort to eminent domain to clear it. The government can simply break the lease on its land, compensate the apartment owners in some minimal fashion, demolish the buildings, and redevelop. Individual unit owners have no recourse.

These are sites fraught with the weight of multiple, conflicting expectations, and they are also sites written and rewritten through daily experience. A 2010 national newspaper poll named KhTZ one of the ten most dysfunctional residential communities in all of Ukraine.[29] Closure of the tractor factory at KhTZ, and the remote location of the *sotsgorod* in relation to the city centre, has left it economically vulnerable, and locals are quick to note that unemployment in this sector of the city is unusually high, as is per capita crime. But negative local perceptions of this experimental site of socialist space-making are difficult to disentangle from pervasive disappointment with the failure of Soviet socialism. A visit to the former tractor factory region on a beautiful summer day, to determine the legibility of the original spatial conditions, refutes these blanket claims of dysfunctionality. The open green spaces between residential buildings are filled with tended flowerbeds. Newly painted wooden play structures see heavy use by the children of the neighbourhood, whose parents and grandparents sit on nearby benches under the shade of now mature trees. Pedestrians moving through the residential precinct—though now joined by vehicles—still enjoy spatial liberation.

I close with three possibilities of how to approach these disappearing remnants of socialist space. In the first case, as with KhTZ, where the model spatial condition persists and yet the economic base is gone, heritage preservation status might offer a second lease of life. Such status would highlight the historical and spatial uniqueness of the site, and might provide justification for restoration funding. Kharkiv is stocked with young, well-educated residents (the city's universities and technical schools are renowned) who are primed to snap up inexpensive live-work spaces in a creative enclave like a revisioned KhTZ. In the second case, as with the first Baku *mikroraion*, which enjoys proximity to the city centre and has adjusted well to its post-socialist circumstances, a strategic plan for growth and change might permit adaptation while ensuring conservation of a large percentage of the landscape for common use.[30] Finally, in extremely attractive development sites such as those under threat in Moscow, the original buildings will invariably be demolished. New buildings could be designed, however, that respect the footprints of the originals or at minimum retain the spirit of the original site plan, replete with high percentages of shared open spaces to be owned and maintained collectively. In all cases, the design of quality exterior spaces for children should be a priority.

Socialist spaces are being dismantled as I write. They will disappear altogether without strong advocacy by historians and design professionals. Together, we must devise strategies that maintain the spirit of openness and collectivity of the residential landscape, while recognizing the exigencies of the economic system in which they now sit. Their spatial uniqueness within the context of the now-capitalist land development regime—fluid and targeted to the needs of a safe childhood, versus enclosed and targeted to those who can pay—is their greatest asset, deserving of preservation for future generations.

Notes

1 If the bill is approved, this demolition programme will displace ten per cent of Moscow's population, or 1.6 million people.
2 A. Baburov and A. Gutnov, *The Ideal Communist City* (New York: G. Braziller, 1971), 27.
3 Smooth space is 'a space of affects, more than one of properties, [and] haptic rather than optical perception'. G. Deleuze and F. Guattari, *A Thousand Plateaus: Capitalism and Schizophrenia* (Minneapolis: University of Minnesota Press, 1987), 479.
4 V.I. Lenin, 'Karl Marx: A Brief Biographical Sketch with an Exposition of Marxism', 1914, https://www.marxists.org/archive/lenin/works/1914/granat/ch04.htm. Used as the opening epigraph in L.M. Sabsovich, *Goroda Budushchego I Organizatsiia Sotsialisticheskogo Byta* (Moscow: Gosudarstvennoe tekhnicheskoe izdatel'stvo, 1929).
5 V.I. Lenin, 'Decree on Land/Peasant Mandate on the Land', 1917, https://www.marxists.org/archive/lenin/works/1917/oct/25-26/26d.htm.
6 For example, Dziga Vertov's film *One Sixth of the World* (*Shestaia chast' mira*) from 1926.
7 E. Widdis, *Visions of a New Land: Soviet Film from the Revolution to the Second World War* (New Haven: Yale University Press, 2003), 3.
8 Vladimir Paperny, *Architecture in the Age of Stalin: Culture Two* (Cambridge: Cambridge University Press, 2002), 34.

9 Sabsovich, *Goroda Budushchego*, 60.
10 Ibid., 15–18.
11 L.M. Sabsovich, *Sotsialisticheskie Goroda* (Moscow: Gosizdat RSFSR 'Moskovskii rabochii', 1930), 41–42, italics original. For fuller discussion of the first Five-Year Plan and Sabsovich's theories, see C.E. Crawford, 'From Tractors to Territory: Socialist Urbanization Through Standardization', *Journal of Urban History* 44, no. 1 (2018).
12 Sabsovich, *Sotsialisticheskie Goroda*, 43, italics original.
13 N.A. Miliutin, *Sotsgorod: The Problem of Building Socialist Cities* (Cambridge, MA: MIT Press, 1974), 65; A. Zelenko, 'Problema Stroitel'stva Sotsialisticheskikh Gorodov', *Planovoe khoziaistvo*, 12 (1929).
14 There are roughly thirty-eight *zhilkombinat* blocks in this site plan. If each held 1,500 residents, this plan would support a *sotsgorod* population of 57,000, a number well in excess of the original demographic target of 36,500.
15 For more on German *zeilenbau* see M. Swenarton, 'Rationality and Rationalism: The Theory and Practice of Site Planning in Modern Architecture 1905–1930,' *AA Files*, no. 4 (1983).
16 *Tsentral'nyi derzhavnyi arkhiv vyshchykh organiv vlady ta upravlinnia Ukrainy* (Central State Archives of Supreme Bodies of Power and Government of Ukraine), TsDAVO, f.5, o.3, d.2085, l.25.
17 Leonid Sabsovich was reproached by name. 'O Rabote Po Perestroike Byta (Postanovlenie Tsk Rkp(B) Ot 16 Maia 1930 Goda)', *Pravda*, 29 May 1930.
18 L.M. Kaganovich, *Socialist Reconstruction of Moscow and Other Cities in the USSR* (New York: International Publishers, 1931), 83.
19 B.A. Ruble, 'From Khruchsheby to Korobki', in *Russian Housing in the Modern Age*, ed. W.C. Brumfield (Washington, DC: Woodrow Wilson Center Press, 1993), 235.
20 N. Khrushchev, 'On the Extensive Introduction of Industrial Methods, Improving the Quality and Reducing the Cost of Construction (December 7, 1954)', *Volume* 21, no. 3 (2009): 26.
21 A. Bronovitskaya, 'Open City: The Soviet Experiment', *Volume* 21, no. 3 (2009): 20.
22 For an excellent discussion of minimum living spaces in Soviet housing, see S.E. Harris, *Communism on Tomorrow Street: Mass Housing and Everyday Life After Stalin* (Washington, DC: Woodrow Wilson Center Press, 2013), chap. 1 and 2.
23 B.R. Rubanenko, *Arkhitekturno-Planirovochnoe Upravlenie. 9-Ĭ Kvartal: Opytno-Pokazatel'noe Stroitel'stvo Zhilogo Kvartala V Moskve (Raĭon Novye Cheremushki)* (Moscow: Glavmosstroi, 1959), 20.
24 R. Mkrtchian, 'Osebennosti Zastroiki Pervogo Mikroraiona V Baku,' *Arkhitektura SSSR*, no. 12 (1964): 31.
25 Ibid., 33.
26 P. Meuser and D. Zadorin, *Towards a Typology of Soviet Mass Housing: Prefabrication in the USSR 1955–1991* (Berlin: DOM Publishers, 2015), 126–7.
27 Mkrtchian, 'Osebennosti', 34.
28 Other extant examples of socialist city/*zhilkombinat* designs can be found at the Stalingrad/Dzerzhinsky Tractor Factory (c. 1929), which sits in present-day Volgograd, Russia, and Uralmash (*Ural'skii Mashinostroitelnyi Zavod*/Ural Heavy Machine-Building Factory, c.1933) in present-day Ekaterinburg, Russia.
29 'Korrespondent Sostavil Top-10 Samikh Neblagopoluchnykh Zhilmassiv V Ukraine', last modified 19 March 2010, http://korrespondent.net/ukraine/events/1058465-korrespondent-sostavil-top-10-samyh-neblagopoluchnyh-zhilmassivov-v-ukraine2010.
30 Positive changes in Baku include lower-level apartments being converted into shops, but there are harbingers of wild development as well, such as infill towers under construction that are plugging some of the shared open spaces.

Bibliography

Baburov, A., and A. Gutnov. *The Ideal Communist City*. New York: G. Braziller, 1971.

Bronovitskaya, A. 'Open City: The Soviet Experiment'. *Volume* 21, no. 3 (2009): 19–25.

Crawford, C.E. 'From Tractors to Territory: Socialist Urbanization Through Standardization'. *Journal of Urban History* 44, no. 1 (2018): 54–77. doi:10.1177/0096144217710233.

Deleuze, G., and F. Guattari. *A Thousand Plateaus: Capitalism and Schizophrenia*. Minneapolis: University of Minnesota Press, 1987.

Harris, S.E. *Communism on Tomorrow Street: Mass Housing and Everyday Life After Stalin*. Washington, DC: Woodrow Wilson Center Press, 2013.

Kaganovich, L.M. *Socialist Reconstruction of Moscow and Other Cities in the USSR*. Williams College Propaganda Pamphlet Series, no. 159. New York: International Publishers, 1931.

Khrushchev, N. 'On the Extensive Introduction of Industrial Methods, Improving the Quality and Reducing the Cost of Construction (December 7, 1954)'. *Volume* 21, no. 3 (2009): 26–34.

Lenin, V.I. 'Karl Marx: A Brief Biographical Sketch with an Exposition of Marxism'. 1914. https://www.marxists.org/archive/lenin/works/1914/granat/ch04.htm.

Lenin, V.I. 'Decree on Land/Peasant Mandate on the Land'. 1917. https://www.marxists.org/archive/lenin/works/1917/oct/25-26/26d.htm.

Meuser, P., and D. Zadorin. *Towards a Typology of Soviet Mass Housing: Prefabrication in the USSR 1955–1991*. Berlin: DOM Publishers, 2015.

Miliutin, N.A. *Sotsgorod: The Problem of Building Socialist Cities*. Cambridge, MA: MIT Press, 1974.

Mkrtchian, R. 'Osebennosti Zastroiki Pervogo Mikroraiona V Baku'. *Arkhitektura SSSR*, no. 12 (1964): 31–6.

Paperny, V. *Architecture in the Age of Stalin: Culture Two*. Cambridge: Cambridge University Press, 2002.

Rubanenko, B.R. *Arkhitekturno-Planirovochnoe Upravlenie. 9-Ĭ Kvartal: Opytno-Pokazatel'noe Stroitel'stvo Zhilogo Kvartala V Moskve (Raĭon Novye Cheremushki)*. Moscow: Glavmosstroi, 1959.

Ruble, B.A. 'From Khruchsheby to Korobki'. In *Russian Housing in the Modern Age*, edited by W.C. Brumfield, 232–70. Washington, DC: Woodrow Wilson Center Press, 1993.

Sabsovich, L.M. *Goroda Budushchego I Organizatsiia Sotsialisticheskogo Byta*. Moscow: Gosudarstvennoe tekhnicheskoe izdatel'stvo, 1929.

Sabsovich, L.M. *Sotsialisticheskie Goroda*. Moscow: Gosizdat RSFSR 'Moskovskii rabochii', 1930.

Swenarton, M. 'Rationality and Rationalism: The Theory and Practice of Site Planning in Modern Architecture 1905–1930'. *AA Files*, no. 4 (1983): 49–59.

Widdis, E. *Visions of a New Land: Soviet Film from the Revolution to the Second World War*. New Haven: Yale University Press, 2003.

Zelenko, A. 'Problema Stroitel'stva Sotsialisticheskikh Gorodov'. *Planovoe khoziaistvo*, no. 12 (1929): 16–31.

PART V

Philosophy of landscape architecture

Knowledge, practice, and education

21

IMAGINARIES IN LANDSCAPE ARCHITECTURE

Ian H. Thompson

It is over two decades since I started the research into underlying value systems in landscape architecture which became the book *Ecology, Community and Delight* (1999). The book was one of the first attempts to examine the motivations of landscape architects and the ends they pursued through their practice. I believe this has proved a valuable framework for analysing and critiquing works of landscape architecture, but, as I found back then when interviewing practitioners, most landscape architects find it hard to articulate a coherent set of values. Practically minded, many were not comfortable being forced to become philosophers, even for an hour or two; others enjoyed it, but readily admitted that it wasn't something they had ever done before. Mulling this over, I've concluded that landscape architects live somewhere between the materiality of their sites, with their leaves and mud, bad drainage and dappled sunlight, and their imaginations, which are stocked, not with abstract ideas, but with complex and compelling 'imaginaries' produced by and shared within the culture. Without wishing to repudiate my earlier work on values, I wish to explore the nature and agency of these imaginaries in this contribution.

What, then, do I mean by an 'imaginary'? Here is an example taken from *The Hobbit*, by J.R.R. Tolkien:

> Morning passed, afternoon came; but in all the silent waste there was no sign of any dwelling. They were growing anxious, for they now saw that the house might be hidden almost anywhere between them and the mountains. They came on unexpected valleys, narrow with deep sides that opened suddenly at their feet, and they looked down surprised to see trees below them and running water at the bottom. There were gullies that they could almost leap over; but very deep with waterfalls in them. There were dark ravines that one could neither jump nor climb into.[1]

On one level, this is a description of an imaginary landscape, but at another level it is a token of a much broader cultural construction which we might call the imaginary of Middle-Earth. Literature is full of imaginary landscapes, but an *imaginary landscape* and a *landscape imaginary* are not quite the same thing. An imaginary landscape might form *part* of a landscape imaginary. A landscape imaginary might include a particular imaginary landscape—it might inspire the creation of imaginary landscapes and indeed real landscapes. In the latter case, we could say that the imaginary had been reified. The imaginary of Middle-Earth is not confined to Tolkien's evocative descriptions. Tolkien drew upon his knowledge of northern European landscapes (with the Alps thrown in), but the films which contributed so much to our collective imaginary were filmed in New Zealand. This is one of the characteristics of imaginaries—they are constantly elaborated, and tend to expand and evolve.

Notice that I am using the word 'imaginary' not as an adjective but as a noun. When I first heard someone speak this way, I recoiled from what seemed an ugly piece of academic jargon, but over the years I have encountered many people, particularly cultural theorists and social scientists, who use the word this way, and have come to find the concept of 'an imaginary' very useful, particularly the version articulated by the Greek psychoanalyst and philosopher Cornelius Castoriadis. Castoriadis says that societies (with all of their laws, institutions, structures and practices) are always based upon a basic conception of the world and the place of humankind within it.[2] There are similarities here with Heidegger's notion of the 'world picture', but Castoriadis's emphasis on creative indeterminacy contrasts with Heidegger's conservative ontology.[3]

Imaginary foundations are often expressed in creation myths, which both reflect and shape behaviour: for example, if you have a myth which states that God placed humankind upon the earth to rule over it, and that other living things were provided for the benefit of humans, then you are likely to develop an exploitative attitude towards nature. So, one use of the term 'imaginary' is to refer to these founding myths. There are also myths in contemporary mass culture, which Roland Barthes explored in his celebrated book *Mythologies*.[4] Social imaginaries represent systems of meanings that govern social structure and practice. They emerge through the interaction of subjects in a society, and they represent an imagined reality in the past, present or future. They do not constitute an established reality, although an imaginary might be part of an emerging reality. The Canadian philosopher Charles Taylor also employs the term 'social imaginary', and he contrasts it with 'social theory', commenting that 'theory is often in the possession of a small minority, whereas what is interesting in the social imaginary is that it is shared by large groups of people, if not the whole society.'[5] John Searle, meanwhile, states that 'the complex structure of social reality is, so to speak, weightless and invisible. The child is brought up in a culture where he or she simply takes social reality for granted.'[6] This remark of Searle's reminded me of garden historian Christopher Hussey's account of growing up at Scotney Castle in Kent and not realizing until adulthood that he had been living in the midst of a landscape contrived on Picturesque principles.[7] He had just assumed it was all natural.

When the emperor Hadrian visited Egypt, he saw the Nile. He held his impressions of the river in his mind, and when he returned to Rome he built the Canopus at his villa near Tivoli to remind him of his travels. In *The Imaginary* (1940), Jean-Paul Sartre uses the term 'quasi-observation' for what goes on when we picture something imaginary. Imaginary objects are, he says, a 'melange of past impressions and recent knowledge'.[8] Our engagement with the world activates our imaginary processes; we combine and recombine things that we have experienced, but because imaginary objects appear to us in a way that is similar to perception, we tend to treat them as real. We might go further and hypothesize that once an imaginary exists, sustained by literary and artistic representations, very soon there will be attempts to realize it, to build it. For example, William Gilpin provided great impetus for the Picturesque movement by giving advice to artists about how to enhance certain qualities in their sketches and paintings, such as by exaggerating vertical dimensions, or by moving things around on the canvas to create more interesting features in the foreground or middle distance. Soon, however, landscape designers such as Uvedale Price, Humphry Repton and John Claudius Loudon were moving earth, water, rocks and plants around to contrive the sorts of Picturesque landscape that fooled the young Christopher Hussey. One could say that these landscapes too were representations, but they were tangible and could be lived among in a way that is not possible for pictures or poems.

Utopias and dystopias

Utopias and dystopias are important types of imaginary with a strong influence upon the kinds of places and societies we seek to create. Le Corbusier's utopian city plans not only form part of a broad Modernist imaginary, but also make use of the centuries-old Pastoral imaginary, illustrating the complex and intertwined nature of imaginaries. In his *Precisions*, he wrote that the lives of the dwellers in his imagined city would be 'set within a Virgilian dream' and that 'nothing will be disturbed—neither the trees, the flowers, nor the flocks and the herds.'[9] There is a canon of utopian writing going back to Plato's *Republic* (c. 380 bce), Thomas More's *Utopia* (1516) and Francis Bacon's *New Atlantis* (1627). It includes such influential nineteenth-century books as Samuel Butler's *Erewhon* (1872) and William Morris's *News from Nowhere* (1890).

A dystopia points out a destination, derived by extrapolating existing trends that people of sound mind would do anything to avoid. Science fiction is full of such futures. Think of movies such as *Soylent Green* (1973), which depicts an overcrowded and polluted world where food is rationed, or *The Day After Tomorrow* (2004), which, however improbably, imagines the consequences of climate change as a new ice age hitting New York. A dystopian imaginary is like a warning sign: do not follow this route. The grim twentieth century produced more dystopias than utopias, including Aldous Huxley's *Brave New World* (1932) and George Orwell's *Nineteen Eighty-four* (1949).

Utopian imaginaries, on the other hand, are like signposts to a better future. Taylor says that a utopia can 'refer us to a way of things which may be realized in some eventually possible conditions; but which meanwhile serves as a standard to steer by'.[10] If we could not imagine, we would be stuck in the real. We would not be able to see possibilities, and that would mean that we had no freedom of choice. James Corner, landscape architect of the celebrated High Line in New York City, finds the term 'imaginary' helpful too. At the end of 'Terra Fluxus', his important essay on landscape urbanism, he calls urgently for new imaginaries, which he usefully explains as 'the collective imagination, informed and stimulated by the experiences of the material world'.[11]

F ailed utopias

When we seek to create perfect places, we inevitably fall short. The philanthropic industrialist Robert Owen (1771–1858), having greatly improved the lives of his employees at his mill in New Lanark, Scotland, then sought to found an ideal socialist community in Indiana. It was called New Harmony, and images of it show the factory set amongst lawns within a large rectangular courtyard of workers' houses, and the whole complex lying within a peaceful and rather English-looking pastoral landscape. The reality was very different. The site was badly prone to flooding, and the inhabitants, belying the settlement's name, were prone to infighting. Established in 1825, it was dissolved in 1829 because of the constant quarrels.

Although literary utopias often include physical descriptions of the built environment, they are generally more concerned with ideal social, political and economic arrangements; but utopian thought has sporadically taken more physical form in the creation of ideal settlements, often informed by Platonic rationalism and a clear geometry. An early example is Palmanova, a fortified town designed by Vincenzo Scamozzi to defend Venice against the Ottomans. Interestingly, despite its bold new design, no one wanted to live there. In 1622 Venice was forced to pardon criminals and offer them free building lots and materials if they would agree to settle the town. This is not an isolated case. The Baroque castle gardens begun in 1709 at Karlsruhe, Baden-Württemberg, Germany, built to the glory of Margrave Karl Wilhelm of Baden-Durlach, were constructed in the shape of a star, with thirty-two avenues radiating from the central hunting tower. In the absolutist spirit of Versailles, the plans for the town were also determined by this strict geometrical pattern. Karl Wilhelm waived taxes for twenty years as an inducement to get people to move in (Figure 21.1).

Perfect societies seem to require perfect people, or at least people who can be bribed or coerced into occupying the rationally determined houses and obeying the rationally conceived laws. The difficulty is that a utopia is imagined to be already perfect, and therefore not open to evolution and change. In order to preserve the status quo, a high degree of coercion is needed. It is not a coincidence that the Utilitarian social reformer Jeremy Bentham, who inspired a range of beneficial social improvements from public parks to public health, also invented the Panopticon, an ideal prison in which inmates could be constantly surveilled. In his book *Discipline and Punish*, Michel Foucault uses the Panopticon as a symbol of the way in which societies order and control their populations.[12] As Gregory Claeys has observed, 'the more urban the ideal, the more highly regulated its vision tends to be.'[13] City planning often involves surveillance and control, and the vision of the city slides towards an efficient but totalitarian dystopia, where any open space is devoted to the ideology of the governing regime.

One person's utopia may be another person's dystopia. William Morris disliked Edmund Bellamy's vision of a society in which technology had removed the need for meaningful work (*Looking Backward*, 1888) so much that he developed an alternative in *News from Nowhere* (1890), one which placed a high value on craftsmanship. New utopias are currently being imagined in

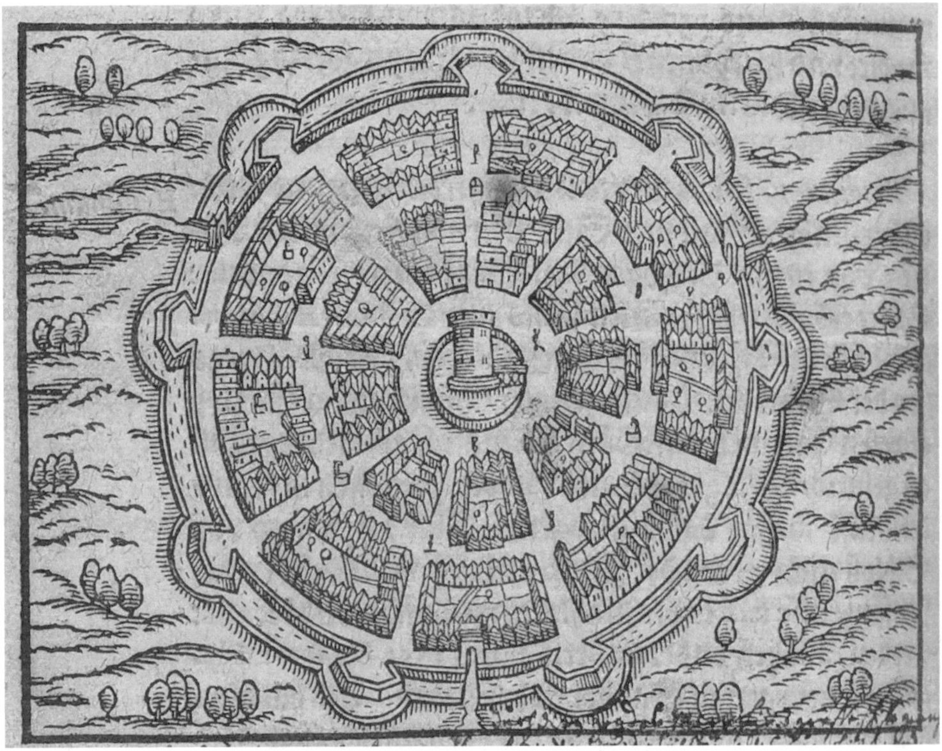

Figure 21.1 Plan of Palmanova, 1610. © Saur, Abraham. Theatrum Urbium: Warhafftige Contrafeytung und Summarische Beschreibung fast aller vornemen und namhafftigen Stätten Schlössern und Klöster (Frankfurt: Richter, 1610).

response to environmental issues, particularly climate change. Landscape architects rarely produce literary utopias, but two stand out. One is to be found in Ian McHarg's *Design with Nature* (1969),[14] where the author imagines a people, the Naturalists, who live close to nature and play a symbiotic role in the management of the earth. A more recent example is the last chapter of *Gray World, Green Heart*[15] written by Robert Thayer, founder of the landscape architecture programme at the University of California, Davis, who imagines that the region around his home city becomes a sustainably managed landscape by the year 2030.

Past, present and future imaginaries

At the present time, we have imaginaries of the past, the present and the future. Every society there has ever been has similarly produced imaginaries of its past, its present and its future. Similarly, all future societies will generate imaginaries of their pasts, presents and futures. The possibilities are mapped in Table 21.1, which could, of itself, suggest an extensive research programme. Though this seems intricate enough, the reality is even more complicated, because societies (or civilizations) produce imaginaries about other societies.

A large part of the interest in studying imaginaries lies in their agency. Although we might be interested in, say, Roman constructions of Egyptian society from a purely historical point of view, as designers and planners we are likely to be most interested in those imaginaries which are active in our culture and thus likely to influence both the sorts of commissions we receive and the sorts of designs we make. Some of these we have inherited from the past (the first column in Table 21.1), though we reimagine and reinterpret them for the present (the second column). The third column in the table is included for completeness, and to show that this process of reimagining will not stop with the present generation, although there is little we can say about it with any certainty.

I do not have sufficient space to explore all the avenues that an interest in imaginaries might open up, but I will briefly discuss two that I find thought-provoking. The first example, the Pastoral, is one with a very long history, and it remains important because it is still very active and because it has the potential to do harm as well as good. The second, which I have called Edgelands, is relatively new, though related to older imaginaries of the Picturesque. The Edgelands imaginary is voguish, influential and relevant to contemporary landscape design practice.

The Pastoral

The Pastoral and the Picturesque are often conflated in writing about landscape, but they should be kept distinct, though they are both aspects of a Romantic sensibility. A pastoral lifestyle is one that involves the herding of livestock according to the seasons and the availability of good pasture. It is thus a life lived close to nature and away from towns and cities. It has been idealized since ancient times as simple and innocent, and contrasted with the complexity and corruption associated with urban life. One version of the Pastoral imaginary is the myth of Arcadia, which is identified with a region of rural Greece. This imaginary was established by a series of poems, the *Eclogues*, written by the Roman poet Virgil in the first century bce, and later elaborated by writers and artists down to the present day. Arcadia became an idyll, a lost rural paradise for which more sophisticated city-dwellers might yearn. The visual arts contributed to the development of this imagined place through such paintings as Poussin's *Et in Arcadia ego* (1637–8), but the imaginary was still going strong in the nineteenth century when Thomas Cole, founder of the Hudson River School, presented his *Dream of Arcadia* (1838), a Romantic vision of the pastoral possibilities he thought were inherent in the American landscape. Closely associated

Table 21.1 Past, Present and Future Imaginaries

	Constructed by people in the past	*Constructed by people in the present*	*To be constructed in the future*
About the past	**Past pasts** Each age looks back to previous ages. For example, the Romans developed imaginaries about the Greeks and Egyptians. We study these using historical methods.	**Present pasts** Contemporary imaginaries about former times. 'Golden Age' myths, pastoral fantasies, etc. These can be negative, e.g. 'the Dark Ages', 'Satanic mills'. Many imaginaries endure from the past. Though 'historical', they can be appropriately studied as elements of our present culture and society.	**Future pasts** How the present (and past) might be constructed by future generations We can try to extrapolate from existing trends—but will probably be wrong.
About the present	**Past presents** Redundant imaginaries which have been superseded by events or otherwise lost their currency. We study these using historical methods.	**Present presents** How we imagine the world and our relations to it at this present time. 'Gaia', for example, is a contemporary imaginary. Contemporary imaginaries slip into the past as they lose their currency.	**Future presents** How future generations will construct imaginaries about their current conditions. Certain to include elements from our present and our past. We can try to extrapolate from existing trends—but will probably be wrong.
About the future	**Past futures** Ways in which the future (including our present) was imagined in the past. Includes both utopian visions and dystopian nightmares. They are redundant in that they have lost any predictive or motivational force. They may, however, provide the researcher with insights into historical conditions, values and preoccupations.	**Present futures** Futurology. How we imagine the future is always from our standpoint in the present. Includes utopian and dystopian visions. These visions may be both predictive and motivating. We investigate these by studying our present culture. Our visions of the future often reveal most about our present.	**Future futures** How future generations will imagine their own futures. We can try to extrapolate from existing trends – but will probably be wrong.

with the Pastoral imaginary, although distinguishable from it, is the Georgic imaginary, derived from another series of poems by Virgil, the *Georgics*, which deal with such themes as viticulture, bee-keeping and the cultivation of olive groves. The difference between the two visions lies in the amount of labour involved. We may envy Arcadian shepherds for the ease of their life, but we admire farmers for their productive husbandry (Figure 21.2).

In England, descriptions and depictions of an ancient Arcadia were an influence upon the rise of the English Landscape School. The designs of William Kent, Lancelot Brown and many

Figure 21.2 Dream of Arcadia, ca. 1838. © Cole, Thomas.

of their contemporaries were influenced by the Pastoral imaginary. The literary critic Raymond Williams argued that the Pastoral was a 'myth functioning as a memory', and that bucolic representations of ideal countryside masked the class conflict, enmity and antagonism which were as real in rural areas as they were in the towns.[16] Here, then, is one way in which the Pastoral imaginary can be regressive.

In terms of eighteenth-century aesthetic categories, the Pastoral can be associated with the Beautiful as identified by Burke in 1758[17] and Hogarth in 1753.[18] William Gilpin's 'Picturesque Beauty' in 1768, however, was 'that kind of beauty which is agreeable in a picture'[19] and was already halfway towards the category of the Sublime, where the human response to the immense dimensions and forces present in nature was one of awe and admiration. Eighteenth-century landscape design moved away from the comfort of the Beautiful and the Pastoral towards the more challenging and rugged Picturesque, and the adjective 'picturesque' has now been watered down so much that it is frequently used as a full synonym for 'beautiful' or 'pastoral' and even as a half-synonym for 'pretty'. In the nineteenth century the Pastoral once again became dominant, at the very moment when the landscape architecture profession emerged from its origins in landscape gardening. The plan that won Olmsted and Vaux the commission for New York's Central Park was the Greensward Plan, a proposal to introduce relaxing countryside scenery into the heart of the city.

Ample research attests to the value of green space in cities[20] and seems to vindicate Olmsted's faith in the restorative virtues of the Pastoral, but we always need to remain cautious and critical when we draw upon the power of a well-established imaginary. One manifestation of the Pastoral tradition is the lawn, which in temperate climates, where grass grows well, is a valuable element for the designer. In America, however, lawn culture became a quest for monocultural perfection, backed up with heavy and environmentally harmful use of chemical weedkillers. Many researchers have called for a change in the cultural norms surrounding grass care.[21] Lawns form part of the Pastoral imaginary, signifying an easy and convenient relationship with nature, but in the United States they also came to represent civic virtues and community cohesion. Keeping one's lawn well maintained was a duty. But manicured grassland is expensive to maintain, and it is thirsty, which is problematic in places where rainfall is low. It is only relatively recently that xeriscaping has offered homeowners in California alternatives to the ubiquitous greensward. A study by Milesi et al. found that American lawns cover three times the area of any other irrigated crop.[22]

A similar case can be made around the worldwide proliferation of golf courses, which again draw upon valorized Pastoral imagery. Golf originated on the short turf of coastal sites in Scotland before being internationalized by British colonialists.[23] It is now a vast worldwide industry, transforming landscapes on a colossal scale. Kim Wheeler and John Nauright noted in 2006 that there were over 25,000 courses worldwide, covering a total area equivalent to the size of Belgium.[24] The authors document both changes in golf course design—involving an idealized version of nature which parallels the perfectionist tendencies of American lawn management—and the spread of golf to new markets in Japan, China and South-East Asia. The grass required for greens and fairways conjures the Pastoral vision, and it is only in exceptional places, such as on some of the courses in Scottsdale, Arizona, that designers have opted to work in ways sympathetic to the local ecology. In many parts of the world, golf development is a threat to local water supplies, so much so that the political scientist Richard P. Hiskes has argued that it represents an incursion upon the 'emergent' human rights of neighbouring communities.[25]

The Pastoral, then, is an imaginary of great scope with far-reaching consequences, for both good and ill. It is hardly likely that landscape designers are going to stop drawing upon it, and with every use the imaginary grows larger and more powerful. It is so familiar that we may not even notice that we are invoking it, but that is the very reason we need to maintain our scrutiny.

Edgelands

The name of my second example of an active landscape imaginary comes from the title of a bestselling book written by contemporary British poets Paul Farley and Michael Symmons Roberts, published in 2011.[26] The poets are fascinated by the indeterminate areas between town and country, which in my student days (the 1970s) were called the 'urban fringe', and which appear in recent planning literature as 'peri-urban areas'. Farley and Roberts list over fifty other terms, including 'dispersed city', 'drosscape', 'dumpspace', 'junkspace', 'splintering urbanism', 'sprawl' and 'stimdross'. Their chapter titles indicate the kinds of places that spark their imagination: 'Cars', 'Paths', 'Dens', Containers', 'Landfill', 'Water', 'Sewage' and so on. Actually, these Edgelands are not confined to the perimeter of urban areas, but also include the interstices: derelict sites, scraps of wasteland, railway corridors through towns, etc. (see Figure 21.3).

Back in the 1970s, such in-between landscapes were vilified as neither good countryside nor good townscape. The general attitude was negative: the Edgelands were vast but unnoticed, unkempt, scruffy and unproductive. The urban fringe needed to be tidied up, prettified, made safe and made productive. The Edgelands are still the place we put things we are not proud of and don't want to live next to: car-crushing facilities, sewage treatment works and sites for Travellers. Sometimes there are heroic bits of infrastructure, such as power stations, waste incinerators and motorway junctions. But there has been a gradual recognition that these places have their positive side. Rather than being a devalued form of countryside or town, they have a character of their own which can be celebrated. Farley and Roberts grew up in the former industrial heartlands of England's north-west, which stretched between Liverpool and Manchester.

Figure 21.3 Weetslade Country Park, with buses. Tyne and Wear, England, 2015. © Thompson, Ian.

They remember these Lancashire Edgelands as a kind of Arcadia, which as children they found both liberating and exhilarating. They set out 'as poets in the English lyric tradition' to praise and celebrate this disregarded landscape. There is a parallel with those poets and artists who were drawn to marginally productive mountainous regions in the late eighteenth and nineteenth centuries, and who brought about a complete revision in the way such places were regarded. In 1727 Daniel Defoe could dismiss the mountains of Cumberland and Westmorland as 'all barren and wild, of no use or advantage either to man or beast',[27] against which we might compare any number of later eulogies, including this one from Charlotte Brontë, who visited friends in Westmorland in 1850: 'dusk as it was, I could feel that the valley and the hills were beautiful as imagination could dream them.'[28]

Farley and Roberts did not conjure the Edgelands imaginary from thin air. A re-evaluation has been underway for some time. In the early 1970s the nature writer Richard Mabey explored the wastelands of that era—crumbling city docks, disused canals, reservoirs, gravel pits and rubbish tips—and discovered that these places were full of species diversity—kestrels hovering above town parks, wildflowers flourishing next to railway lines, fox cubs playing on a motorway embankment, etc. He called this *The Unofficial Countryside.*[29] In 2002 Jennifer Jenkins edited a book called *Remaking the Landscape: The Changing Face of Britain.* In it there was a chapter by Marian Shoard called 'Edgelands', and this is where, at least in British culture, the name for this sort of landscape was coined.[30]

The first sustained literary attempt to capture the essence of the idea was Iain Sinclair's *London Orbital*, an evocative account of a walk around the M25. The tone, as shown in this passage, is rather bleaker than that adopted by Farley and Roberts:

> We drive in two cars, over the railway line, through caverns of brightly coloured containers, under the A13, past breakers yards and out onto the marshes. Unlisted,

> this is one of Europe's great roads. Drainage channel on one side, landfill on the other. Filthy lorries, trucks, vans trying to shove you into the ditch. A stench of unbelievable complexity: necrotic, polluted, maggoty, piscine. Magnificent. London, animal, vegetable and mineral, rotting into the ground.[31]

Sinclair is frequently associated with the revival in the 1990s of psychogeography, a critical artistic practice first developed by Guy Debord and the Situationists in the late 1950s. Farley and Roberts complain that this school merely uses the Edgelands as 'a backdrop for bleak observations on the mess we humans have made of our lives, landscapes, politics and each other'.[32] Their own writing about the Edgelands is wistful, pensive and elegiac, as in this description of a scrapyard:

> These are the automotive equivalents of the Paris catacombs. Mass graves in orderly array, but above ground, exposed to the elements. In these yards you find rusting car cadavers piled three, five, eight high, towering columns of ex-cars leaning on each other for support. Stripped of all that's worth taking—alloys, radios, lights—all windows smashed, these are a record of our conduct. Who was taken for a ride in these? Who bought and sold them, thrashed them on the motorways?[33]

The psychological process whereby this scene of destruction became a subject for contemplation was prefigured by those eighteenth-century aesthetes who were drawn to (what seemed to them) wild and unproductive mountains. Dr John Brown, an Oxford don, wrote these lines about the Vale of Keswick in 1755:

> Horrors like these at first alarm,
> But soon with savage grandeur charm,
> And raise to noblest thought the mind.[34]

Farley and Roberts find sublimity on landfill sites, richness on motorway verges and grandeur in power stations. The imagination can soar in the Edgelands, as shown in their meditation upon cooling towers:

> Cooling towers stand as deceptively still as mills once did, but are full of busy, watery activity. When a curious child on a train points and asks his father what a cooling tower is, and he replies, 'That's where they make the clouds,' the imagination is plumbed in again to an ancient sense of water spirits, the modern connected to the mythical and fabulous.[35]

The largest coal-burning power station in Britain is at Drax in Yorkshire, and it produces so much ash that from the 1970s onwards a landscape architecture practice, Weddle Landscape Design, has been employed to sculpt it into pleasant-looking hills. Accommodating industry and infrastructure in the landscape, hiding, screening, ameliorating and disguising, have been lucrative aspects of the professional workload, but the shift towards the Edgelands aesthetic poses questions and introduces a tension. In the 1970s the urban fringe was presented as problematic. Something had to be done about it, and indeed one of the local authorities I worked for employed a management team specifically to improve this sort of landscape, looking after footpaths, repairing gates and stiles, providing signage and screening eyesores. But there is another way of 'doing something about' a despoiled landscape, which is to alter perceptions of it, to revalue it, to find virtues in it, to redescribe, relabel, rename.

Changing perceptions has been part of landscape architecture since the time of Olmsted, but it has usually involved a Pastoral makeover. The first prominent signal of the arrival of a new aesthetic was Seattle's Gas Works Park, conceived by Richard Haag and opened in 1975. When Haag realized that the gas plant was the last of its kind in the United States, he persuaded both the authorities and the public that the old industrial structures should be kept, in all their rusty glory. It was a significant departure in terms of land reclamation, which up until that point had usually involved the erasure of all that went before. In the 1980s, the German landscape architect Peter Latz developed a similar approach while working on plans for the Bürgerpark in Saarbrücken, which was to be built on the derelict site of a coal dock bombed during the Second World War. Latz found a way of working which valued the historical traces left upon the site as well as the accrued ecology of the wastelands, and this approach served him well for what would become his most celebrated work, the creation of the Landschaftspark Duisburg-Nord upon the site of a former Ruhr steelworks, realized between 1990 and 2002. The Landschaftspark has become emblematic of a certain approach to landscape design, as has New York's High Line, designed by James Corner Field Operations in collaboration with architects Diller Scofidio + Renfro and plantsman Piet Oudolf (realized 2009–14) (see Figure 21.4).

In such work, landscape architects are both responding to and helping to shape the emerging imaginary, but the imperative 'to do something about' often creates a tension between the Edgelands aesthetic and older notions of managed public space. This becomes evident if one compares images of High Line as it is now with the photographs taken by Joel Sternfeld of the derelict elevated railway before its transformation, when it bore comparison to a ribbon of lonely prairie threading through the city. Ironically it was its liminal, Edgelands character that led to the campaign for its preservation, but this character was bound to change when it became a must-see tourist attraction.

Figure 21.4 Latz + Partner, Landschaftspark Duisburg-Nord, Germany. © Schäfer, Peter.

Deception and delusion

I have suggested that we have a tendency to try to realize our imaginaries, but this can be stated the other way around, giving agency to the imaginaries. These visions reciprocate by tending to shape our actions. This fact, of itself, is aesthetically and ethically neutral. There is nothing intrinsically wrong about actualizing an imaginary. However, there are undoubtedly both helpful and unhelpful imaginaries; there are idealistic imaginaries and ethically dubious imaginaries. There are imaginaries which spring to some extent from veridical observation of the world and others which are ill-informed or downright deceptive. It is part of the tumbleweed nature of imaginaries to lose touch with the realities on the ground—this is part of their creative promise, but also of their capacity to mislead. Jacques Lacan said that the imaginary was 'the field of images and imagination, deception and lure'.[36] It is important to keep in mind the possibility that the field of the imagination is a place where we can be tricked, deceived or deluded. It is also difficult to control an imaginary. A useful metaphor for thinking about imaginaries is to see them as giant snowballs or tumbleweeds which gather material as they roll around, growing ever larger and gathering a momentum of their own. All of this suggests caution. It is important to keep the gap between the imaginary and the real in mind.

Shaping the imaginary

On the other hand, imaginaries are not optional, in the sense that societies cannot avoid creating them. They are a society's collective dreams. I also hope that I've shown how they can work for both good and ill. Cultural producers of all kinds—writers, painters, film-makers, photographers, designers and the rest—are responsible for our imaginaries, but generally in the way that hill-walkers might add a stone or two to a trailside cairn to indicate the way. It is a shared responsibility, but it is still a responsibility. To push the metaphor a bit further, some paths lead to dangerous cliffs, while others take us to places of safety and delight. Of course, to the hard-pressed landscape architect struggling with client demands, budgetary constraints and tight deadlines, it might seem that all this talk of imaginaries is somewhat nebulous and hardly related to their everyday activities, but I would argue that this is not so. As the recent flood of smart-city, green-city and future-city imagery suggests, the call to contribute to an evolving imaginary can be an invigorating challenge. If we believe that design is an ethical activity, then critical attention to the sorts of imaginaries we create, foster and sustain is an imperative.

Notes

1 J.R.R. Tolkien, *The Hobbit* (London: George Allen and Unwin, 1937), chap. III.
2 Cornelius Castoriadis, *The Imaginary Institution of Society* (Cambridge, MA: MIT Press, 1997).
3 Angelos Mouzakitis, 'Autonomy and Authenticity. On the Aporetic Nature of Time and History: Castoriadis—Heidegger', *Critical Horizons* 7, no. 1 (2006).
4 Roland Barthes, *Mythologies* (London: Jonathan Cape, 1972).
5 Charles Taylor, *A Secular Age* (Cambridge, MA: Belknap Press of Harvard University Press, 2007), 172.
6 John R. Searle, *The Construction of Social Reality* (London: Penguin, 1996), 4.
7 Christopher Hussey, *The Picturesque: Studies in a Point of View* (London and New York: G.P. Putnam's Sons, 1927).
8 Jean-Paul Sartre, *L'Imaginaire: Psychologie phénoménologique de l'imagination* (Paris: Gallimard, 1940), 90.
9 Le Corbusier, *Precisions: On the Present State of Architecture and City Planning* (Cambridge, MA: MIT Press, 1991).
10 Taylor, *Secular Age*, 161.
11 James Corner, 'Terra Fluxus', in *The Landscape Urbanism Reader*, ed. Charles Waldheim (New York: Princeton Architectural Press, 2006), 32.

12 Michel Foucault, *Discipline and Punish* (London: Penguin, 1991).
13 Gregory Claeys, *Searching for Utopia. The History of an Idea* (London: Thames & Hundson, 2011).
14 Ian L. McHarg, *Design with Nature*, (Garden City, NY: Natural History Press, 1969).
15 Robert L. Thayer, *Gray World, Green Heart: Technology, Nature and the Sustainable Landscape* (New York: John Wiley and Sons, 1994).
16 Raymond Williams, *The Country and the City* (London: Chatto and Windus, 1973).
17 Edmund Burke, *A Philosophical Enquiry into the Sublime and the Beautiful* (Oxford: Oxford University Press, 1998).
18 William Hogarth, *The Analysis of Beauty* (Menston: Scolar Press, 1971).
19 William Gilpin, *An Essay on Prints* (Cadell and Davies: London, 1802).
20 Helena Nordh, Caroline M. Hägerhäll and Terry Hartig, 'Urban Nature as a Resource for Public Health', in *The Routledge Companion to Landscape Studies*, eds. Peter Howard, Ian Thompson and Emma Waterton (London: Routledge, 2013).
21 See, for example: Herbert Bormann, Diana Balmori and Gordon T. Geballe, *Redesigning the American Lawn: A Search for Environmental Harmony* (New Haven and London: Yale University Press, 2001); Paul F. Robbins, *Lawn People: How Grasses, Weeds, and Chemicals Make Us Who We Are* (Philadelphia: Temple University Press, 2007); Ted Steinberg, *American Green: The Obsessive Quest for the Perfect Lawn* (New York: W.W. Norton and Co., 2006).
22 Cristina Milesi et al., 'A Strategy for Mapping and Modeling the Ecological Effects of US Lawns', *Journal of Turfgrass Management* 1(2005): 83–97.
23 See: Erik Jönsson, 'The Nature of an Upscale Nature: Bro Hof Slott Golf Club and the Political Ecology of High-End Golf', *Tourist Studies* (2015); Robert Price, *Scotland's Golf Courses* (Edinburgh: Mercat Press, 2002).
24 Kim Wheeler and John Nauright, 'A Global Perspective on the Environmental Impact of Golf', *Sport in Society* 9, no. 3 (2006).
25 Richard P. Hiskes, 'Missing the Green: Golf Course Ecology, Environmental Justice, and Local "Fulfilment" of the Human Right to Water', *Human Rights Quarterly* 32, no. 2 (2010).
26 Paul Farley and Michael S. Roberts, *Edgelands: Journeys into England's True Wilderness* (London: Jonathan Cape, 2011).
27 Daniel Defoe, *A Tour Through the Whole Island of Great Britain* (Exeter: Webb and Bower, 1989), 195–6.
28 Letter, Charlotte Brontë to James Taylor, 6 November 1850.
29 Richard Mabey, *The Unofficial Countryside* (London: William Collins Sons and Co., 1973).
30 Marian Shoard, 'Edgelands', in *Remaking the Landscape: The Changing Face of Britain*, ed. Jennifer Jenkins (London: Profile Books, 2002).
31 Iain Sinclair, *London Orbital* (London: Penguin, 2003), 503.
32 Farley and Roberts, *Edgelands*, 9.
33 Ibid., 13.
34 John Brown, *Description of the Lake and Vale of Keswick* (Newcastle, 1767).
35 Farley and Roberts, *Edgelands*, 188.
36 See: Dylan Evans, *An Introductory Dictionary of Lacanian Psychoanalysis* (Routledge: London,1986), 84.

Bibliography

Barthes, Roland. *Mythologies*. London: Jonathan Cape, 1972.

Bormann, Herbert, Diana Balmori and Gordon T. Geballe. *Redesigning the American Lawn: A Search for Environmental Harmony.* New Haven and London: Yale University Press, 2001.

Brontë, Charlotte. Letter to James Taylor, 6 November 1850.

Brown, John. *Description of the Lake and Vale of Keswick.* Newcastle, 1767.

Burke, Edmund. *A Philosophical Enquiry into the Sublime and the Beautiful.* Oxford: Oxford University Press, 1998.

Castoriadis, Cornelius. *The Imaginary Institution of Society*. Cambridge, MA: MIT Press, 1997.

Claeys, Gregory. *Searching for Utopia. The History of an Idea.* London: Thames & Hundson, 2011.

Corner, James. 'Terra Fluxus', in *The Landscape Urbanism Reader*, edited by Charles Waldheim. New York: Princeton Architectural Press, 2006.

Defoe, Daniel. *A Tour Through the Whole Island of Great Britain.* Exeter: Webb and Bower, 1989.

Evans, Dylan. *An Introductory Dictionary of Lacanian Psychoanalysis.* Routledge: London,1986.

Farley, Paul and Michael S. Roberts. *Edgelands: Journeys into England's True Wilderness.* London: Jonathan Cape, 2011.

Foucault, Michel. *Discipline and Punish.* London: Penguin, 1991.

Gilpin, William. *An Essay on Prints.* Cadell and Davies: London, 1802.

Hiskes, Richard P. 'Missing the Green: Golf Course Ecology, Environmental Justice, and Local "Fulfilment" of the Human Right to Water.' *Human Rights Quarterly* 32, no. 2 (2010).

Hogarth, William. *The Analysis of Beauty.* Menston: Scolar Press, 1971.

Hussey, Christopher. *The Picturesque: Studies in a Point of View.* London and New York: G.P. Putnam's Sons, 1927.

Jönsson, Erik. 'The Nature of an Upscale Nature: Bro Hof Slott Golf Club and the Political Ecology of High-End Golf.' *Tourist Studies* (2015).

Le Corbusier. *Precisions: On the Present State of Architecture and City Planning.* Cambridge, MA: MIT Press, 1991.

Mabey, Richard. *The Unofficial Countryside.* London: William Collins Sons and Co., 1973.

McHarg, Ian L. *Design with Nature.* Garden City, NY: Natural History Press, 1969.

Milesi, Cristina, C. D. Elvidge, J. B. Dietz, B. T. Tuttle, R. R. Nemani, and S. W. Running. 'A Strategy for Mapping and Modeling the Ecological Effects of US Lawns.' *Journal of Turfgrass Management* 1(2005): 83–97.

Mouzakitis, Angelos. 'Autonomy and Authenticity. On the Aporetic Nature of Time and History: Castoriadis—Heidegger', *Critical Horizons* 7, no. 1 (2006).

Nordh, Helena, Caroline M. Hägerhäll and Terry Hartig, 'Urban Nature as a Resource for Public Health.' In *The Routledge Companion to Landscape Studies*, edited by Peter Howard, Ian Thompson and Emma Waterton. Abingdon: Routledge, 2013.

Price, Robert. *Scotland's Golf Courses.* Edinburgh: Mercat Press, 2002.

Robbins, Paul F. *Lawn People: How Grasses, Weeds, and Chemicals Make Us Who We Are.* Philadelphia: Temple University Press, 2007.

Sartre, Jean-Paul. *L'Imaginaire: Psychologie phénoménologique de l'imagination.* Paris: Gallimard, 1940.

Searle, John R. *The Construction of Social Reality.* London: Penguin, 1996.

Shoard, Marian. 'Edgelands.' In *Remaking the Landscape: The Changing Face of Britain*, edited by Jennifer Jenkins. London: Profile Books, 2002.

Sinclair, Iain. *London Orbital.* London: Penguin, 2003.

Steinberg, Ted. *American Green: The Obsessive Quest for the Perfect Lawn.* New York: W.W. Norton and Co., 2006.

Taylor, Charles. *A Secular Age.* Cambridge, MA: Belknap Press of Harvard University Press, 2007.

Thayer, Robert L. *Gray World, Green Heart: Technology, Nature and the Sustainable Landscape.* New York: John Wiley and Sons, 1994.

Tolkien, J.R.R. *The Hobbit.* London: George Allen and Unwin, 1937.

Wheeler, Kim and John Nauright. 'A Global Perspective on the Environmental Impact of Golf'. *Sport in Society* 9, no. 3 (2006).

Williams, Raymond. *The Country and the City.* London: Chatto and Windus, 1973.

22

WHOSE CITY IS IT?

Public space as agent of change in marginalized settlements in Buenos Aires

Flavio Janches

Informal urbanism is outpacing all other forms of urban development. One third of the entire urban population, nearly one billion people, live in informal settlements around the world. With no possibilities for integration or urban mobility, this condition of marginalization is strengthened at the frontiers of urban space, where social inequities and segregation are exacerbated. Today, this marginalization is getting worse. The problem relates not only to the lack of access to income, but also to the social stigma that marks those unable to integrate into formal systems of urbanization and modernization. At the same time, however, informal urbanism can generate survival strategies that help to consolidate unplanned processes that may further social interaction, collective identity, and cultural integration. This common or shared ability to integrate is vividly expressed through daily experiences of actual or potential shared urban public spaces, thus emphasizing collective identities that honour the already existing—that which is already there. Using public space as a tool for urban transformation is therefore a way to improve socio-territorial integration in these marginalized environments. This chapter is an inquiry into the problem of urban marginalization, and it also aims to provide design parameters for sustainable social and spatial urban integration, drawing on research in landscape architecture/urban design carried out in informal settlements in Buenos Aires. It does so by recognizing the collective and inclusive capital inherent in public space, and discusses the potential role of landscape architecture/urban design and its inherent agency in a more inclusive understanding and experience of the city, driven by social and spatial urban integration.

The 'otherness' of informal urban settlements

Shared social life has significant value as a relational dimension amongst different people, be they socially, ethnically, or culturally diverse. It is through such interdependency with and awareness of the 'other' that a community expresses its sociocultural identity. According to Argentinean anthropologist Néstor García Canclini, this separation or distance from the other (understood in terms of cultural interaction) defines a set of social processes of production, circulation, and consumption of social signification.[1] In his book *Diferentes, desiguales y desconectados: Mapas de la interculturalidad* (*Different, Unequal and Disconnected: Maps of Inter-Culturality*), García Canclini relates these ordinary and quotidian interactions to a unique way of building cultural significance.[2] 'Difference' therefore represents the potential for an urban edification process, since it implies a way of establishing socio-territorial standing that brings with it the ambiguity of stigma together with the richness of collective urban life. As a physical expression of shared community values, urban space is not just a significant intrinsic feature; it also acquires significance extrinsically, that is to say in interactions, confrontations, and discriminations in areas that lack urban space.

In the case of Buenos Aires, this system of diverse urban spaces with specific meanings for specific groups is recognizable in the dialogue among its multiple *barrios* or neighbourhoods. Through ordinary and extraordinary events, and through their spatial particularities, Buenos Aires' *barrios* promote alternative processes for social participation and a sense of belonging within the community. According to anthropologist Ariel Gravano, the *barrios*' potential ability to integrate enables the delimitation of a dialogue with the other through a spatial definition of 'us'.[3] Thus, he argues, the *barrios*' inherent spatial condition 'brings along with it a particular way of inhabitation and collective living that highlights the existing positive values of primary systems of relationship such as tradition, authenticity, belonging, solidarity, and ideals'.[4] However, the process of differentiation can easily turn from definitions of otherness into real actions of marginalization of the different. Argentinean sociologist Javier Auyero's work on poverty and marginality (with added connotations of political ethnography and urban violence) recognizes that this spatial condition of difference also acts as 'an interaction process of discrimination and hostility, as a place for social discrimination'. For Auyero, these marginalized spaces then 'cease being those recognized as *other*, and enter into processes of radical segregation that can even culminate in racial stigmatization of people and places.'[5] Understood in this way, cities can in effect also establish separation or distance from the 'different'. In his book on social fabric, Brazilian architect and urbanist Vinicius Netto recognizes this behaviour of urban space to separate and generate spatial segregation and social distance. For him, 'urban space is usually seen as both the *materialization* and the *medium of segregation*.'[6]

In the city of Buenos Aires, informal settlements are one of the strongest manifestations of this social and physical segregation. This is evident in the precariousness of their territory, which is a result of both economic polarization and social rejection. Buenos Aires' informal settlements, locally called *villas*, are defined by their infrastructural deficiencies as well as by the stigma associated with their territories and populations. Mabel Nélida Giménez and María Elena Ginobili from the Universidad Nacional del Sur in Argentina explain the pejorative connotation of these informal environments as based on stereotypes produced by other sectors of society:

> The *villa* is for others a hidden, chaotic, and dangerous no-man's territory. It represents a shameful place, a permanent symbol of social inequality. The poor are those who live in certain neighborhoods and, considering their good moral values, deserve to be helped. But the *villeros* (those who live in the *villas* or slums) are the disreputable poor: those living in conditions of poverty, but above all those living outside the law [...] and because of this they are the ones who are suspected, discriminated, and segregated from society.[7]

Nevertheless, despite being places self-defined by informality and the absence of lawful intervention, the *villas* also embody a socio-territorial structure that supports an internal system of social and cultural relationships, both within and among its communities, just like in the formal city. The shared public spaces of these informal settlements could offer this marginalized sector of society an opportunity to build their own collective: based on their own informal rules, their everyday life, their coexistences, their cohesion, their tolerance, and their particular way of being part of the city.

It is specifically this demeaning condition that presents itself as an empowering duality when reinterpreted by the inhabitants themselves: a condition that offers the potential for real and symbolic appropriation. On the one hand, and for a particular social group, the *villa* or slum is a place of fear (due to the high rate of violence and crime associated with it), but on the other hand, for its inhabitants, it is exactly 'the space where their own civil power could be exercised'.[8] This sociocultural behaviour of appropriation, made possible by the inhabitants' organizational ability and collective identity, enables the inhabitants themselves to adapt to living in precarious conditions, to build and preserve boundaries with other neighbourhoods, family, and friends, and to take advantage of resources provided by outside organizations (governmental or otherwise).[9] Their informal systems of mutual aid and reciprocity are thus key aspects of their survival strategy—particularly given that it is through this 'social capital' (in Pierre Bourdieu's sense) that the networks of institutional relationships (of material and symbolic exchange) are defined.[10] These resources, which are integral to the inhabitants own culture, are highly useful for them to preserve the continuity of family life, and are directly expressed in their use of public space and in the materialization of their housing.

The survival strategies adopted by the inhabitants of the *villa* also range from the symbolic and cultural appropriation of *villero* identity to the embodiment of this identity in cultural products and community-based institutions or associations. Soccer clubs are a significant example. The practice of soccer in a *villa* club demonstrates how an everyday event or activity may be transformative, in that it may effectively be a tool to reaffirm community inclusion and belonging. In the context of a club, soccer helps neighbours to meet and interact around a common activity. This facilitates more complex forms of organization and the reinforcement of group or collective identity. Apart from soccer clubs, other key community-based institutions, such as mothers' groups or networks and neighbourhood committees, as well as religious, political, or charity associations, enhance the bonds and fabric of social relationships. This is why the loss of a *villa*'s social solidarity networks, produced mainly by changes in their habitat, is so traumatic for the *villero*. Community-based activities are generally consolidated in shared spaces, and are characterized mainly by their ability to build representative cultural spaces of social cohesion. This is why conceiving *incumbent upgrading* strategies (where there is no displacement) that build upon these informal tactical actions (to enhance these spaces for further inclusion and belonging) could be a possible way to stimulate, through a network of new or existing public spaces, the growth of planned and spontaneous modes of socio-territorial development and integration.

Public space as a catalyst for socio-territorial urban integration

As indicated above, differences—where social agreements need to be negotiated—play a double role. On the one hand, the overlapping of (differences in) meanings separates people according to particular identities; on the other hand, it improves the city's ability to bring about greater sociocultural integration, thanks to the diversity of experience. Dialogue between differences is therefore a key element for transformative and integrative urban strategies. As French philosopher Henri Lefebvre has pointed out, difference can describe urban lifestyles, but it can also enable the city to reinvent itself.[11] Dutch political scientist Maarten Hajer and urban sociologist Arnold Reijndorp also hail the importance of differences in order for 'an antidote to stereotyping and stigmatization'[12] to emerge, through the presence of and confrontation with the 'other' and the other's cultural manifestations. The question is, then, how can urbanism, landscape architecture, and urban design, as disciplines, generate these dialogues, in order to avoid the fragmented, marginalized, and stigmatized urban conditions of informal settlements? One alternative way is by generating an interdependent system of community or collective domains that foster a dialogical urban synergy between cultural references and symbolic identifications. Open urban spaces, as places of meeting and interaction between people and events (ordinary and extraordinary), are the physical expression of this potential integrative process between differences. According to Lefebvre, the production of these urban spaces can be understood in three related dimensions. He calls them '*formants or moments* of the production of space' and characterizes them as 'doubly defined and consequently doubly named as well. On the one hand, there is the triad of *spatial practice, representation of space, and spaces of representation*; and on the other, *perceived, conceived, and lived space*.'[13] In this way, the complexity of the lived city, for example, comprising relations 'both *legible and illegible, visible and invisible*',[14] identified by multiple sociocultural signs and symbols, could be defined by a heterogenic and interdependent network of places for intense cultural interactions. This understanding of the ability of urban space to organize or define cultural change recognizes the link between people and site, and therefore, in my view, becomes an opportunity for an integrative open-space design intervention and, even more, a method for socio-territorial urban integration.

Various international examples show how the use of systems of public space are a key element for negotiating differences when consolidating social agreements for urban integration processes. The urban transformation initiated during the 1980s in Barcelona, Spain, clearly demonstrates the use of public space as part of an urban vision focused on integration and non-exclusion. This strategy of 'making city in the city' was based on the positive impact that the construction of public space at all scales had on the degraded environment. The methodology of city construction was based on a polycentral model that re-evaluated the public sphere: as urban habitat, as quality of life, and as a dialectical space between city and neighbours.[15] Another intervention providing compelling evidence of the power of public spaces (as places for social cohesion and integration) in the recovery of degraded areas of the city can be found in the recent experience in Medellin, Colombia. Since 2004 the municipality of Medellin has developed a new model of public space management. Through cultural, educational, and social development, a network of urban initiatives addresses the physical, social, and institutional dimensions of the city. It involves actions on public spaces, community infrastructure, housing programmes, and support for institutional projects. A third internationally recognized example is the Amsterdam playground network, planned by the municipality of Amsterdam, the Netherlands, and designed by architect Aldo van Eyck in the post-war years. Taking into consideration the ability of public playgrounds to achieve a connection between people and urban place, Van Eyck's project established an urban transformation and integration process for post-war Amsterdam through activities accessible to people of all ages and from different cultural backgrounds. Making an intervention of a polycentric, interstitial, and participative character, the network consists of approximately 700 playground areas that strengthened the sense of the collective and the community in the devastated socio-territorial context of post-war Amsterdam.[16]

In relation to this idea of the playground as an opportunity to foster dialogue and enable integration within a community, a recent example is the Villa Tranquila project in Buenos Aires. The project was studied in different academic institutions (Architecture Academy of Amsterdam, Harvard University Graduate School of Design, and Universidad de Buenos Aires) and supported by the Dutch PlaySpace Foundation and the Municipality of Avellaneda. The proposal consisted in building shared community space with the aim to integrate the *villa*, both within its own diverse communities and with the rest of the city. To capitalize on the existing sense of community, BJC Architects and Max Rohm Architect, devised an urban design strategy defined not just by specific architectural interventions, but also by its potential future modifications, expansions, and adaptations depending on the further growth of the *villa* and its possibilities for urban integration. The novelty of the strategy was that it defined scenarios and ways of programming, rather than providing fixed guidelines for specific planning stages. The final result was thus defined according to the way the community adapted the proposal to their own expectations and daily habits as the project evolved. This allowed not only for changes in places where specific projects had been built, but also for adaptation to unforeseen transformations within the surrounding areas. Based on what García Canclini called 'micropolis', this project reflects how adaptive and flexible models of urban intervention offer a possible alternative path towards socio-territorial integration for areas and societies struggling with ongoing marginalization processes.

Public space design components for socio-territorial urban integration processes

Existing socialization networks, dynamics of daily life, and cultural signification are key parameters for any incumbent upgrading urban strategy driven by socio-territorial integration. Drawing on the experience of Villa Tranquila in Buenos Aires and its ability to generate social integration and cohesion, the aim here is to present premises that could guide others in this particular urban approach to the socio-territorial integration of marginalized communities. One possible transformative process, based on the continuation of evolutionary synergies, is to modify existing trends of segregation by following an interdependent system of daily spatial dialogues, negotiations, and agreements. This enables a process that is based on facilitating physical connection, permeability, and interaction. It is therefore a process that is imbued with diverse and integrative public interactions. It entails multiple cross-use territorial intersection networks, both within a neighbourhood and with the rest of the city. These interdependent forms of territorial appropriation could improve opportunities to strengthen sociocultural cohesion through multi-actor and multiscale networks related to urban space. With the appropriation of 'what is available', this network of complementary activities can intensify and provide greater interaction and connection. It can also stimulate unforeseen dynamics to enhance neighbourhood identity and a sense of inclusion and belonging among inhabitants.

One of the challenges is to analyse the various public spaces through an existing sociocultural framework, that is to say an understanding of everyday life experiences and collective activities, in order to organize and establish the coherence of a strategy. Quotidian places have the potential to support integration due to their ability to open alternative forms of socio-territorial arrangements with others;[17] therefore agreements need to be forged to reach a common strategy. Future concentration and integration of economic, cultural, sports, and other recreational activities is a way to reorganize an unbalanced distribution of local development and infrastructure services, and to empower these collective activities and urban services through equitable distribution and access for all. In doing so, it addresses the communities' particular modes of shared and integrated urban life. It is equally important to consolidate access to each of these public activities and their connection with similar or complementary activities. Furthermore, it is possible to motivate transformation in both defined public spaces and the in-between areas defined as 'going-to' or 'coming from' these main activities. In this way, improving mobility and accessibility between community micro-environments also helps to avoid the existence of (inner and external) urban barriers, and promotes a permanent integrated and sustainable *unslumming*—in Jane Jacobs' positive understanding of the concept, as supporting incumbent upgrading, not displacement—as an urban process in continuous transformation. It is thus an adaptable and flexible urban space strategy that also constitutes a tactical proposal, because it aims to modify defined urban problems and also anticipates unforeseen reactions, whether stemming from social or environmental conflicts, when working with the community it most directly affects.

Another example of this strategy of socio-territorial integration for informal settlements is the work done in Buenos Aires for the area known as Comuna Ocho, where BJC Architects worked together with the Secretary of Habitat and Inclusion for Buenos Aires (a government agency of the Buenos Aires City Government). Today, this area, located on the south-west edge of the city, has multiple socio-territorial conflicts deriving from the coexistence of many physical barriers (mainly produced by the existence of several informal settlements with different community backgrounds), disconnected social housing projects, and a high rate of environmental degradation. The area also lacks community facilities, mixed uses, land regulation, and access to any physical or social infrastructures.

The Comuna Ocho strategy had two goals: to define an integral and integrated urban transformation process, and to devise a method that could be replicated in places with similar socio-territorial conflicts. In this case, the strategy framework was defined by four starting premises. First, an understanding of existing socio-spatial experiences, given that the opportunity for integration rests in the quotidian life and cultural signification of the neighbourhood itself. Second, the development of a new or existing public space network, given that the opportunities for community interaction depend on the coexistence of spaces for shared use. Third, a recognition of the instability of the socio-territorial integration process, while also recognizing its adaptability and flexibility as the key to consolidating a transformative process through time. Fourth, an understanding of community participation as an inclusive process, given that negotiated agreements may ground not only the design and construction of urban spaces, but also the long-term sustainability of the transformative process.

Therefore, exploring community consensus was the starting point from which the informal inner organization could be modified and transformed into multiple integrative urban structures. The value of the diagnosis helped to determine the transformative opportunities in each of the components analysed, and also to define the general coherence of the intervention proposal. The main challenge was to analyse the area from a territorial point of view, as well as in relation to its sociocultural, economic, and institutional potential for growth and development. Social networks, everyday life experiences, and cultural signification thus provided the greatest opportunities for the transformative strategy to influence change in conditions of social marginalization and physical fragmentation. According to this criterion for analysis, the strategy was defined by two main urban systems that defined territorial occupation: public space and movement dynamics. The combination of both systems allowed the achievement of (sociocultural) programmes and (physical) projects that led towards an inclusive and integrative urban process. Through the improvement of public spaces, the strategy aimed to empower the existing sociocultural character. It did so by organizing, through multiple spaces for meeting and interaction, a balanced distribution of activities and urban services. Besides optimizing the mobility and access system, the strategy also aimed to integrate these designed spaces of social interaction within a new multilayered network that not only allowed the improvement of connectivity, but also avoided the fragmented organization of the inner and external city connections.

Based on existing sociocultural conditions, the integral socio-territorial transformative strategy was guided by three main components. First, a multifunctional and multiscale centrality network; second, a multimodal mobility and accessibility system; and third, axes for programming agreements.

The result was a multidimensional public space network that articulates both a new territorial distribution of collective sociocultural activities, and an integrative and exchange connection system that articulates interactions between neighbourhoods within Comuna, and between Comuna and the city. The transformation of the area into this new polycentral system was a way to promote the integration of the area, from both community and government perspectives. This integration was made possible through a balance of proposals that related to accessibility and mobility, as well as to sociocultural activities that involved institutional, community, and economic considerations. Given the potential of each of the proposals as part of the interdependent system, the adaptable and flexible method enabled not only an understanding of the social and physical transformation strategy, but also a consideration of design possibilities with alternative tactical implementations. Finally, this way of understanding a possible socio-physical transformation, which takes into account an existing community's sociocultural character, is also determined by time as a fundamental design parameter. Time allows for possible modifications according to unforeseen socio-territorial, political, or economic conditions that arise throughout the process.

This specific strategic proposal for Comuna Ocho aimed to consolidate an urban development and transformation process through multiple actions that would enable social and territorial integration. The sequence of interventions, with differentiated and complementary scales of projects and programmes, reinforced the existing potential for regeneration and integration within and among the communities of Comuna. Although the strategy in general seeks transformation through specific activities in particular spaces, the main goal of the socio-territorial integration process is to strengthen existing or potential networks of socialization. As such, the strategy can support social behaviour in favour of sociocultural recognition and collective identification, given that communities identify with their urban spaces and participate in their potential for change. While each intervention deploys specific tactics according to the diverse particularities of a place, they are all intent on enhancing both existing and potential synergies for socio-spatial integration, as well as the stability and permanence of its transformative process. As mentioned, the other fundamental characteristic of this strategy was its adaptability to support possible social, economic, and cultural changes generated by the transformation process itself. This condition of flexibility allowed the strategy to organize the process around the community's own expectations, given that proposed programmes and projects inevitably impacted on their actual reception in desired future directions. From this perspective, the proposed actions were also a way to generate collective ideas that responded—in many cases unconsciously—to the demands and desires of each community involved. Having understood the local and subjective values, each project could build, in an action/reaction dynamic, community symbols and an identity for Comuna Ocho and each of its participating neighbourhoods.

As a working model for the socio-territorial integration of Comuna Ocho, the strategy also generated guiding principles that, when understood as components, could be replicated in other contexts with similar conflicts and conditions of social exclusion and territorial fragmentation. The components that define this strategy are:

- The importance of understanding the social, spatial, and urban systems that organize the daily life and experience of the local communities.
- The formulation not of an unmodifiable physical model, but of a network of sociocultural activities or events that can promote, through their evolution and interaction, a process for continuous urban transformation.
- The recognition that the strength of community identity, and a sense of inclusion and belonging amongst the community and their environment, is grounded in public activities or events, and in places or urban spaces with symbolic power.
- The acknowledgement that public space is at the same time the tool and the purpose of the strategy, given that it is through the intensity of sociocultural collective activities that new forms of interaction, negotiation, and integration can best be articulated.

- The recognition that generating a flexible process, adaptable to unforeseen economic, social, or political limitations, must be inherent to the socio-territorial integration process, because it is the only way to consolidate the permanence of an iterative process.

This strategic approach is defined by its ability to articulate and promote urban spaces as spaces of sociocultural experience. The main goal of this urban intervention model is thus to promote continuous community involvement and development. It does so not only through the quality of new physical public spaces, but also—as Jordi Borja, the Catalan urban geographer, points out—through the intensity of the social relationships they facilitate, their potential to create and strengthen group interactions, and their ability to encourage symbolic identification, expression, and cultural integration between people and place.[18] This brings us back to our initial question: whose city is it? This sustainable social and spatial urban integration strategy necessarily entails a recognition of the significance of the collective in public urban space, and the potential role of landscape architecture/urban design in an ongoing process of socio-territorial integration. In this context, design has agency in changes towards an iterative, and more inclusive, understanding and experience of *our* city.

Notes

1 Néstor García Canclini, *Diferentes, desiguales y desconectados: Mapas de la interculturalidad* (Barcelona: Gedisa, 2014).
2 García Canclini, *Diferentes.*
3 Ariel Gravano, *Antropología de lo barrial: Estudios sobre producción simbólica de la vida urbana* (Buenos Aires: Espacio, 2003), 12.
4 Ibid., 42.
5 Javier Auyero, 'Claves para pensar la marginalización', in *Parias urbanos: Marginalidad en la ciudad a comienzos del milenio*, ed. Loic Wacquant (Buenos Aires: Manantial, 2007), 20–1.
6 Vinicius Netto, *The Social Fabric of Cities* (New York: Routledge, 2017): p. 17 (original italics).
7 Mabel Giménez and María Ginobili, 'Las villas de emergencia como espacios urbanos estigmatizados', *HAOL* 1 (2003): p. 77.
8 Alicia Ziccardi, quoted in Beatriz Cuenya, *Programa de Radicación e Integración de las Villas y Barrios Carenciados de Capital Federal* (Municipalidad de la Ciudad de Buenos Aires: Programa de las Naciones Unidas para el desarrollo, 1993), 11.
9 Mario Margulis, 'Las villas: Aspectos sociales', in *Hacia la gestión de un hábitat sostenible*, eds. Juan Manuel Borthagaray, María Adela Igarzabal de Nistal, and Olba Wainstein-Krasuk (Buenos Aires: FADU-UBA, 2006), 35–6.
10 Alberto Bialakowsky and Cristina Reynals, 'Hábitat, conflicto social y nuevos padecimientos', in *Social del hábitat y neoliberalismo*, ed. Seminario Internacional Producción (Montevideo, 2001) (working paper, n.p.).
11 Christian Schmid, 'Networks, Borders, Differences: Towards a Theory of the Urban', in *Implosion/Explosion: Towards a Study of Planetary Urbanization*, ed. Neil Brenner (Berlin: Jovis Verlag, 2014).
12 Maarten Hajer and Arnold Reijndorp, *In Search of New Public Domain* (Rotterdam: NAi, 2001), 13.
13 Schmid, 'Networks', 73.
14 Henri Lefebvre, *The Urban Revolution* (Minneapolis: University of Minnesota Press, 2003), 46.
15 Jordi Borja and Zaida Muxi, *El espacio público: Ciudad y ciudadanía* (Barcelona: Electa, 2003).
16 Liane Lefaivre, 'Space, Place and Play', in *Aldo van Eyck: The Playgrounds and the City*, eds. Liane Lefaivre and Ingeborg de Roode (Rotterdam: NAi, 2002).
17 Margaret Crawford, John Chase, and John Kaliski, *Everyday Urbanism* (New York: Monacelli Press, 1999).
18 Borja and Muxi, *El espacio público.*

Bibliography

Auyero, Javier. 'Claves para pensar la marginación'. In *Parias urbanos: Marginalidad en la ciudad a comienzos del milenio*, edited by Loic Wacquant. Buenos Aires: Manantial, 2007, pp. 9–29.

Bialakowsky, Alberto, and Cristina Reynals. 'Hábitat, conflicto social y nuevos padecimientos'. In *Social del hábitat y neoliberalismo*, edited by Seminario Internacional Producción. Montevideo: 2001. Working paper for an international seminar, n.p.
BJC Architects. 'Plan urbano integral'. In *Modelo Estratégico de Intervención Unidad Territorial de Intervención Urbana, UTIU 8 Norte working Report*. Buenos Aires: 2014. N.p.
Borja, Jordi, and Zaida Muxi. *El espacio público: Ciudad y ciudadanía*. Barcelona: Electa, 2003.
Crawford, Margaret, John Chase, and John Kaliski. *Everyday Urbanism*. New York: Monacelli Press, 1999.
Cuenya, Beatriz, *Programa de Radicación e Integración de las Villas y Barrios Carenciados de Capital Federal*. Municipalidad de la Ciudad de Buenos Aires: Programa de las Naciones Unidas para el desarrollo, 1993.
García Canclini, Néstor. *Diferentes, desiguales y desconectados: Mapas de la interculturalidad*. Barcelona: Gedisa, 2004.
Giménez, Mabel, and María Ginobili. 'Las villas de emergencia como espacios urbanos estigmatizados.' *HAOL* 1 (2003):75–81.
Gravano, Ariel. *Antropología de lo barrial: Estudios sobre producción simbólica de la vida urbana*. Buenos Aires: Espacio, 2003.
Hajer, Maarten, and Arnold Reijndorp. *In Search of New Public Domain*. Rotterdam: NAi, 2001.
Janches, Flavio. *Public Space in the Fragmented City: Strategy for Socio-Physical Urban Intervention in Marginalized Communities*. Buenos Aires: Nobuko, 2012.
Janches, Flavio and Rohm Max. *Urban Interrelations*. Buenos Aires: Piedra, Papel y Tijera, 2012.
Lefaivre, Liane. 'Space, Place and Play'. In *Aldo van Eyck: The Playgrounds and the City*, edited by Liane Lefaivre and Ingeborg de Roode, Rotterdam: NAi, 2002, pp. 24–47.
Lefebvre, Henri. *The Urban Revolution*. Minneapolis: University of Minnesota Press, 2003.
Margulis, Mario. 'Las villas: Aspectos sociales'. In *Hacia la gestión de un hábitat sostenible*, edited by Juan Manuel Borthagaray, María Adela Igarzabal de Nistal, and Olba Wainstein-Krasuk, 33–49. Buenos Aires: FADU-UBA, 2006.
Netto, Vinicius. *The Social Fabric of Cities*. New York: Routledge, 2017.
Schmid, Christian. 'Networks, Borders, Differences: Towards a Theory of the Urban'. In *Implosion/Explosion: Towards a Study of Planetary Urbanization*, edited by Neil Brenner. Berlin: Jovis Verlag, 2014, pp. 67–81.

23
KHÔROGRAPHOS
Space-scripting

Michael Tawa

Two questions frame this inquiry: how can the city enable citizens not merely to live, but to live well; that is, not only to find their way in a utilitarian sense, but also in a profounder, more enduring and engaging sociocultural sense? How can the city provide not only the physical amenity necessary to a liveable environment, but also the mental indexes and contexts fundamental to human life, to life that connects person to community, and that connects both person and community to place? Pivotal here is how the city might enable the construction of situated identities: individual citizens and collectives that are collocated in space and time, in relation to particular kinds of spatialities and temporalities.

Evidently, this prompts a series of questions around history and of memory. How might the city make possible multiple kinds of recollection—remembering one's own situatedness or displacement, one's own personal history, but also the histories and archaeologies of place? How might the city enable the recuperation of such human and territorial registers? How might the city itself become an apparatus or enabling infrastructure for such recuperation?

One way of thinking about history, memory and recuperation in relation to the city is through the theme of narrative. What stories does a city tell? How does it recount those stories? What spatial tactics are mobilized and deployed to narrate them? How are they received, inscribed in and retained by urban fabrics? What role do they play in how the city constitutes itself as a place? In other words, what relationships might be construed and built between remembrance as a foundational human need, narrative as a foundational human construct, and the spatial and material fabric of cities—their gardens, streets, squares and buildings—in order to choreograph and script civic life and civil society at the intersection of human, urban and environmental dimensions of place?

While the inquiry will here be situated within a Western framework, cross-cultural references will also be made—for example, to Chinese landscape practices and the concept of Country in Australian Aboriginal culture—so as to provisionally scope out the importance of narrative to place; to signal parallel, resonant indices that can challenge and enrich Western perspectives; and to serve as a reparative counterpoint to contemporary urban trends pointing to a generalized virtualization and disembodiment of civic life.

If, with the promulgation of virtual communities, the urban commons can no longer be situated within the physical coordinates of particular concrete spaces, and its formation can no longer take place within the temporality of particular times—and if this were to be considered problematic or risky for human-being—what agency and critical role might be taken up by practitioners of the designed environment? Would a practice of resistance to these inevitable trajectories be possible? How might it be framed and played out? How could landscape, urban and architectural design enhance the capacity of the city to act as a common ground and receptive support for choreographing political life? In what ways might the city counteract the virtualization of experience and its disaggregation from material contexts? Are designed environments capable of enabling deep engagement with place?

To conceptually frame these questions, I have drawn from numerous philosophical and cultural sources in what might appear to be a rather syncretic manner. My general approach is to bring together as diverse a range of material as is prompted by the thematics of a question, but to do this in such a way as to foreground parallels and attend to emergent possibilities arising in the interstices and overlaps. My goal is not to produce any kind of systematic veracity founded on incontrovertible evidence, but rather, as a designer attending to the tectonic possibilities of assemblage, to produce consilient neighbourhoods of sense that resonate without necessarily amounting to any singularity. I also make extensive use of etymology. Again, this is not geared to legitimation or validation—I am not after originary meaning. Likewise, I am more interested in how multiple meanings, strangely imbricated in words, can be unpacked to form semantic landscapes that can be traversed, or semantic threads that can be woven in multiple ways to produce multiple registers of sense around the key themes and problematics of the inquiry. In that sense I proceed slowly and tentatively, as a bricoleur caught up in a process of making, fabricating and confabulating. Ultimately this is the work of storytelling—whether scripted through cultural, architectural, landscape or urban practices.

A major contrast in what follows is between the city as a virtual environment for civic life, mediated by computational systems and informatics, and the city as a concrete socio-spatial and socio-temporal environment whose actual, formal and material presence contributes to the constitution and practice of that civic life. My contention is that virtualized civic life runs the risk of excluding and forgetting the circumstantial, of devaluing the earth, of promoting a kind of suspended, disengaged existence—and that carries significant implications for civic life, as it does for a public domain that might nourish it. If, with ubiquitous computing, one can be anywhere at any time, then one is effectively nowhere, and the situational can then cease to matter.

Between virtual and actual there is an ambiguous borderline. For that reason, I have drawn several themes from post-structural philosophy to help mobilize a counterargument that avoids oppositional thinking. Gilles Deleuze and Félix Guattari's couplet territorialization/deterritorialization is useful to broach the double gesture of being in place and of displacing oneself simultaneously and interminably—whether virtually or concretely.[1] Likewise, Giorgio Agamben's pivotal themes of indiscernibility and indeterminacy challenge dominant oppositional thinking that pits city against country, human against animal, architecture against landscape, inside against outside, private against public, and so on. In this respect, what is interesting is less the definitive borderline and non-communication between such antinomies—between city and country, for example—than the porosity of their interface and their transactional, choreographic capacity.

The motif of *khôra*, space—in Plato, then in Heidegger and Derrida—is useful to conceptualize space as radical receptivity and capacity for affordance, rather than in terms of its normative conception as emptiness and deficiency. Such a concept of space—not void, but full of as yet unexploited or unexercised possibility—can be clarified through a recurrent concept in Giorgio Agamben's thinking: that of potentiality or latency.[2] The question then arises as to how such implicit potentiality might be incorporated into designed environments so as to become readily available and mobilizable.

Another critical motif is the tight relationship between space and narrative, between place and storytelling. Spaces are known and remembered in relation to what took place there; to the events that render them memorable and remain there, harboured and suspended, awaiting recollection. Derrida's reading of *khôra* in relation to receptivity, naming, inscription, memory and narrative[3] triggers important questions for the city and its urban and landscape settings. How can *khôra*—as the collective, interstitial zones of a city: cavity, room, courtyard, garden, public square, park, natural reserve or 'wilderness'—be construed so as to function mnemonically, that is, as a setting that enables recollection and recuperation of the sociocultural, urban, political and environmental memories of place, together with their future projections and reconstitutions? How can the landscape within, around and beyond the city be conceptualized, framed and constructed to provide the kind of radical receptivity that Derrida ascribes to *khôra*?

The collective spatial practices of a civil society might be spoken of as choreography: a dramatization or playing-out of place engagement, together with the stories and memories inscribed there. The coinage *khôrographos* then brings together Platonic and Derridean *khôra* with the notion of 'scripting' (Greek *graphein*)—in the theatrical, cinematic or more accurately kinematic sense of organizing how a story is to unfold: in this case, how place is made to unfold in the mind and in the embodied or kinaesthetic experience of the citizen. From 'script' and 'narrative', themes lead to allied notions of 'thread', 'weave', 'chord' and 'yarn', returning to the spatial cognate 'yard' as that girded or bound enclosure that delineates the *khôra* as site of potential engagement and recollection. In that sense, the urban commons, the public realm, becomes an apparatus, device or infrastructure that retains the memory of what may have been erased, demolished, forgotten or overlooked, and makes it available for recuperation.

Choral works: City and landscape

The archaic Greek city is founded on one tactic: territorial appropriation—that is, land grab. The ancient city—the *polis*—furnishes that tactic's enduring socio-spatial, political model by way of a boundary wall that traces a foundational architectonic gesture—the spacing-out and articulation of inside and outside, private and public, culture and nature. Two distinct regions, brought into relationship through a rupture that assures communication between them: the city gate. *Polis* is from Greek *pollus*, many, and the etymon *PEL, fold. Hence the city is a kind of pullulation in the landscape—a pall, maybe even an appalling irritation.[4]

The *polis*—city-state and basic unit of political sovereignty—comprises two key components: *astu*, the city proper as a formal assemblage or agglomeration; and *khôra*, agricultural land, hinterland and wilderness. The *astu* is the cultic, political, jurisprudential and economic centre of the *polis*. Its structure comprises an *acropolis*, or fortified hill; an *agora*, or public square and market; *temenai*, temples; and the *boulterion*, or place of the council. In this figuration, the city is conceived over and against the country—*astu* over and against *khôra*; and country is, literally, what counters the city, its counterpart.[5] The *polis* is the political ensemble of the city and its territory, remotely apprehended as an objective entity.[6] Seen proximately and from the inside by its own citizens (*politai*), the city is *astu*. Considered in opposition to its surrounding territory, the *agros*, *astu* is a site of the sophisticated, astute life of civilized society.

This macrocosmic set-up is paralleled by a microcosmic correlate at the scale of the individual house, the *oikos*. This is how Aristotle describes it:

> The natural unit established to meet all man's daily needs is thus *oikos*. [...] Then, when a number of *oikiai* are first united for the satisfaction of something more than day to day needs, the result is the village [*kome*]. [...] Finally the ultimate partnership, made up of numbers of villages and having already obtained the height, one might say, of self-sufficiency [i.e. *autarchia*]—this is the *polis*. It has come into being in order, simply,

that life can go on; but now it exists so as to make that life a good life. [. . .] So from all this it is evident that the *polis* exists by natural processes, and that it is natural for a man to live in a *polis*; he who is *apolis* [city-less, apolitical] is, by nature and not by chance, a being either degraded or else superior to man.[7]

The diremption between city and country is explicit in Ambrogio Lorenzetti's *Effects of Good Government in the City and in the Country* (Figures 23.1 and 23.2). The Platonic references of this work's antithesis, *Bad Government and the Effects of Bad Government in the City*, which Lorenzetti

Figure 23.1 Effects of good government in the city and in the country, Palazzo Publico, Siena, 1338. © Lorenzetti, Ambrogio. Wikimedia Commons.

Figure 23.2 Effects of good government in the city and in the country, Palazzo Publico, Siena, 1338 (detail). © Lorenzetti, Ambrogio. Wikimedia Commons.

painted opposite the former, are evident. The figure representing bad government shares the name Tyranny with Plato's 'worst city', corrupt Atlantis. In this companion work, the *urbs* or *astu* is forcefully delineated from the countryside or *khôra* by a monumental city wall and gates.

On one side of the wall, civil (or, in the case of *Bad Government*, uncivilized) society—whose conduct is tightly framed within the streets and alleyways of a dense urban context. On the other side, agricultural lands—appropriated, fully marked out and rendered productive (or, in the case of *Bad Government*, inoperative).

Plato's paradigmatic cities provide a useful indication of the relationship between *astu* and *khôra* within the *polis*. Plato's cities are not principally urban or architectonic figures; their major register and function are political and ethical. Each city is a geometrical, spatial and territorial assemblage corresponding to one of four political systems that Plato seeks to either valorize or devalue in line with the principles of his philosophy. These are—in order from the first to the last, from the most just to the least just—timocracy (Athens), oligarchy (Kallipolis), democracy (Magnesia) and tyranny (Atlantis).[8] In paradigmatic Athens,[9] the number of householders or holdings must be fixed at 5,040 'hearths'. Each holding is twofold—with one part near the centre of the village and another near the boundary:

> And they must also divide the twelve sections of the city in the same manner as they divided the rest of the country; and each citizen must take as his share two dwellings, one near the center of the country the other near the outskirts (τοῦ μέσου καὶ τὴν τῶν ἐσχάτων). Thus the settlement shall be completed.[10]

In this way citizens dwell between and actualize two forces: the centralizing attraction of a civic and domestic communitarian polity, and the individualizing distraction of a choretic, peripheral and centrifugal deterritorialization. Interesting here is not the opposition between *astu* and *khôra* in the *polis*, but the manner in which the dual set-up between centre or middle (*meson*) and periphery (*eschaton*), between town and country, between city and landscape, mobilizes a transactional, alternating relay that passes between them to constitute the rhythm of civic life.

Khôra, khortus, hortus: From chasm to garden

The Greek word for space is *khôra*, from the etymon *KHA, gape, gap, chasm, yawning abyss—cognate with *GHER, grasp, enclose. The common interpretation of the word is 'inhabited region, territory or land'[11]—but the range of allied terms and their implications for urban and architectural tectonics is considerable: cave, cube, cup, cap, cape, scape, shape, escape, landscape, capital, coping, chapel, chorus, court, cohort, yard, garden, horticulture, *hortus conclusus*, yawn, gawp, gape, chasm, *khaos*, guttural, gullet, goal, glut, grasp, garth.[12] Throughout this resonant lexicon, the general sense emerges of an open, interstitial region in the midst of a given circumambience—a clearing in a forest; a cave in a mountain; a public square in a city; a garden in a town; a courtyard in a house; a cavity in a wall.[13] *Khôra* is an undifferentiated site of production—what Plato referred to as the 'receptacle' (*khôra*) and 'nurse of all becoming' that receives all bodies (*panta dekhomenes somata phuseos*) and 'provides room for all things that have birth';[14] or, architectonically, the room that makes life possible; the porosity that alleviates mass and gives prospect; the doors that render architecture operative and the fissures that welcome deterritorialization.

In the loose-leaf paratext to his 1993 *Khôra*, Jacques Derrida noted that the book is the first chapter of a trilogy around the aporia of naming—specifically in terms of what 'the Timaeus names *khôra* (locality, place, spacing, emplacement) as that "thing" which appears to "give place"—without ever *giving* anything'.[15] The self-contradiction of *khôra* is that it functions as a generalized matrix of becoming that is always-already unaffected by or unidentified with whatever it gives place to. Hence *khôra* is both full (of what it makes possible) and empty (of any specific possibility that might arise in it). Derrida reads *khôra*—here in the demeanour of Socrates—as radical receptivity, as '*receptacle of everything* [*pandekhes*] that will henceforth inscribe itself':[16]

> His speech receives, in the event itself, more than it gives. As such he is *ready* and *prepared*, disposed to *receive* everything that will be offered to him. [...] Once again the question returns: what does *receive* mean. What does *dekhomai* mean. [...] *Dekhomai*, which will determine the relation of *khôra* to everything that it is not and that it receives (it is *pandekhes*, 51a), plays out across an entire scale of sense and connotation: receive or accept (a deposit, a salary, a gift), welcome, gather (*recueillir*), if not await.[17]

This receptivity of *khôra* determines it as a 'place of inscription',[18] identified with the receptacle (*ekmageion*) of becoming—the imprint-carrying medium that is 'always ready to receive' the stamp or seal of inscription,[19] as an 'always virginal'[20] wax tablet marked with 'inerasable characters'.[21] Echoing the gift of Mnemosyne (memory), mother of the Muses,[22] Derrida then moves from *khôra*, through *ekmageion* to the mnemonic and technical functions of writing and the archive—that is, to the scripting and inscription of narratives.

Yard/yarn

As that which receives and gathers, space (*khôra*) aligns with a key domiciliary zone: the courtyard. A dual sense is evident in this word 'courtyard' through the conjunction of the etymons *KHA, gape, gap, chasm and *GHER, grasp, enclose, gird. Cave, square, garden, yard—these are all species of porosity,[23] sequestered portions of spatial extension, mobilized and made habitable through gestures and tactics of appropriation. Such gestures are cosmogonic, clearly—since what is at stake is the creation of a world from what is not yet one: *kosmos* from *khaos*, form/essence from matter/substance, culture from nature, civilization from savagery, the tamed from the wild, humanity from animality, the political life (*zoon politikon*) from bare life (*bios*). A further sense is relayed by the connection between *khôra* and the girdle or limit that assures it—*khorde*, chord, yarn. If the association of yarn-twisting (thread-making) and yarn-spinning (telling stories) originates among sailors in the early nineteenth century, it is because weaving, splicing, binding, naming and measuring are apposite tactics germane to narrative and *khorography*.[24]

The close relationship between narrative, inscription and space—and the pivotal role they play in the constitution of place—can be gleaned from the Chinese landscape tradition. In Cao Xueqin's *The Story of the Stone*, a new garden—Prospect Garden—is described and its various features named.[25] The garden's spatial organization and the naming of its diverse vistas and elements show an intricate interdependence between narrative, rhetorical, and spatial or *khorographic* registers. Here, naming is not solely descriptive but allusive and *poietic*[26]—that is, naming produces and fabricates. Given in relation to traditional texts, the 'best name' is one with the most hidden, indirect, intricate and *ingenious* sequences of allusion. For example, the main mountain is named Little Censer 'after the famous Censer Peak in Kiangsi'. The name refers to another mountain, and doubles it in miniature. To make the allusion less direct, another strategy is to name it 'after the line in Chang Jian's poem about the mountain temple: *A path winds upwards to mysterious places*'. Here the name does not only refer to a mountain. It also defers to a path, leading elsewhere.

The various tropes of deferral involve play and pun by shuttling between explicit texts and allusive references. This peripatetic practice[27] rehearses a poetic tradition while construing the garden's spatial organization and scripting or choreographing its experience. As much a rhetorical landscape as an arrangement of mountains, streams, pavilions, bridges and plants, the garden provides a context for remembrance where sequence and juxtaposition of elements can trigger recollection. But the garden is also a *world*; and its design a practice of world formation. Landscape elements and settings—each with its own associations, propensities, ambiences and affects—are brought into strategic assemblage. Places are especially significant where a *conjunction* of conditions literally provokes or compels activities—'sipping tea', 'playing the *qin*', 'seeking inspiration', 'burning incense for sweet smell'. Each vista or prospect has a distinctive name that frames such conjunctions—for example, 'pear tree blossom in springtime rain', 'rushes in the winter snow'. Hence the garden is not merely a place of reflection or aesthetic meditation. It is a mnemonic apparatus for recuperating, recollecting and producing culture through the choreographing of courtly life; it is a civilizing memory-machine.

In Australian Aboriginal concepts of place, narrative is central to the constitution and reconstitution of subjectivity, situated through traditional kinaesthetic practices of storytelling and walking Country.[28] These practices are paralleled in Western Desert painting typical of the Ngaanyatjarra region around Patjarr and the Warburton Ranges. In *Yankaltjunkunya* (Figure 23.3), the Ngaanyatjarra painter Pulpuru Davies reiterates a well-known story: two totemic animals, emu and turkey, pursue each other across a land they are at the same time *producing*.[29] The landscape we see in the painting registers traces of demiurgic peregrination in the same way that the physical landscape functions as a mnemonic apparatus of memorialization—much like a book, library or archive. The land in turn conscripts human beings into an acculturated being-with-Country, comprising a diverse set of comportments, responsibilities and practices.

The painting functions across many registers. It is a map of the land that identifies specific locations, historically significant places or mythically significant events. Practically, *Yankaltjunkunya*

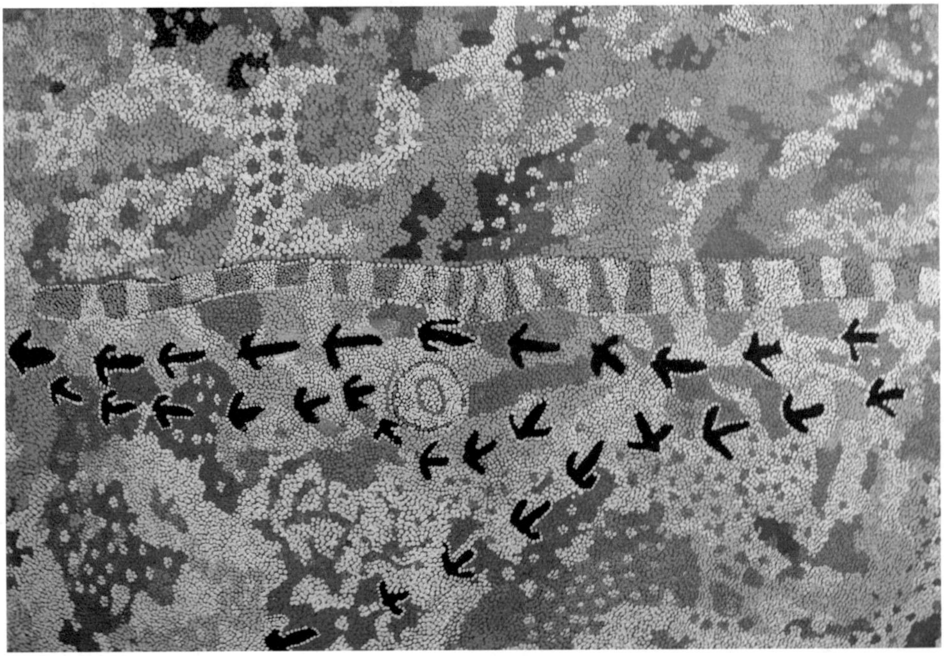

Figure 23.3 Yankaltjunkunya, 2000. © Davies, Pulpuru. Photograph by Michael Tawa.

functions as an aide-memoire for the landscape itself, and a trigger for the process of narrative production, storytelling and cultural sustainment. The delineation of components in the painting is left loose so that each narration might function as a distinctive interpretation and adaptation of a generic story. As some totems are shared across territorial borders between different groups, one painting or one narration will only ever represent one part of a much larger story within a constellation of narrative responsibilities. In that sense stories never totalize into singular authorized versions, but always remain open to interminable repetition, reiteration and variation. Worthy of retention here is the intimate connection assured by the painting between physical and mythical landscapes, between these and the purely graphic or cartographic traces in the work, and between the patterns of everyday living practices and the cosmographical tracery of world formation that the painting registers. The painting serves as an *enabling framework* that opens indefinite narrative and interpretative variations, but it is also a veil that dissimulates the whole story: a story that cannot be recounted to strangers—strangers to culture, but also those strangers to women that are their men, who possess their own forms of representation. It is precisely as a site of veiling-unveiling that *Yankaltjunkunya* draws its narrative force.

As a map, what matters in the painting is not its cartographic or artefactual dimensions, but its choreographic or space-scripting register. Free of perspectival depth, its overlaid patterns and strata foreground an archaeology of place—at once ecological, geological, social, spatial and mythical. It marks an inventory of ecosystems: familial and totemic lineage and association; circuits of peregrination and ceremony; cosmogonic trajectories; geographical patterns and forms. As such, the painting in fact produces the places it depicts and places the people who produce it. This performative or presentational rather than representational function determines stratifications not focused on aesthetics but on the *ethical* factors that ground community. This is the painting's depth and potency—not a formal or aesthetic spacing of layers, but a means of arraying Country as a resonant field in the midst of which one's being-in-common emerges.

Recuperation

If the classic *polis* is characterized by interiority—a bounded *astu* over against the expanse of a circumambient, appropriated *khôra*—the contemporary city must be its radical other, characterized by an extreme interpenetration of *astu* and *khôra* where interiority and exteriority, boundary and access, have become indiscernible and indeterminate. Countering the uni-centric, physically defined territories and textures of classic cities—with their city walls, gateways, districts, temples and marketplaces—contemporary cities are polycentric, rhizomatic networks of relays, with ambiguous boundaries that must be interminably negotiated and recast. Computational technologies are making possible ubiquitous, simultaneous presence in multiple locations at multiple times—or at least a simulacrum of presence—through a virtual evaporation of the circumstantial and the particular. Such virtualization transposes the *politeia* from concrete space and time to a non-place that is everywhere and at all times interchangeable. Yet being everywhere simultaneously also means being nowhere; or at least it means not being where one is, being always-already elsewhere, always-already absenting oneself so as to remain interminably indeterminable, always-already in touch so as to remain interminably out of reach. Alternatively, it means *being-divided*, so that where one is can be definitively uncoupled from the locational specificity, spatio-temporal coordinates and orientations of a physical place: the situated human body, the body of the earth, the body politic. At risk is a devaluation of place-embeddedness and time-embeddedness that in turn furnishes the grounds for a generalized sequestration from, disinterest in and disavowal of context and environment—consigning the latter to forgetfulness and the *politeia* to the disingenuous, surreptitious designs of a pervasive attention economy.

This kind of *khorographic* amnesia weighs heavily on the possibility of public space—of a public domain that is structurally and genuinely open to the unencumbered formation of communities, of situated beings-in-common. The advent of conurbations further erodes prospects, exacerbating the erasure of centres and boundaries in favour of extensive multi-nodal urban networks that permeate so many *terrains vagues*. As public domains are increasingly privatized, shifting from concrete to virtual spatialities and temporalities further disables human beings' capacities to engage deep connections to situated place. The formation of publics and communities—particularly publics and communities of resistance—is now largely practised within the virtual dimensions of social media. As Jordan Geiger contends in his 2015 survey *Entr'acte*,[30] citizens can no longer be considered subjects of the *polis* but rather actors and 'actants', with the capacity to form, dissolve and reform as necessary. On the other hand, the virtual realm is resolutely under the control of vested interests within the post-capitalist economy and its apparatuses. In this context, the conditions of freedom and enslavement within physical and virtual public domains become indistinguishable. Keller Easterling's afterword to *Entr'acte* is a sobering advance warning about the vigilance necessary to counter novel, incessantly reformulated dangers emergent 'in all concentrations of authoritarian power and all obstructions of information whether digital or spatial'.[31]

What, then, might constitute the conditions for a counteractive civic space of resistance, for an urban commons that might unclench and up-build a circumstantially engaged polity? A useful theme in this regard might be the idea of recuperation; as might be Derrida's reading of *khôra* as radical receptivity. The difficulty is to adapt and frame such ideas *tectonically* so they can function productively in the programming and design of landscapes, cities and buildings. What renders a public space or an urban garden recuperative? What renders it receptive? What kinds of zonings, boundaries, thresholds, levels, scales, frameworks, porosities, densities, transitions, nodes, relays, interactions, continuities, discontinuities, aspects, prospects, inclinations, declinations, revelations, occlusions and infrastructures might be incorporated as the fabric of a public domain so that it becomes capable of receiving life lived in its variegated complexity? How can the material conditions of designed environments receive, register and indicate the memories of a place and its future prospects? How can such recuperation not merely recollect and memorialize but also *produce* culture?

Contrary to Derrida's reading, receptivity in the context of design cannot be delivered through absence of determination, minimal specificity or open-endedness. Receptive places must be designed, which means that they must be designated, articulated, measured and materialized. Resilience—that is, the capacity to rebound, to be genuinely flexible and adaptable to multiple registers—demands of designed environments that they consist of multiple overlaid and simultaneous propensities, suitabilities and proclivities in order to give purchase to political life. Yet in any particular place—a landscape, a garden, a courtyard, a house, a street, a city—these propensities, while present and always available, must be retained in a state of latency, since it is in the density of unrealized potential, rather than in actualized potential, that *khôra* affords the highest degree of receptivity and gives access to the highest degree of recuperation.

In the designed environment, three *scales* of recuperation might be ventured: the recuperation of what has been definitively lost—to be managed by practices and apparatuses of recollection, mourning and grief; the recuperation of what has been erased, but leaves traces—to be managed by practices and apparatuses of mapping and assemblage; and the recuperation of what has been concealed, but is recoverable—to be managed by practices and apparatuses of procurement, modification and reconstitution. Likewise, three *registers* of recuperation might be ventured: the recuperation of environmental conditions that have been occluded—such as topography, hydrology, climate, ecology, species, processes and systems; the recuperation of political and sociocultural indices that have been overlooked—such as indigenous and non-indigenous, settler and immigrant histories, stories, practices and events associated with place; and the recuperation of designed environments that have been demolished or altered beyond repair—such as constructed landscapes, urban fragments, infrastructures and buildings.

The gathering of these scales and registers of recuperation implies a particular kind of practice in which research and assemblage alternate within a reiterative process to build texture and thickness into the environment being designed. Such an environment can then function as a score or script, open to multiple kinds of performance by the citizens who occupy and use it. The diverse registers that constitute the design do not totalize functionality to a singular script, to an explicit name. Rather, and because those registers exist in consilient discrepancy, the script will always lend itself to interpretation, and every interpretation will always be original. In such a scenario, design becomes a question of space-scoring or space-scripting—that is, of *khôrographos*: setting up and framing spatio-temporal settings so as to afford multiple, implicit possibilities of composition and infiltration, combination and performance. Playing these out, the citizen is drawn into a twofold process of recuperation: retrieval of a spatio-temporal order (external, objective) that at first appears ambiguous and unresolved, but that in time slowly gathers and begins to make sense; and retrieval of an individual engagement with place (internal, subjective) that pivots on personal memory and recollection.

By scripting space-time in such a way as to enable this kind of recuperation, design can promote deep, embodied and mnemonic engagement with place. At one level, such engagement counters the virtualization of being-in-common that characterizes contemporary political life. At another level, it keeps open the possibility of personal and collective investment in place through recollection and storytelling. As a *khorographic* practice, design shifts its field of operation from aesthetic and formal concerns to ethical concerns, from tectonics and technological tactics to issues of tact, solicitude and care. The implied infrastructural function of designed environments, or their equipmental status as apparatuses of affordance, must mean—as for every perfect instrument, whose function should be exercisable seamlessly and without impediment—that the ideal state of a designed environment can only lie in its effective withdrawal and disappearance: its own obliviousness and forgetting.

Notes

1 Gilles Deleuze and Félix Guattari, *A Thousand Plateaus: Capitalism and Schizophrenia* (London: Continuum, 1988).

2 Giorgio Agamben, *Potentialities: Collected Essays in Philosophy* (Stanford: Stanford University Press, 1999). See also Agamben, *The Open: Man and Animal* (Stanford: Stanford University Press, 2004).

3 Jacques Derrida, *Khôra* (Paris: Galilée, 1993).

4 Related words include Greek *pollos*, abundance; *polemos/pelomai*, warfare, polemics; *hoi polloi*, the masses; and Latin *pollutio*, defilement, contamination.

5 See Jean-Christophe Bailly, *Le Champ Mimétique* (Paris: Editions du Seuil, 2005), 228:

> The city conceives and constructs itself as an ordered and geometric clearing in the midst of a forest of branching and fleeing signs which continue to carry a dimension of dread. […] Greek religion can be understood in its ensemble and in its contrasts as the effort to render compatible these two distinct universes, that of the city entirely given to measure, and that of a world sensed as immense, dangerous and unmapped (*inarpenté*).

6 Michel Casevitz, Edmond Lévy and Michel Woronoff, '"Astu" et "polis," essai de bilan', *Lalies* 7 (1985).

7 Aristotle, *Politics*, 1253a. See also Jacques Derrida, *The Beast and the Sovereign, Volume 1* (Chicago: University of Chicago Press, 2011), 461.

8 All references to Plato are drawn from *The Collected Dialogues of Plato*, eds. Edith Hamilton and Huntington Cairns (Princeton: Princeton University Press, 1961). Plato's *Republic*, Books 8 and 9

describe the characteristics of each city, together with the corresponding dispositions of their citizens; see 543d–5c; 547e. See also Book 5 of Plato's *Laws*, 745b–e. Aristotle gives his version in *Politics*, Books VII and VIII.

9 Plato, *Critias*, 109b–12c.

10 Plato, *Laws* 745e: ἐσχάτων *(eschaton)* is from the etymon *EGHS, *ex-*, and can mean last, furthest, uttermost, most remote, out of/beyond, external to.

11 Susan E. Alcock, 'The Essential Countryside: The Greek World', in *Classical Archaeology*, eds. Susan E. Alcock and Robin Osborne (Chichester: Blackwell Publishing, 2012).

12 See Latin *cavus*, hollow; *chorus* and Greek *khoros*, circular dance, dancers, enclosed dancing ground; *khortos*, pasture, and Latin *hortus*, garden; Greek *khaos*, abyss, gape, what is vast and empty.

13 Martin Heidegger defines *khôra* as 'that which separates itself, deflection of all particular things, what effaces itself, that which thereby precisely admits something other and "makes room" for it (*Platz macht*)'; and *khaos* as 'the yawning (*das Gähnen*), the gape, that which rents itself in two. We understand *khaos* in close connection to an original interpretation of the essence of *aletheia* as the self-dilating/opening abyss (cf. Hesiod, *Theogony*)' Derrida, *Khôra*, 101–2.

14 Plato, *Timaeus*, 51e–52b. See Keimpe Algra, *Concepts of Space in Greek Thought* (Leiden: E.J. Brill, 1995), 79.

15 Derrida, *Khôra*, 1, 3 (my translation). For the formative agency of the paratext as something 'off-frame' yet pivotal, see Gérard Genette, *Paratexts: Thresholds of Interpretation* (Cambridge: Cambridge University Press, 1997).

16 Derrida, *Khôra*, 61.

17 Ibid., 62.

18 Ibid., 68.

19 Ibid., 69.

20 Ibid., 75.

21 Plato, *Timaeus*, 26b–c.

22 Plato, *Theaetetus*, 191a–e.

23 Architectural theorist Benoît Goetz develops the idea of porosity, using Walter Benjamin's writings on Naples to suggest that the anfractuous and perforate urban fabric is fundamental to enabling the communitarian, hospitable life of a vital polity. See *Théorie des maisons: L'habitation, la surprise* (Paris: Éditions Verdier, 2011), 113–36.

24 There are parallels in the Sanskrit *sutra*, which can variously mean thread, textile, suture, sew—and text, manual, aphorism, scripture, rule for ritual conduct. The central space of a Hindu temple, where the deity is installed, is called the *garbhagrha*, womb-house—or, more accurately, seed (*garbha*)-grasping (*grha*). For the sexual symbolism of this couplet see Stella Kramrish, *The Presence of Shiva* (Delhi: Motilal Banarsidass, 1988).

25 Cao Xueqin, *The Story of the Stone* (Harmondsworth: Penguin, 1986), 324–52.

26 In the sense of Greek *poiesis*, from *poiein* meaning to make, produce, articulate, versify.

27 David L. Hall and Roger T. Ames. 'The Cosmological Setting of Chinese Gardens', *Studies in the History of Gardens & Designed Landscapes* 18, no. 3 (1998); John Makeham, 'The Confucian Role of Names in Traditional Chinese Gardens', *Studies in the History of Gardens & Designed Landscapes* 18, no. 3 (1998): 187–210.

28 See my 'Place, Country, Chorography: Towards a Kinaesthetic and Narrative Practice of Place', *Architectural Theory Review* 7 (2002) and *Agencies of the Frame: Tectonic Strategies in Cinema and Architecture* (Newcastle upon Tyne: Cambridge Scholars Publishing, 2010), 57–9.

29 The definitive painted version by Pulpuru Davies, *Yankaltjunkunya* (1991), is in the archive of the Warburton Arts Centre, http://www.warburtonarts.com/english/collection-form.html (accessed 26 November 2016).

30 Jordan Geiger, *Entr'acte: Performing Publics, Pervasive Media, and Architecture* (New York: Palgrave Macmillan, 2015).

31 Ibid., 215.

Bibliography

Agamben, Giorgio. *Potentialities: Collected Essays in Philosophy*. Stanford: Stanford University Press, 1999.

Agamben, Giorgio. *The Open: Man and Animal*. Stanford: Stanford University Press, 2004.

Alcock, Susan E. 'The Essential Countryside: The Greek World'. In *Classical Archaeology*, edited by Susan E. Alcock and Robin Osborne, 124–32. Chichester: Blackwell Publishing, 2012.

Algra, Keimpe. *Concepts of Space in Greek Thought*. Leiden: E.J. Brill, 1995.
Aristotle. *Politics*. Perseus Digital Library. Last accessed 1 May 2017. http://www.perseus.tufts.edu/hopper/text?doc=Perseus:text:1999.01.0058.
Bailly, Jean-Christophe. *Le Champ Mimétique*. Paris: Éditions du Seuil, 2005.
Cao Xueqin. *The Story of the Stone*. Harmondsworth: Penguin, 1986.
Casevitz, Michel, Edmond Lévy and Michel Woronoff. '"Astu" et "polis," essai de bilan'. *Lalies* 7 (1985): 279–85.
Coomaraswamy, Ananda K. *Essays in Early Indian Architecture, Volume 1*. New Delhi: Indira Gandhi National Centre for the Arts, 1992.
Deleuze, Gilles, and Félix Guattari. *A Thousand Plateaus: Capitalism and Schizophrenia*. London: Continuum, 1988.
Derrida, Jacques. *Khôra*. Paris: Galilée, 1993.
Derrida, Jacques. *The Beast and the Sovereign, Vol. 1*. Chicago: University of Chicago Press, 2011.
Geiger, Jordan. *Entr'acte: Performing Publics, Pervasive Media, and Architecture*. New York: Palgrave Macmillan, 2015.
Genette, Gérard. *Paratexts: Thresholds of Interpretation*. Cambridge: Cambridge University Press, 1997.
Goetz, Benoît. *Théorie des maisons: L'habitation, la surprise*. Paris: Éditions Verdier, 2011.
Hall, David L., and Roger T. Ames. 'The Cosmological Setting of Chinese Gardens'. *Studies in the History of Gardens & Designed Landscapes* 18, no. 3 (1998): 175–86.
Kramrish, Stella. *The Presence of Shiva*. Delhi: Motilal Banarsidass, 1988.
Makeham, John. 'The Confucian Role of Names in Traditional Chinese Gardens'. *Studies in the History of Gardens & Designed Landscapes* 18, no. 3 (1998): 187–210.
Plato. *The Collected Dialogues of Plato*, edited by Edith Hamilton and Huntington Cairns. Princeton: Princeton University Press, 1961.
Tawa, Michael. 'Place, Country, Chorography: Towards a Kinaesthetic and Narrative Practice of Place'. *Architectural Theory Review* 7 (2002): 45–58.
Tawa, Michael. *Agencies of the Frame: Tectonic Strategies in Cinema and Architecture*. Newcastle upon Tyne: Cambridge Scholars Publishing, 2010.

24

TOWARDS NEW RESEARCH METHODOLOGIES IN DESIGN

Shifting inquiry away from the unequivocal towards the ambiguous

Kathryn Moore

Research in design, as in any subject, is 'a process of investigation leading to new insights, effectively shared'.[1] This simple definition makes you wonder why it is apparently so difficult to undertake, why design research remains 'highly contested',[2] weighed down with 'confusion and controversy'[3] and still fraught with the misunderstandings and misconceptions that Durling[4] identified more than a decade ago, when it was a relatively young, emergent discipline.

Arguing for a fundamental shift in the way we think about how we perceive, to deal head-on with the 'one big mistake' that Ryle derides as 'the absurdity of the Official Doctrine',[5] this chapter traces the contentious nature of design research directly back to the dichotomy between body and mind to which Ryle refers. Proposing that research in design should not really be any different from research in any other discipline does not suggest that design research is just another form of '"problem solving" or "information processing"'. It does tell us, however, that groundbreaking research in or through design is absolutely achievable, and that it can be critical, rigorous and 'brilliant in idiosyncratic freewheeling ways'.[6] The point is that research in design does not necessarily have to be scientific, and neither does it need to be based on rationalist views about the nature of intelligence, emotions, facts and values.

This chapter outlines a central premise that fundamentally redefines the relationship between the senses and intelligence,[7] and that has far-reaching consequences for our understanding of language, intelligence, meaning, the senses and subjectivity. A pragmatic and holistic approach to consciousness has been used as a tool to examine and reconceptualize the epistemology, pedagogy and function of design, and it is used here to re-evaluate some of the assumptions underlying practice-based research inquiry and research through design. Set here within the context of landscape architecture, it also has implications for other art and design disciplines, architecture, philosophy, aesthetics and education more generally.

Philosophical underpinning

This chapter is underpinned by research taking a pragmatic line of inquiry into the perceptual realm, drawing in particular on the work of the American pragmatists including William James, John Dewey, Hilary Putnam and Richard Rorty. Rorty brought pragmatism back into focus after several decades of decline with the publication *Philosophy and the Mirror of Nature.*[8] The radical redefinition of perception on which this chapter is based is presented in my book *Overlooking the Visual: Demystifying the Art of Design.*[9]

Since its emergence as an intellectual movement in the latter part of the nineteenth century, pragmatism's main thrust has been to question and debunk the metaphysical basis of disciplines. Cutting across the 'transcendental empiricist distinction by questioning the common presupposition that there is an invidious distinction to be drawn between kinds of truths',[10] pragmatism sets itself against the traditions of analytical philosophy, including those of language, evolutionary psychology, ecopsychology and phenomenology, which currently underpin much of design discourse. Analytical philosophy has many guises, but from the point of view of the pragmatist, they all share the idea that there is a distinctively philosophical method of analysis that can be used to get to the bottom of problems about the mind, knowledge, meaning, truth and so on.[11] Pragmatism also challenges evolutionary psychology with its 'central premise' that 'there is a universal human nature' and its belief that 'this universality exists primarily at the level of evolved psychological mechanisms, not of expressed cultural behaviours'.[12] It questions the cognitive psychologists for whom the 'challenge lies in explicating the universal rules that govern perception'.[13] Countering the argument that there is a collective subconscious or human memory, the bottom line is that pragmatism suggests there are no predetermined end points and no universal truths to measure up to, even in vision and perception. The aim of pragmatism, far from finding universal truths, Rorty explains, is to undermine the reader's confidence in 'the mind' as something about which one should have a 'philosophical view', in 'knowledge' as something about which there ought to be a 'theory' and which has 'foundations', and in 'philosophy' as it has been conceived since Kant.[14]

Overlooking the Visual develops the argument that, from a pragmatic perspective, designing is an iterative, complex process involving researching, testing, redefining, refocusing and expressing ideas in a particular medium. It is the synthesis and analysis of a plethora of information to make propositions for the future. To do it well takes aesthetic skill, artistic sensibility, expertise and judgement as well as technological know-how. Currently, the predicament in which design research finds itself is that no matter how scientific or phenomenological the process is, some part of it is thought to involve a deeply mysterious and unique act that lies beyond investigation, separate from intelligence, entangled in creativity, the mind's eye and the subconscious, engaging with universal truths and essences, archetypes and visual modes of thinking.

Although it might make design seem special and alluring, this introduces a fundamental weakness, a conceptual void at the heart of the process that compromises research inquiry as much as it does design pedagogy. *Overlooking the Visual* puts forward the radical idea that the problem stems from theories of perception. Descartes was largely responsible for maintaining the perceptual myth, and we are still suffering the consequences.

The sense datum theory of perception

Intensely nuanced and variable, the general picture we have of the perceptual process is that it depends on a sensory mode of thinking that somehow intervenes on our behalf to organize various inputs in order to serve intelligence—'a disastrous idea that has haunted Western philosophy since the seventeenth century'.[15]

According to legend, absorbing the plethora of information that surrounds us, the sensory interface sifts, crystallizes or in some other way processes it all, acting as a mediator between us and the world, and as an interlocutor between what are taken to be different conceptual realms or ways of knowing or thinking. The interface is called a number of things—the haptic, the experiential, the visual, creativity, the black box, the genius loci, the mind's eye, even a kind of understanding that lies just beneath intelligence. However, it is characterized, and this changes with the times, the details of the process remain shrouded in mystery.

A raft of rationalist beliefs and practices supports the premise. These include the idea that there are different ways of thinking (for example, visual or verbal, emotional or rational) and the idea that there are pre-linguistic modes of knowing, such as deep-seated structures in our brains, primeval yearnings or subconscious memories. Such ideas are dependent on the concepts of universal truth, independent logic and determinate facts. These are the 'real truths' 'in here', 'out there', 'somewhere', for us to find if we are clever enough or sensitive enough. Candidates for these truths include 'God, the material or "brute act" world, rationality in general or logic in particular' and 'the set of eternal values'.[16]

It is widely recognized that if we remove the assumption that there are different types of truth, this dismantles the idea that there are different kinds of reasoning or separate modes of thinking. As Fish remarks, this rationalist tradition has been consistently undermined over the last century, but it has proved 'remarkably resilient and resourceful'[17] mainly because these distinctions are so deeply embedded in our culture that they have become part of Western common sense.[18] The sense datum theory of perception is not only dependent on but also constantly reinforces the distinctions made between reality and appearance, pure radiance and diffuse reflection, intellectual rigour and sensual sloppiness, absolute and relative, nature and convention, body and mind.

Despite all the postmodern rhetoric, concepts such as visual thinking, intuition, language, emotions, artistic sensibility and design expertise remain imbued with the fundamental Cartesian distinction between mind and body, between facts and values, real truth and mere opinion—the consequence of a damaging metaphysical duality that has slipped under the intellectual radar, disguised in visual and perceptual theories. It is rarely recognized that this is just one way of understanding the world.

The impact of these dichotomies

It is difficult to exaggerate how much the general understanding of intelligence is dominated by the false notion of a sensory interface, and by the difficulties this notion creates. It lies at the heart of the common idea that art involves a different conceptual framework from science and requires a different mode of thinking; that art is a pleasurable pastime, whereas science is a serious endeavour; that it is possible to forget all you know in order to appreciate fully a piece of music, a painting or the landscape, embracing the sensuality of the experience with a clean slate, uncontaminated by knowledge or rationality. This is why, despite so much evidence to the contrary, we still characterize scientists as cool, detached and unencumbered by emotion, and artists as passionate, subjective and slightly deranged; why we think decisions can be made on the one hand intuitively, without knowledge, and on the other hand objectively, without value judgements. More generally, it skews the way intelligence is defined or what counts as valid knowledge, and gives a prejudicial and narrow view of the role of language.

Educationally this is disastrous. For example, at the centre of aesthetic experience, the sensory mode of thinking is what students are expected to reap the benefits of if they are to be in any way successful. But the fundamental dichotomy between body and mind enshrined in theories of perception actually creates insoluble puzzles within aesthetics that inevitably spill into design discourse.

Aesthetics, almost more than any other discipline, is dependent on the idea of universal truth. It does not seem to matter whether the notion of truth is approached from a transcendental or empirical perspective: in aesthetic theory, what really count are the universal superstructures that are thought to stand outside culture but also act to underpin and unite our responses. In the attempt to identify these universals we are supposed to set aside all reason, opening ourselves without reservation to what is outside of us in order to sense something 'other'. However, from a pragmatic perspective, asking anyone to step outside of what he or she knows and to sense significance or beauty as it really is, without the encumbrances of knowledge and culture, is as pointless as it is ridiculous.

Then, most damaging of all, running through a whole range of design theory is the highly pejorative attitude towards the visual, underscoring the contention that, whatever it is that determines our responses, it is certainly not 'merely' visual. The visual may well be acknowledged as a component, but it is also thought to be a distraction. The physical, material qualities of place are thus edged out of the frame because an appreciation of such things is considered too subjective or ephemeral.

As a result, society generally speaking has lost the art of critical looking. Through long-term neglect and discrimination, we no longer have the confidence, the appetite or even the language to talk about appearances. It is abundantly clear, however, that we live and work in a visual, spatial medium. It is both pretentious and foolhardy to think we can manipulate that medium without knowing the implications of what we are dealing with. Undervaluing the cultural and social criteria of appearances disables our attempts to understand the impact of the made environment on our quality of life.

Within design we habitually think that it is possible to reconcile what are by definition irreconcilable opposites such as visual thinking with verbal thinking, creativity with intelligence; that it is possible to switch on one's creativity (can we ever switch it off?), that it is possible to 'graze the senses' and engage our intuition by drawing, in the belief that, the hand captures what neither the eye nor language can grasp. We expect novice students to sense the genius loci or sense of place without thinking, to leave their ideas at the studio door and rely on their intuition or subconscious, without using language. We seek what lies beneath the surface of what we see, without understanding what is in front of our noses. More broadly, this habitual thinking shapes current disciplinary and institutional silos and hierarchies.

Ramifications

Underpinning the distinction made between different conceptual frameworks such as science and art, this oddly enduring duality leaves us with a narrowly defined view of intelligence and rationality: language is seen as linear and logical, while emotions and intuitions are subjective, irrational and inexplicable. Belief in a sensory interface that is supposedly making decisions on our behalf means that whether research is undertaken from an empirical or transcendental perspective—or indeed anywhere between the two—it cannot escape the clutch of metaphysical concepts based on objective, universal truth or subjective, hidden essences. Yet either position is nothing more than a camouflage for all sorts of agendas that are poorly articulated and open to abuse, essentially opinion masquerading as self-evidence. The process of designing is thought to be a special case, because it supposedly straddles these two conceptual realms.

It leaves us with a number of fairly predictable scenarios: on the one hand, it translates into efforts to identify primeval, subconscious yearnings and recognitions, the invisible, or what lies beneath the surface; on the other, it calls for research to be neutral, simple, clean and objective, with replicable analyses based on hard facts and incontrovertible truth. Caught on the horns of this dichotomy, an awful lot of time is spent developing increasingly complex and elaborate strategies to build bridges or gateways between what are characterized as the emotional, intuitive aspects of design and the logical side that deals with practicalities and language. Similarly, trying to understand the creative possibilities of a 'confusion of thought and perception',[19] working out how we can synthesize thinking in images with thinking in words, as well as how we might teach such a skill, has become a preoccupation. In their attempts to patch up the division between the senses and intelligence, either by stressing the close proximity of the two or even by claiming to reverse the usual rationalist bias, researchers are forever looking for new ways to freshen up old propositions, for example by focusing on the aesthetic nature of scientific thought or language and the consensual, rational basis of poetic discourse. In contrast, speculation about what is actually perceived is negligible, even though, as Ingold[20] observes, this is almost certainly a far more significant question to ask.

Notwithstanding the substantial and impressive body of research dealing with an array of historical, contextual and technological issues as esoteric, practical, obscure or technical as you could wish, the picture remains pretty much the same. At the critical point, when it tries to address the designing part of design, significant chunks of the process go missing; they slip away into an arcane, sensory netherworld. The spatial, conceptual and visual skills needed to generate form, to express ideas through materiality, the nuts and bolts of understanding why things look the way they do given the time, place and context, are hardly ever addressed (see *Overlooking the Visual* for a detailed analysis of this problem).

The rule of rationalism

The extraordinary success of the scientific paradigm has led to our being practically transfixed by the idea that research has to adopt a scientific methodology, maintaining at all times a neutral objectivity, even though time and again it has been shown that the design process does not sit easily within it[21]—and incidentally, neither do many other disciplines, including, paradoxically, the sciences. Over the years, warning bells have been sounded about the validity of the rationalist doctrine. Railing against the proclivity for empiricism at the turn of the twentieth century, James said that the devotion to science was so overwhelming that it was, to all intents and purposes, a religion: 'Our children, one might say, are almost born scientific.'[22] The 'cult of the fact', Hudson argues, remains almost impassible,[23] and we are still 'dazzled', Midgley says, by science.[24] Support for such dissenting views, particularly from within the scientific community itself, has made a bit of a dent in science's otherwise copper-bottomed reputation, and design research is not alone in emerging from decades of analytical, logical, inductive reasoning, number-crunching, longitudinal studies and so-called objective analysis. Many agree with Cross that it is 'no longer necessary to turn design into an imitation of science; neither do we have to treat design as a mysterious, ineffable art'.[25] But despite a slow migration away from explicitly scientific systems of inquiry such as those espoused by McHarg—led by, among others, the existential 'happenings' organized on the beaches of California by Halprin,[26] phenomenological explorations of topophilia by Yi-Fu Tuan[27] and (the) concepts of placelessness of Edward Relph[28] the promise of certainty and truth offered by hard scientific methodology is difficult to resist. One alarming piece of evidence, for example, discussed at the Forty-Sixth World Council of the International Federation of Landscape Architects (Rio de Janeiro, 2009) and the General Assembly of the European Federation for Landscape Architecture (Brussels, 2009), is the growing number of university departments requiring teachers of design to hold a scientific PhD. Both meetings voted unanimously to urge funding bodies, universities, ministries of education and professional organizations to address the decline in knowledge and expertise this is causing as a matter of urgency, on the basis that it is damagingly prescriptive and will do enormous harm to the future development of the discipline.

More insidious, however, is the fact that even when a scientific methodology is not explicitly adopted, the underlying rationalist principles are just too sticky to peel off—evident, for example, in the belief that we are '"getting closer to the way things really are" or "more fully grasping the essence of…" or "finding out how it really should be done"'[29] (Kuhn, quoted by Rorty). Such beliefs underlie attempts to find descriptions of the world as it really is, and are clearly exposed in the notion that it is possible to gather practical, utilitarian hard facts, remote from the 'muddy, painful and perplexed' world of personal experience,[30] as well as the idea that these facts can be separated from values or that values can be added on after the facts have been established. Lurking in the background is a residual, deep-seated dependence on universal conceptions that are beyond all doubt, impermeable and implacable. The divided consciousness remains absolutely fundamental.

The alternative

The alternative is to avoid altogether the 'obsolete and clumsy tools' that distinguish 'between absolutism and relativism, between rationality and irrationality, and between morality and expediency'.[31] If we adopt an interpretative view of perception, the whole metaphysical edifice built on the flawed conception of a sensory mode of thinking comes tumbling down. Rather than argue that we should recognize the intelligence of perception, we should be seeing that perception *is* intelligence. Such a belief unlocks a major part of the debate and disengages aesthetics, the visual, creativity and many aspects of consciousness from primitive bodily ways of knowing. It becomes disentangled from psychology and uses a fresh, common-sense approach, bringing materiality back into the picture.

From a pragmatic perspective, it follows that all thinking, whether in the arts or the sciences, is therefore interpretative and metaphorical; neither uses a special kind of reasoning. Essentially, this is to say that we think the same way no matter what we happen to be thinking about. In understanding emotions or equations, formulae or artistic responses, we interpret, reinterpret, judge and try to make sense of our feelings because there is simply no other way to make sense of what we see, to make sense of the world. Just because we are looking at a painting does not mean we are thinking in pictures, or that when we are reading a book we are thinking linguistically. Whatever grabs our attention or catches our eye, no matter what gets us thinking, we always get to think about it by the same route, through language. There are no exceptions, no special cases, no ifs, ands or buts. Language binds us, separates us: it quite literally defines us.

Recognizing that both perception and language are interpretive removes a blindfold. This recognition is the final radical shift that enables us to understand one of the most obscure aspects of the whole design process: that of generating form. It demonstrates the indivisibility of ideas, theory, expression and technology in practice, making us realize that it is as impossible to design without concepts as it is to talk without a tongue. Sensible discussions can emerge about the making of informed, imaginative and often difficult design decisions, making it clear that there is nothing magical or mysterious about the process.

The impact this makes in the studio is in many respects quite simple—it is a matter of consistently asking 'why' things look as they do, what ideas are being worked with, what they look like, and why they are appropriate given the site, project brief and context. What spatial principles are being worked with, and how are ideas being expressed—whether at a strategic or a detailed level? It means asking for an explanation as to why a particular kind of materiality is involved (light, shape, form, texture) and what quality of experience is being designed for what kind of user. This way of teaching is about encouraging students to plunge wholeheartedly into the visual, spatial world, working with ideas, space, form and materiality. It is also about adopting strategies to discourage the habitual or clichéd decisions that students often fall back on (presumed to be instinctive or intuitive, but actually learned) and instead pushing and challenging students to go further. This develops their confidence in knowing which line of inquiry is a good one to follow, supports risk-taking, and challenges both faculty and student-held preconceptions. Above all else, to teach in this way is *not* to presume that anything visual or spatial or conceptual is self-evident. It *is* to enter an ambiguous world.

Embracing ambiguity

Not only does this give us a means of dealing with spatial, visual information that is artistically and conceptually rigorous, but we can also reject the idea of universal, inviolable truth without necessarily being sucked into the argument that the only alternative is to believe everything is relative and dependent on a point of view. But moving the purpose and methodology of design inquiry into such potentially ambiguous areas requires taking on board what may seem at first to be a number of contradictory propositions. For example, apart from recognizing the slippery quality of language and the interpretative nature of facts, we have to accept the rationality of emotions. We do not switch modes of thinking. Pragmatism focuses unequivocally on knowledge within a particular medium rather than any notion of innate, generic skill, suggesting that all perceptions, observations and analyses (even the most scientifically based) are ambiguous, flexible and open to interpretation.

Offering 'a middle way between reactionary metaphysics and irresponsible relativism', as Putnam asserts,[32] and redefining the relationship between the senses and intelligence means that essentially there is no need to choose one or the other. This releases us from the endless debate between positions that are natural or cultural, scientific or artistic, theoretical or practical, value-laden or quantitative. Collapsing the visual, intelligence, language and many other elements of consciousness into a holistic concept of perception takes the supernatural element of design theory and education out of the equation. It also reveals that, far from masking design ability or creativity, concepts and language actually allow us access to the arts, in both their making and criticism.

Profoundly changing the epistemological basis of design leaves us no option but to engage with ideas at every stage of the process. Even the most intimate, seemingly mystical elements of design are based on knowledge and knowledge alone. The bottom line is that, as individuals or as a community, in any study, design or otherwise, we are constrained or liberated by the language and concepts we have at our disposal. There is no other way of knowing, no other kind of meaning to uncover, no 'genius loci' to give us a nudge in the right direction. Neither the site nor what lies beneath, within or without it, nor even the fears and desires of our prehistoric ancestors, can speak to us beyond what we know.

There is no way to operate with the presumed objective neutrality of a so-called scientific approach. We need a healthy measure of scepticism to deal with the hard facts enshrined in regional spatial plans, perennially used to justify the economic imperative for new roads, the distribution of new settlements, how big they should be, or the cost the market will stand in terms of quality housing or town centre development. The evidence of the impact of such quantitative factual decisions is all too clear: you just need to look around any town or city. Or compare today's transport, housing and agricultural policies with those of fifteen years ago. Were all those experts just plain wrong back then, or were they simply working under different circumstances and with a different set of values—different ideals?

Re-evaluating fundamental assumptions about objectivity inevitably has an impact on research and pedagogy, especially in teaching aspects of the design process. For example, rather than simply aiding the mechanical or practical part of a project, we have learned to see technology as a means of understanding how far materials might be pushed or manipulated in order to

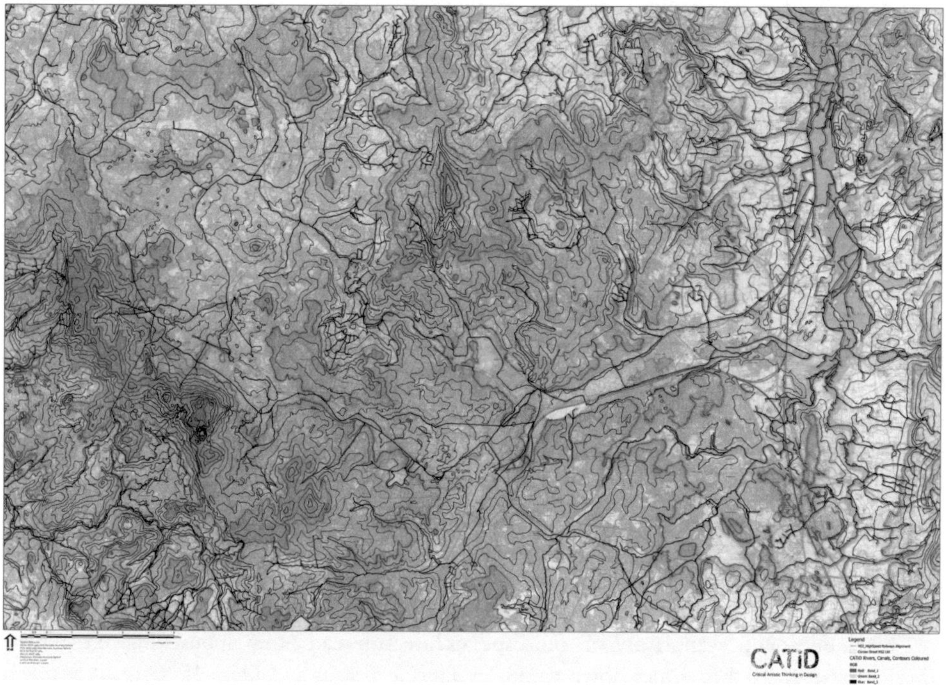

Figure 24.1 Drawing as research and research through design, 2017. Exploring the materiality the West Midlands, United Kingdom, this drawing, one of a sequence of studies, re-discovers a vast, hidden landscape that has been largely overlooked and undervalued for many, many decades. © Moore, Kathryn.

express ideas with style and confidence. Similarly, drawing, rather than just a technique offering access to intuition, somehow kicking part of the brain into the long grass, should be valued more as a way of working things out, exploring ideas and speculating on the possibilities. As an investigative tool, drawing is hard to beat (Figure 24.1).

All of this prepares the ground for a fresh artistic and conceptual approach to design, establishing it as a holistic, critical endeavour. From this perspective, any and every part of the design process becomes accessible to investigation. It also makes clear that the limits of our inquiries are governed only by our knowledge, nothing more or less. Responsibility for understanding what sense we make of the world is thus handed back to us. The driest, most reductive statistical equation or number-crunching analysis is as full of presumptions and preconceptions as any ephemeral, instinctive response. Look at the debates relating to climate change and it is easy to see how open to interpretation the facts can be, let alone finding any consensus as to what is an actual fact and what is not.

What makes a good researcher?

We should recognize that what is considered to be clear and rigorous research is absolutely contingent upon the knowledge, values and opinions of those who judge it. This explains why Swaffield and Deming[33] find that what is valued in research is shaped by academic location, the educational background of academics, and the particular approach of editors and reviewers. Those undertaking research effectively enter a lion's den, as work can easily end up in the hands of someone with a conflicting agenda, an entirely different view of the world. So, as supervisors, reviewers and editors, our role is to be informed and make judgements from a position of knowledge and experience, aware of our prejudices, preconceptions and desires. The hard part is to recognize what these are and then have the courage to put them to one side if necessary: not trying to gauge how closely the work measures up to our own ideas, but being open and pragmatic enough to appreciate what might be an entirely different way of understanding things, aware that there is no single model for good work in any academic discipline. It also means that we can even be, as Richard Rorty suggests, more relaxed about whether we have a rigorous research methodology, or whether the work produces knowledge rather than mere opinion.[34] This is the embodiment of objectivity.

Precisely what excites or appeals to us will depend on our inclinations and temperament. James distinguishes between those who are tough-minded and those who are tender-minded,[35] whereas Rorty suggests that a more apt divide is between 'those busy conforming to well-understood criteria for making contributions to knowledge [and] people trying to expand their own moral imaginations'.[36] And this is precisely the point. Rather than staying within the safety of fixed disciplinary parameters, and in order to overcome a long period of technological stagnation, we need to be more aggressively expansive, appropriating and operating confidently, making connections between disciplines, linking theory and practice, ideas and form; evaluating the ethical, aesthetic, ecological and artistic value of physical and imagined environments, with the explicit purpose of investigating how this knowledge can be used directly to inform design.

The richness and complexity of landscape architecture can often appear difficult to capture, and essentially this comes down to our own conceptions and ideas. If we really want to fully articulate the way we experience the world, there can be no room for the dry bureaucratic talk that squeezes the life out of any debate about place and space. We need a better set of descriptions. It is not as though we are stuck for ideas, or indeed for words. There is a wealth of literature and research, with evidence scientific, academic and anecdotal—imaginative narratives to inspire and show us things we had not noticed in the world. The real skill of a designer or a researcher is in using the information to capture these narratives and/or create new ones through good investigative digging, and then explicating the work in such a way that it fires the imagination. Obviously, this is not just about language and language alone. The narratives, the words, must be made real, supported by a demonstration of their spatial implications. When we are dealing with the transformation of a place, it is not only the understanding of new ideas that enables us to adjust to new circumstances and possibilities, but also the convincing and appropriate evidence of their expression in physical form. Within both research and design, if we steer clear of the safe options, we can begin to fill the conceptual void by talking seriously about ideas and their function in quality design. Whereas designers tend to express ideas and experience in space, form, light and texture, the researcher relies more on words and a wide range of images. To practise either as a designer or as a researcher requires different kinds of expertise, calls on quite different kinds of contextual information (including literature, editorial objectives and peer expectations), and has different motivations and outputs. But the real skill in both is to shape the experience of those with whom we wish to engage, either as readers or people who use the spaces we create.

What makes a skilful researcher? Being objective enough from a position of knowledge to realize, for example, that some of your initial assumptions are erroneous. Being brave enough to reconceptualize basic beliefs again and again, to work things out without having a preconceived idea as to what the results are going to be. Having the insight to bring to bear new ideas and understandings that can enlighten and inform. Discovering something that effects real change and affects policy, people's lives, the way we teach. Making research that makes a difference.

Conclusion

This perspective gives us a new imperative to improve the academic credibility of design research and design itself. Ditching the metaphysical baggage that weighs down most current theories of perception enables us to demystify the art of design, teach the generation of form, connect spatial strategies to real places, and develop ways of working that not only encourage but also demand the expression of ideas—the ideas that are fundamental to the design process. Changing the focus of how we think of landscape from technology towards ideas, seeing it as both a cultural and natural resource and a physical and abstract entity that has economic and social value, and looking at the experience people have of their physical environment as well as making the vital connections between governance, culture, health and economics: these steps go some of the way to providing a viable new platform from which to deal holistically with the rural and the urban, wilderness and human-made, the most treasured and memorable as well as the unloved and degraded. They set a new agenda for both design and research to bring fresh insights that will shape the future of our environment. That is not to say that research and design are equivalent. Researcher and designer are governed by different protocols, agendas and ambitions. Each is judged by different criteria, criteria that, as Swaffield and Deming show,[37] are absolutely contingent on the social and cultural milieu.

Whether one is a researcher or a designer, there is inevitably a degree of anxiety when old certainties are challenged and the interpretative, transient nature of everything we believe to be true finally dawns on us. Shifting any inquiry away from the unequivocal towards the ambiguous is perhaps one of the most difficult aspects of this paradigm. It is not just another way of saying that anything goes, but suggests rather that work must be judged against different criteria. Truth is contingent, beliefs change; nothing is set in stone. And it is this flexibility that gives us such a great opportunity. If we have the confidence to move away from the central hard core of scientific assumption and methodology, there is a real chance to develop new approaches, make connections across and between disciplines, and erase rigidly drawn boundaries delineating and distinguishing practice from theory. The old Cartesian duality is a house of cards … time to blow it down.

Notes

1 Higher Education Funding Council for England, 'REF 2014 – Research Excellence Framework', 2011.
2 S. Swaffield and E. Deming, 'Research Strategies in Landscape Architecture: Mapping the Terrain', *Journal of Landscape Architecture* 11 (Spring, 2011): 43.
3 N. Cross, *Designerly Ways of Knowing*, Board of International Research in Design (Basel, Boston and Berlin: Birkhäuser, 2007), 126.
4 D. Durling, 'Discourses on Research and the PhD in Design', *Quality Assurance in Education* 10, no. 2 (2002): 79–85.
5 G. Ryle, *The Concept of Mind* (London: Penguin, 1990 [1949]), 17.
6 Richard Rorty, *Philosophy and Social Hope* (London: Penguin, 1999), 178.
7 See K. Moore, *Overlooking the Visual: Demystifying the Art of Design* (Abingdon: Routledge, 2010).
8 Richard Rorty, *Philosophy and the Mirror of Nature* (Oxford: Blackwell, 1980).
9 See Moore, *Overlooking the Visual.*
10 Richard Rorty, *Consequences of Pragmatism* (Minneapolis: University of Minnesota Press, 1982), xvi.
11 Louis Menand, ed., *Pragmatism: A Reader* (New York: Vintage Books, 1997), xxxii.
12 J.H. Barkow et al., eds., *The Adapted Mind, Evolutionary Psychology and the Generation of Culture* (New York, Oxford: Oxford University Press, 1995 (first published 1992)), 5.
13 Gabriella Goldschmidt, 'On Visual Design Thinking: The Vis Kids of Architecture', *Design Studies* 15, no. 2 (April 1994).

14 Rorty, *Consequences of Pragmatism*, 7.
15 H. Putnam, *The Threefold Cord: Mind, Body and World* (New York, Chichester and Surrey: Columbia University Press, 1999), 43.
16 S. Fish, *Doing What Comes Naturally: Change, Rhetoric, and the Practice of Theory in Literary and Legal Studies* (Oxford: Duke University Press, 1989): 342–3.
17 Ibid., 345.
18 Rorty, *Philosophy and Social Hope*.
19 Nicholas Davey, 'The Hermeneutics of Seeing', in *Interpreting Visual Culture Explorations in the Hermeneutics of the Visual*, eds. I. Heywood and B. Sandywell (London and New York: Routledge, 1999).
20 Tim Ingold, *The Perception of the Environment* (London and New York: Routledge, 2000).
21 See G. Broadbent, *Design in Architecture, Architecture and the Human Sciences* (Letchworth, Hertfordshire: David Fulton Publishers Ltd, 1988 [1975]), 321.
22 W. James, *Pragmatism*, ed. Bruce Kuklick (Indianapolis and Cambridge: Hackett Publishing Company, 1981 [1907]), 20.
23 L. Hudson, *The Cult of the Fact* (London: Jonathan Cape, 1976 [1972]).
24 M. Midgley, *Science and Poetry* (London and New York: Routledge, 2001), 59.
25 N. Cross, *Designerly Ways of Knowing*, 126.
26 See Peter Walker and Melanie Simo, *Invisible Gardens: The Search for Modernism in the American Landscape* (MIT Press, 1994).
27 Y.-F. Tuan, *Topophilia* (New York: Columbia University Press, 1974).
28 Edward Relph, *Place and Placelessness* (London: Pion Limited, 1976 (third reprint 1986)).
29 Kuhn, quoted by Rorty, *Philosophy and Social Hope*, 187.
30 James, *Pragmatism*, 23.
31 Rorty, *Philosophy and Social Hope*, 44.
32 Putnam, *The Threefold Cord*, 5.
33 Swaffield and Deming, 'Research Strategies in Landscape Architecture'.
34 Rorty, *Philosophy and Social Hope*, 181.
35 James, *Pragmatism*, Lecture 1.
36 Ibid., 127.
37 Swaffield and Deming, 'Research Strategies in Landscape Architecture'.

Bibliography

Barkow, J.H., Cosmides L. and Tooby J., eds. *The Adapted Mind: Evolutionary Psychology and the Generation of Culture*. New York, Oxford: Oxford University Press, 1995 (first published 1992).

Broadbent, Geoffrey. *Design in Architecture: Architecture and the Human Sciences*. Letchworth, Herts.: David Fulton Publishers Ltd, 1988 (first published in 1975).

Cross, Nigel. *Designerly Ways of Knowing*. Board of International Research in Design. Basel, Boston, Berlin: Birkhauser, 2007.

Davey, Nicholas. 'The Hermeneutics of Seeing'. In *Interpreting Visual Culture Explorations in the Hermeneutics of the Visual*, edited by I. Heywood and B. Sandywell, 3–29. London and New York: Routledge, 1999.

Durling, David. 'Discourses on Research and the Phd in Design'. *Quality Assurance in Education* 10, no. 2 (2002): 79–85.

Fish, S. *Doing What Comes Naturally: Change, Rhetoric, and the Practice of Theory in Literary and Legal Studies*. Oxford: Duke University Press, 1989.

Goldschmidt, Gabriella. 'On Visual Design Thinking: The Vis Kids of Architecture'. *Design Studies* 15, no. 2 (April 1994): 158–74.

HEFCE. 'Ref 2014 Research Excellence Framework'. *Higher Education Funding Council for England*, 2011.

Hudson, Liam. *The Cult of the Fact*. London: Jonathan Cape, 1976 (first published 1972).

Ingold, Tim. *The Perception of the Environment*. London and New York: Routledge, 2000.

James, William. *Pragmatism*, edited by Bruce Kuklick Indianapolis. and Cambridge: Hackett Publishing Company, 1981 (first published in 1907).

Menand, Louis, ed. *Pragmatism: A Reader*. New York: Vintage Books, 1997.

Midgley, Mary. *Science and Poetry*. London and New York: Routledge, 2001.

Moore, Kathryn. *Overlooking the Visual: Demystifying the Art of Design*. Abingdon: Routledge, 2010.

Putnam, Hilary. *The Threefold Cord: Mind, Body and World*. New York, Chichester, Surrey: Colombia University Press, 1999.
Relph, Edward. *Place and Placelessness*. London: Pion Limited, 1976 (third reprint 1986).
Rorty, Richard. *Philosophy and the Mirror of Nature*. Oxford: Blackwell, 1980.
Rorty, Richard. *Consequences of Pragmatism*. Minneapolis: University of Minnesota Press, 1982.
Rorty, Richard. *Philosophy and Social Hope*. London, England, New York, USA, Victoria, Australia, Toronto, Canada and Auckland, New Zealand: The Penguin Group, 1999.
Ryle, Gilbert. *The Concept of Mind*. London: Penguin Books, 1949 (reprinted 1990).
Swaffield, Simon, and Elen Deming. 'Research Strategies in Landscape Architecture: Mapping the Terrain'. *JOLA* (Spring 2011): 34–45.
Tuan, Yi-Fu. *Topophilia*. New York: Columbia University Press, 1974.
Walker, Peter, and Melanie Simo. *Invisible Gardens: The Search for Modernism in the American Landscape*. MIT Press, 1994.

25

A CONVERSATION ON EDUCATION

Ellen Braae and Henriette Steiner

with Samantha L. Martin-McAuliffe, Anne Bordeleau, Torben Dam, Lilli Lička, Alan Tate, Tom Nielsen, Inge Bobbink, David Grahame Shane, Catharina Dyrssen, Maggie Roe, Tao DuFour, Gini Lee, Anders Busse Nielsen, and Catherine Dee

Landscape architecture and landscape architecture research—the heart of this volume—are both objects and agents of study, each on their own terms. The growing academization of architecture and design educational programmes (for example, in the Bologna process in European educational institutions), and the increasing emphasis on quantifiable research outputs from educational institutions, raises the important question of how these strands will interact and develop in the future. As we began to collate the chapters of this volume, we found that many contributions touched on the relationship between knowledge and practice, but they mainly concentrated on the role of the practitioner, and on the dialogue between architectural practice and scholarly reflection. We realized that something was missing: a reflection on the role and potential of architectural education as a specific institutional setting where much of the exchange between research and practice not only takes place but actually develops in deeply interdependent ways. Most university-based researchers also contribute to landscape architecture education, and this makes it even more important not to draw dividing lines between research, practice and education in landscape architecture, but rather to lay bare their intrinsic overlaps and interdependencies.

In this final chapter, we would like to zoom in on the institutional framework that defines the education of landscape architects and hence conditions the way the contours of the discipline are drawn, in both practice and research. The chapter offers a discussion of how current challenges in practice and research come together in landscape architecture education. It mainly focuses on bachelor's and master's curricula rather than PhD programmes, although a discussion of the latter would also be highly relevant. For now, we consider how university educators best can prepare tomorrow's landscape architects—as researchers and as practitioners. How might we understand the hinges between education, research and practice and the way they depend on one another? It was against this background that we decided to approach fourteen significant educators from different university institutions around Europe and North America. These are people who are highly engaged in educational questions and who represent different corners of the discipline, whether studio, history, theory, or adjacent fields such as urban planning or art.

Our original aspiration was to bring these individuals together for a conversation, but physical distance ultimately made it impossible to gather everyone in the same room. Rather than completely abandoning the idea, we instead opted for a more distributed and 'virtual' form of discussion by asking everyone the same set of questions, making the chapter itself a vehicle to link together perspectives on educational matters from educators in different national and institutional contexts. Nevertheless, we still insist on calling the outcome 'a conversation', as we would like this chapter to outline possible places from which to start a multifaceted discussion.

The framework of discussion we have proposed to the contributors hinges on the different forms of embedded 'knowledge' related to the discipline, on what we see as the recurrent paradox of a globalized profession working in distinctive local contexts, and on the relationship between ethics and aesthetics. Our contributors chose to approach these themes in various ways, emphasizing different aspects. The contributors' wide geographical spread across Europe and North America encompasses whole continents with highly diverse climatic, natural and cultural traditions. Rather than trying to align this diversity in terms of the contributors' frames of reference, we have decided to allow the differences and commonalities alike to materialize in the following pages, to provide a starting point for continued discussion and exchange, for new conversations to develop and blossom.

Nevertheless, it was particularly important for our general purposes to encourage the contributors to reflect on how landscape architecture can be positioned in the wider educational field and on the different kinds of knowledge and scientific tradition currently sustaining the discipline. These reflections are fundamental to emphasizing how landscape architects, or indeed landscape architecture students, may regard their position in the future. Landscape architecture today is becoming increasingly professionalized as a quasi-global profession—not least through the successful expansion of academic research efforts within the discipline. This includes the kinds of effort to which this volume as a whole contributes, and this chapter thus also serves to broaden our understanding of the institutional structures within which we operate. Indeed, education may be one of the most important realms for the engagement and transmission of knowledge that moves back and forth and connects practice and research in intricate ways—with the added benefit that education remains a dialogical and personal setting for exchange, not least in studio-driven design programmes. However, we asked contributors to reflect on the tensions and challenges the profession currently faces, and on what we can do to prepare new generations of landscape architects who are now part of the educational system or will be so in the future. We also considered it important to inspire a discussion of how the aesthetic and ethical dimensions of landscape architecture potentially impact on the discipline, and of the role they should play in future landscape architecture education.

Each contributor's written answer to our questions is set out in turn in the following pages. We hope the discussion will continue—in the mind of the reader, with peers, researchers and practitioners, and beyond disciplinary confines. We sincerely thank Samantha L. Martin-McAuliffe, Anne Bordeleau, Torben Dam, Lilli Lička, Alan Tate, Tom Nielsen, Inge Bobbink, David Grahame Shane, Catharina Dyrssen, Maggie Roe, Tao DuFour, Gini Lee, Anders Busse Nielsen, and Catherine Dee for their time, commitment and courage in entering this experiment. They have done so with precisely the courage to enter unknown constellations and territories that we all require of our landscape architecture students every day.

Samantha L. Martin-McAuliffe (University College Dublin, Ireland)

In 1366, the Italian scholar and poet Francesco Petrarca (Petrarch, 1304–74) climbed Mont Ventoux (Ventosus, 'windy'), near Avignon, with the express purpose of seeing the view from its summit, an undertaking virtually unheard of at the time. A local shepherd tried in vain to dissuade him from the ascent, citing the dangers of exposure, yet Petrarch persisted, and his account can be read as a physical as well as spiritual expedition: 'What you have experienced so often today in the ascent of this mountain, certainly happens to you as it does to many others in their journey toward the blessed life [...] At the summit is both the end of our struggles and the goal of our journey's climb.'[1] What is remarkable, however, about his description is the deeply personal transformation that unfolds on the mountaintop. The climate and especially the view overwhelm him, and his vulnerability is laid bare—revealed—by the yawning expanse of the panorama: 'I stood there like a dazed person.' Over the course of a single afternoon, Petrarch had experienced an entirely new way of contemplating topographical scale.

The ascent of Mont Ventoux is often cited as a paradigmatic moment in the development of the picturesque in landscape architecture: an early humanist aesthetic of visualizing not only immense scale, but also perspective, prospect and, ultimately, dominion over the landscape. Yet there is more to be gleaned from this passage in a pedagogical sense. Rather than being confined to an episode in the history and theory of designed landscapes, it should also be seen as something which continues to exert an ethical dimension insofar as it influences the way we question human intention and especially agency. In deliberately spurning the shepherd's warnings, Petrarch gained exposure to something else—the ineffable. The geographer Donald Meinig, in his consideration of landscape as aesthetic, foregrounds an understanding of landscape that is situated beyond what is quantifiable or objective: 'Landscape becomes a mystery holding meanings we strive to grasp but cannot reach.'[2]

A defining characteristic of landscape architecture today is a complex and variegated pedagogy that requires a mastery not only of design and planning principles, but also of essential concepts in ecology, horticulture, hydrology, soil chemistry and food systems, among others. Yet, at the same time, landscape architects must recognize the ambiguity inherent to their discipline, in particular the effects of climate change and exposure to major ecological events. The point to take from this is that in addition to tackling what is measurable and empirical, the discipline of landscape architecture must also find new ways to experience the unknowable. In the design studio, we describe outlook and scale in cardinal coordinates and Cartesian terms. Perhaps the present challenge is how to write a brief that also fosters, or at least encourages, the revelation of something which is beyond description. To some degree, Petrarch's experience and account from Mont Ventoux still informs the practice of landscape architecture, but this does not preclude the possibility that a new paradigm of seeing the landscape will be forthcoming.

Anne Bordeleau (University of Waterloo, Canada)

The term 'landscape', when placed in front of 'architecture', both challenges and requalifies architecture. On the one hand, it challenges the potential dichotomy between the idea of durable built constructions and the cyclical movements of nature. On the other hand, it requalifies the scope of architectural practice to embrace larger spatial and temporal territories—often involving projects that develop over hectares and decades. With this in mind, we can appreciate how students in landscape architecture should be most aptly positioned to move beyond traditional dualities such as culture and nature, or human and nature. At the same time, the field is the most likely to fall prey to the desire to assert the autonomy of a project over ever-larger territories. The fragile and thoughtful equilibrium that the field must maintain rests on traditions of scientific investigation of the surface and crust of the earth—the geologists', geographers' and geometers' mappings, along with the biologists', botanists' and mineralogists' studies—coupled with the necessary awareness of the rich and tangled political, economic, social and cultural questions that any material or site holds.

The tensions between the local and the global touch the very foundations of landscape architecture, as it typically sets out to first map and then respond to the conditions already existing on-site. As noted by James Corner,[3] landscape is intrinsically linked with the idealization and imaging of nature, and current modes of visualization have opened up new avenues by enabling the representation of landscape in different phases of growth. However, representation always implies a certain form of jurisdiction of the architect over nature. Hence, while presumably intended to privilege architecture's context and locality over its autonomy and potential universality, mapping can sometimes result instead in the subjugation of a larger territory that can be near or far. This impact is even more dramatic when paired with the ambition to project landscape architecture as a practice that changes over time and participates in larger processes with effects anticipated over five, ten, twenty-five or fifty years. As landscape architects seek to overcome contemplative distance, and to favour the agency and projective quality of architecture as landscape—setting their projects within the larger processes (natural, political or infrastructural) already at work in the designated site—our education must also enable them to remain mindful of what can be neither entirely known nor fully controlled, so that they resist blurring all boundaries, whether spatial or temporal, actual or imagined.

Aesthetic and ethical questions are pervasive and particularly visceral in the discipline of landscape architecture. Whether we consider the analytical tool of choice—mapping—the unalienable material quality of the sites of intervention, or the sociopolitical conditions that underpin any production of landscape, landscape architecture education seems to offer a prime field in which to address the inevitable interplay between ethics and aesthetics. As we dwell on the mesmerizing beauty of maps, and on our ability to delaminate and apparently comprehend complex territorial processes across space and time, J.B. Harley and Denis Cosgrove[4] remind us that maps have a subjugating tendency and invite us to carefully reflect on our presumed power as map makers. And, as Jane Hutton eloquently traces in her studies of reciprocal landscapes, the matter we work with and live with is laden with sociopolitical and cultural dimensions that force us to confront questions of exploitation, human labour or commodification, necessarily bringing a form of material ethics to bear on a project's material aesthetics.[5] Finally, the scale of interventions in landscape architecture inevitably raises questions tied to the environment, sustainability and even responsive technologies, concerns in relation to which our solution-driven society must constantly revisit how it envisions the terms of human-to-nature relations, with all of their ethical and aesthetic implications.

Torben Dam (University of Copenhagen, Denmark)

In Europe there are approximately 110 landscape architecture programmes, originating in arts, architecture, technology, engineering, agriculture and horticulture. Landscape architecture students have to navigate this sea, which is often described in terms of 'the inextricable relationship between natural science and the humanities' despite the differing lineages.[6] The would-be landscape architectural student has to create clever and skilful landscape architecture, balancing the expanding amount of research-based knowledge with landscape architectural competences in a reflexive way. One could argue that landscape architecture education programmes globally mirror the knowledge, skills and competences commonly used or considered important in landscape architectural practice, combined with the possibilities offered by their disciplines of origin—somehow merging other fields and disciplines into a heterogeneous landscape architecture education profile that is practical and solution-oriented, like the profession itself. The profile deriving from these components varies from institution to institution and from time period to time period. In essence, all these educational variations face the same challenge: how to correlate many (but not all) issues and still achieve a satisfactory landscape architectural result. Quality in landscape architecture assists the navigation of educational programmes and helps to prevent researchers and students being lured into a singular discipline. Landscape architecture education programmes thus include many kinds of knowledge and scientific tradition, and are rich in variety, yet very alike from a quasi-global perspective. In practice, it is the students that take on the task of gluing the components of an educational programme together. So how do such composite programmes maintain the quality of landscape architecture?

One might say that all educational programmes can work well as long as the combination of themes in focus and integrated research are seen as parts of a structure where the conception of landscape architecture is centre stage. Despite the general focus on knowledge in our cerebral society, it is rarely the kind of knowledge associated with making, e.g. the design process in landscape architecture. A simple change of prepositions linked to knowledge spells out the difference: knowledge *of* can be labelled as rational, science-oriented and reliable information that can be reused in other contexts; knowledge *for*, on the other hand, relates to situated places, specific problems (including tradition and aesthetics) and times. Landscape architecture design integrates both kinds of knowledge as foundings and findings.

At the University of Copenhagen, we see design as a means to explore and reflect on emerging fields of concern within practice, and we sustain students in their critical reflection on educational knowledge, global precedents and external practice. Design is thus not a scientific activity per se. Rather, it is considered an activity in its own right and a sound support for research and teaching.

The role of future landscape architecture education will still be to mediate related disciplines in such a way that landscape architecture glues together the students' competences. Future landscape architecture education needs critical reflection to operate with the increasing research, global outreach and demanding professional practice. Moreover, we should dare to communicate clearly about the mysterious and wonderful uncertainty intrinsic to landscape architecture and design, as spelled out in the notes of the eleveth-century Japanese garden design book "sakuteiki". Sakuteiki insists on the designer´s own mental clarity in a poetic and indeterminate way that will probably provoke many Westerners. On stone-setting, he says: 'When you are making up your mind how many stones to use and where to place them, be guided by the lie of the land as well as your own passing mood.'[7]

Lilli Lička (Universität für Bodenkultur Wien, Austria)

Landscape architectural education is part of a more general creative and cultural education. Its core object, however—the landscape—is formed by natural processes, by construction and engineering. Disciplines which relate to the making of landscape in a physical as well as social sense overlap, and their boundaries are becoming increasingly blurred. Interdisciplinary work is always important, but (like landscape architecture practice) it also requires a high degree of discipline-specific knowledge and self-assuredness from the contributors. In the future, landscape architecture education and practice alike will have to be explicit about the core competences required to interpret, analyse, develop and design landscape, and to be able to connect and collaborate with disciplines in the arts, humanities, sciences and technical professions.

Many have stated that we live in the age of the Anthropocene. Human influence concerns the whole planet, and goes beyond its climatic state. The rules of the market are probably the most global influence, as is shown in large landscape firms that act all over the globe and whose designs are becoming decreasingly site-specific—particularly on waterfronts and in corporate landscapes, but even in urban developments and large parks. To properly prepare students for global challenges, their awareness of these mechanisms must be raised, and their ability to conduct critical analysis must be developed. Academic research covers both global mechanisms and local specificities, especially when it comes to issues such as sites and identities.

The seemingly contradictory dichotomy between the local and the global must be dissolved, but at the same time a thorough examination of their interdependence is urgent. Landscape architecture projects and questions, however, are related to specific sites and landscapes, which is why local conditions also form a starting point for a global understanding.

Ethics are required as a general basic attitude for any responsible academic, including landscape architects, scholars and teachers. Education therefore has to point out specific issues and details where ethical considerations are essential. These range from cultural and societal impacts, questions of justice and sustainability, and the relationship with nature, right down to materials and products. Ethics are a pervasive issue throughout the majority of courses and subjects within education, and they must not replace the core content of landscape architecture. This core content includes artistic skill and aesthetic sensibility. These seem to have become subordinated to practical, scientific or sociological research questions, and we risk underestimating them in landscape architecture education. As a design practice, landscape architecture in its full relevance overlaps with the arts. Imparting design skills requires practice, as well as a knowledge of cultural history and an understanding of the cultural underpinnings of art and design forms.

Alan Tate (University of Manitoba, Canada)

Landscape architecture touches on all aspects of human activity in outdoor space. This makes it an extraordinarily broad discipline. Whatever else may be said about the discipline, its central role remains mediation between humans and nature. As such it draws on ecology, horticulture, soil science, abiotic materials and their durability, and comprehension of human society and human interventions (over time) in natural processes.

Landscape architects should view their future role(s) in the context of the UN Sustainable Development Goals[8] and through promoting the role of urban and rural landscapes in providing ecological and health services to all forms of life on earth. The mantra of thinking globally and acting locally is reflected in the practice of landscape architecture at scales ranging from the site to the city.

A particular challenge in this respect is the ability to recognize fully the existence of local peculiarities in biotic and abiotic materials and the importance of local customs and practices. And there is a danger in believing that 'an expert is someone from out of town'. So what can we do to prepare the new generation of landscape architects who are still part of the educational system? We should expose students to the appropriate questions to ask rather than letting them believe that there are standard answers to standard questions.

Design as a human activity occurs within aesthetic and ethical realms. Aesthetic and ethical sensibilities should inform all design decision-making. And they should be fundamental to landscape architecture education—in the same way that concerns for human health, safety and pleasure underpin all aspects of the discipline.

Tom Nielsen (Arkitektskolen Aarhus, Denmark)

At Aarhus School of Architecture, we teach landscape architecture together with urban design and urban planning in a sort of hybrid urban landscape education. Urbanization processes take place in both denser and less dense areas, and they are related through both physical and non-physical connections. Understanding these systems, and acquiring relevant specific tools and competences such as different forms of mapping, representation and analysis, is what we emphasize in our education—and it is important for other landscape architects too. Landscape architecture and landscape architects play a big role in the development of urbanized areas.

Figure 25.1a,b,c,d FABRICATING WILDERNESS explored the transforming ecology of the city of Milwaukee's post-industrial landscapes through the development of a landscape design proposal and temporary site installation. The design–build research project was initiated within the framework of the Architecture Fellow & Distinguished Visiting Design Critic Studio at the School of Architecture and Urban Planning at the University of Wisconsin–Milwaukee. The project involved experimentation with fabrication processes, employing analogue and digital methods in casting a series of prototypical concrete panels. The panels were envisioned as elevated and horizontally oriented, supported on a structural grid, constituting a parallel ground through and upon which landscape processes of 're-wilding' could be facilitated and experienced. The project was temporarily installed in Milwaukee's Inner Harbor brownfield site along the industrial shoreline of Lake Michigan. Tao DuFour, with student research assistants Travis Nissen, James Sequenz and Kelly Yuen. Figure 25.1d photo: Kelly Yuen.

Figure 25.1a,b,c,d Continued.

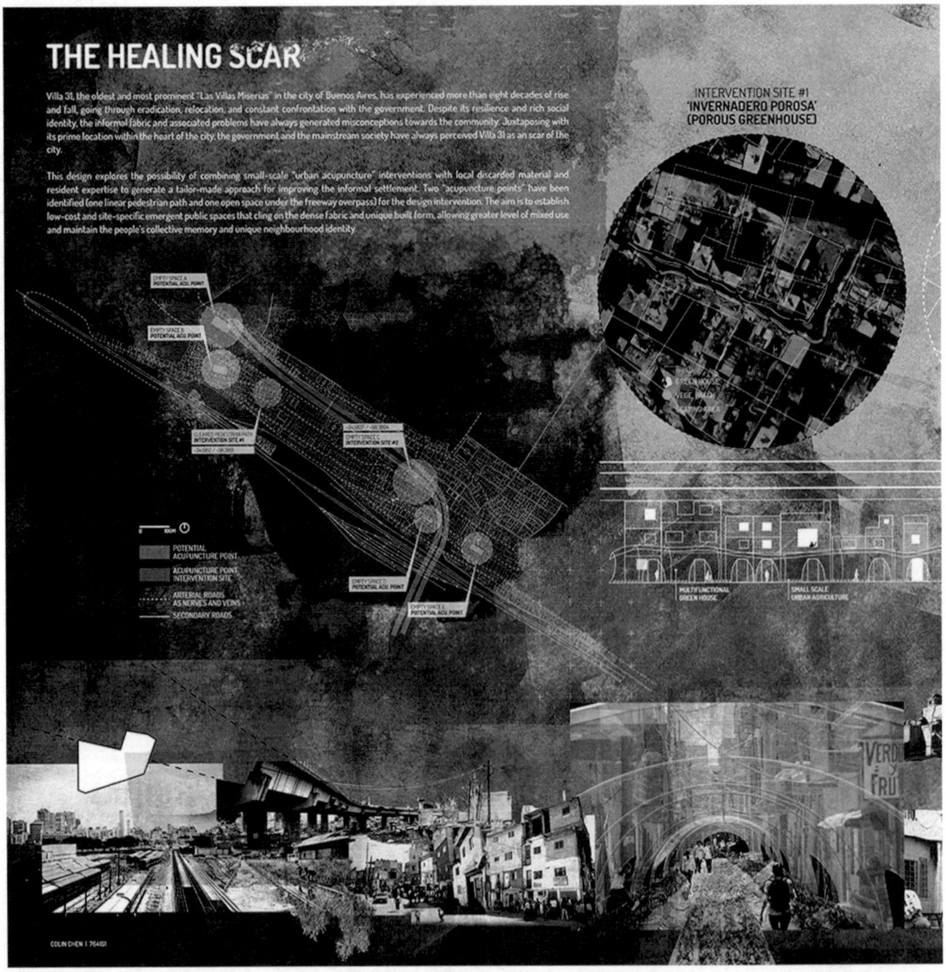

Figure 25.2 This design explores the possibility of combining small-scale 'urban acupuncture' interventions with local discarded material and resident expertise to generate a tailor-made approach for improving the informal settlement of Villa 31, the oldest and most prominent of 'Las Villas Miserias' in the city of Buenos Aires. Two 'acupuncture points' have been identified (one linear pedestrian path and one open space under the freeway overpass) for the design intervention. The aim is to establish low-cost and site-specific emergent public spaces that cling onto the dense fabric and unique built form, allowing greater level of mixed use and the maintenance of people's collective memory and unique neighbourhood identity. (Author: Haifan Chen, recent graduate of Master of Landscape Architecture, University of Melbourne.)

PROBLEMS AS INSPIRATIONS
Villa 31's growth depends on spontaneous and unregulated buildings. Large amount of construction material normally lie in front of the completed buildings, creating additional obstacles on the narrow streets and generating hazardous environment. This prominent issue can somehow be turned into a design opportunity, by clearing the streets and utilising the leftover material as existing resources to create public spaces in Villa 31. As most of the local residents are also experienced farmers and construction workers, their skills and knowledge will contribute to the additional functions of the newly created public spaces.
INTERVENTION SITE #2
'ESPACIO RECICLADO'
(RECYCLED SPACE)
BEACH SOCCER / SAND PLAY
CONCRETE MOUND
ROPE NET
PLASTIC SHEET
SCRAP METAL
TIMBER
VEGE. SEEDLINGS
SITE #1
ROPES
CEMENT
SAND
TIMBER
SITE #2
SEATING AREA
ARTURO ILLIA HIGHWAY OVERPASS
VILLA BEACH SOCCER
CONCRETE PLATFORM
SLIDE & CLIMBING NET
CONCRETE MOUND
0
10m
VILLA
31
70 AÑO
DE HISTOR

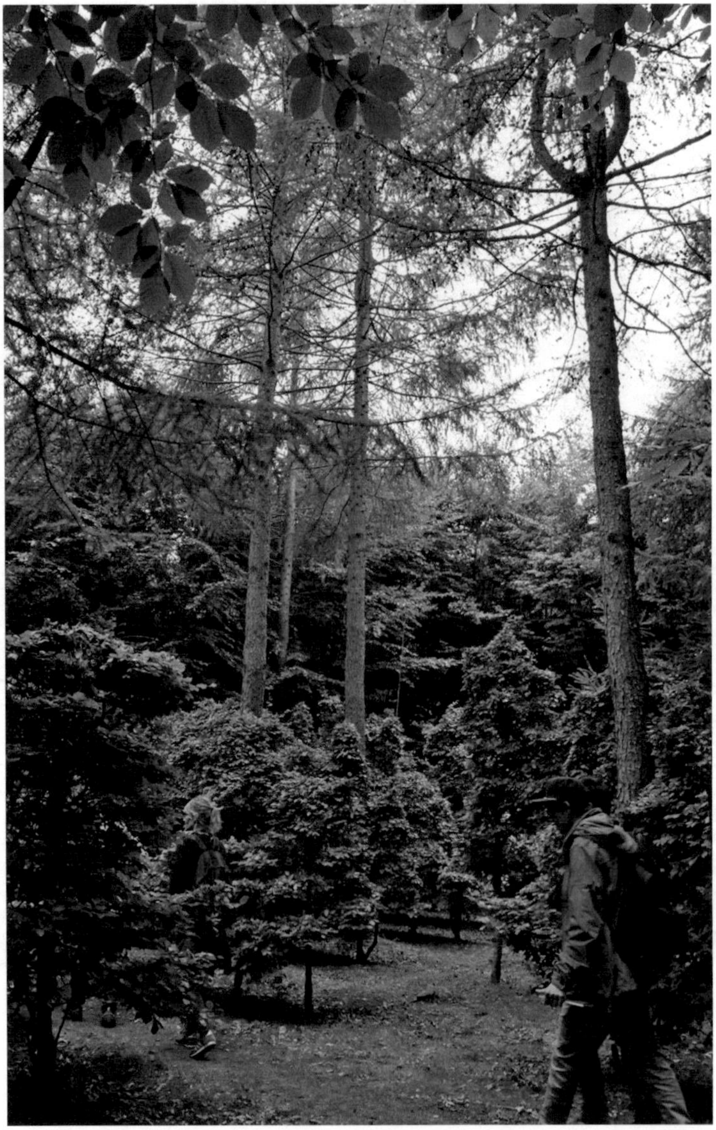

Figure 25.3 Alnarp landscape laboratory. Recycling the residues of making landscape over time. The wood path, called *Erik's cat walk*, is made of Larix x europelis that had served as nurse trees for the neighboring beech plantation. Anders Busse Nielsen. Photo: Bjørn Wiström

Figure 25.4 Alnarp landscape laboratory functions as a didactic learning space that allows for the embodiment of knowledge. Here students make full scale tests of design-directed management action as part of the Dynamic Vegetation Design course. Anders Busse Nielsen. Photo: Bjørn Wiström

Figure 25.5 Alnarp landscape laboratory is a 1:1 platform where researchers, teachers and students can meet and cooperate on the development and testing of new design concepts for establishing and managing urban landscapes at large (like *Lin's place* shown on the photo). The location at the doorstep of the Swedish university of agricultural sciences campus allows for the replacement of indoor lectures and representations of landscape with *in-situ* lectures. Photo: Anders Busse Nielsen

'Sculpting Light: Experiments in Not-Green Studios' There is no division between my teaching studio, research, and art and design practice. I conduct experimental studios with postgraduate students in *Tree Urbanism* and in the productive interface between contemporary and conceptual art and landscape architecture. Teaching approaches, design methods and media are both informed by prior research, and reciprocally, elucidate next stage research questions and aims. The purpose of teaching studio is to change ways of thinking, through ways of doing. Illustrated in these photographs are student responses to a number of workshops exploring landscape architecture not as "green", but as the (spatial and conceptual) medium of *light*.

Figure 25.6 *Frottage* applied as fixed percentages of forest, grove and glades to structure new urban form. Catherine Dee

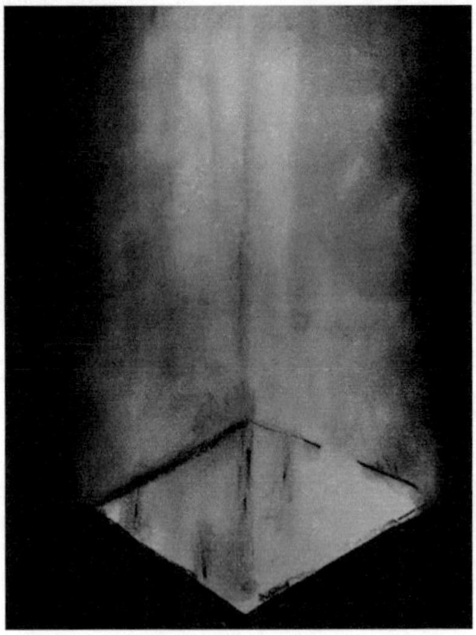

Figure 25.7 Drawing exemplifying the use of light to articulate volumetric space. Catherine Dee

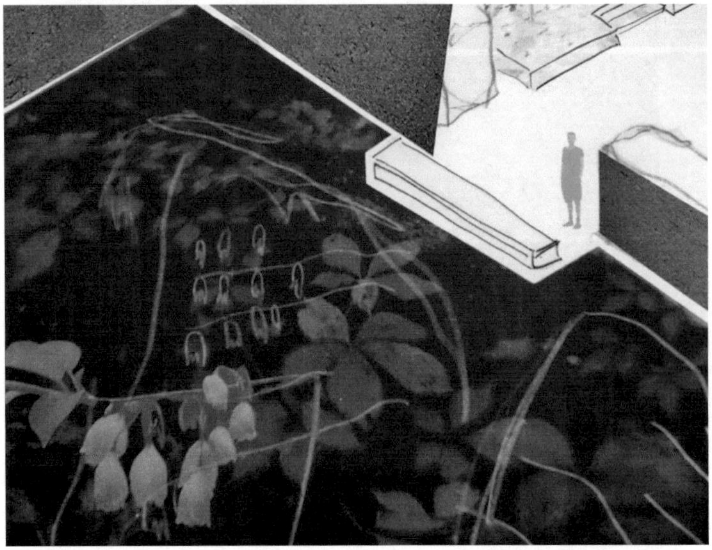

Figure 25.8 The dark flowers and heavy blocks of a sylvan garden. Abrupt and modulated blacks, greys, and whites are atmospheric *and* definitively structural. Catherine Dee

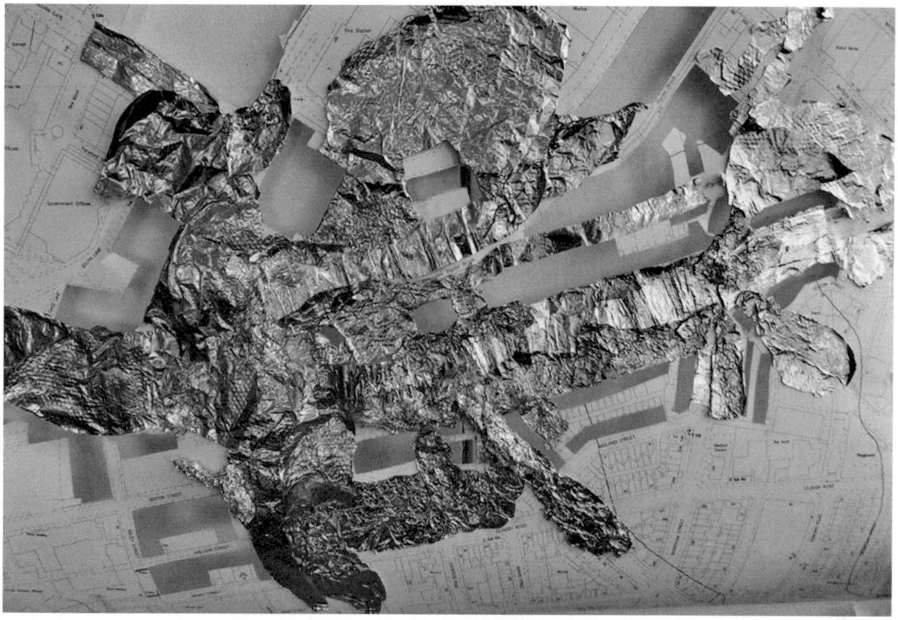

Figure 25.9 Shiny media, 'shatters' urban form; so to 'swim' in tree light. Catherine Dee

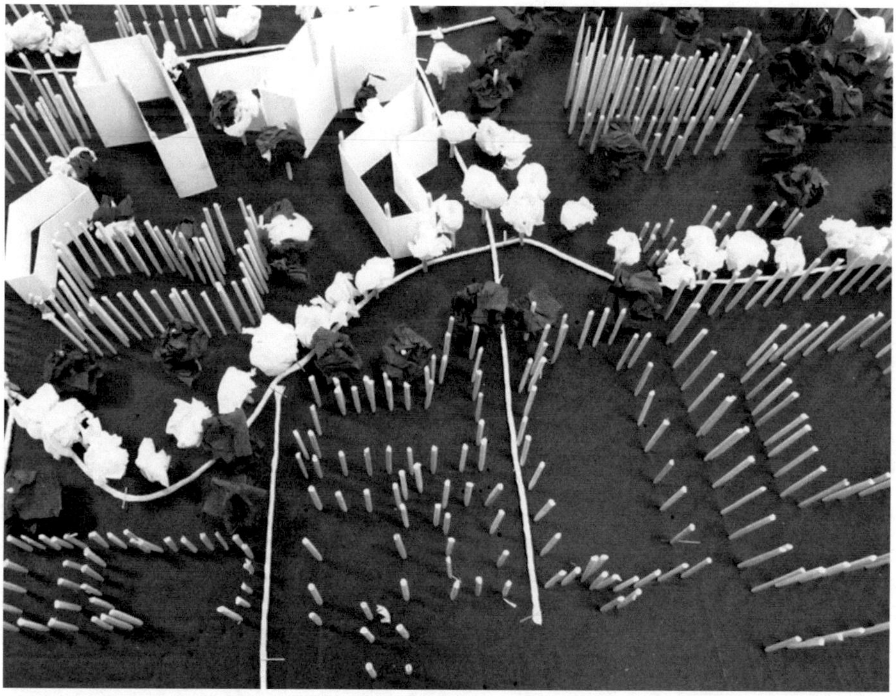

Figure 25.10 'Transparent' buildings and reduced-colour models reveal 'unseen' indoors-outdoors gradients and structure. Catherine Dee

Traditionally, the architect is more focused on the buildings and the programme; the urban designer on the urban area, the neighbourhood and the urban renewal plan; and the urban planner on the more systemic aspects. The connective structures that go beyond the buildings and the urban area (e.g. the neighbourhood) thus tend to get lost from view. In that sense, landscape architecture as a discipline goes well beyond trees, grasses and water management. Hence we regard landscape architecture as entangling built-up and non-built-up areas in urbanized territories that are tightly interwoven with urbanization processes on different scales and levels.

You may say that people are very different, coming from different cultures, but humans are all part of the same global culture. Everyone uses smartphones, for instance, which affects how we move through the city today. But landscape conditions, plants, water or soil conditions—those are still pretty local. Landscape architects have to be experts in their own local contexts. It is very beneficial for us here in Aarhus that we have students from abroad, because they ask questions that are very basic, for instance, about the local context, that we and their fellow students then have to try to answer. In this way, they make implicit knowledge explicit.

Importantly, all teaching here revolves around project work supported by teachers, engineers, other students, small workshops etc., so we do not teach courses, being very much in the *beaux arts* tradition.

Aesthetic and ethical discussions greatly influence each other, in my opinion. There is a long tradition of discussion that has tried to keep them separate, but that does not work at all. The two discussions are intertwined. There are ethical dimensions to what most aesthetic things should do, and vice versa. There are extremely important ethical-aesthetic discussions right now related to sustainability, the sustainability of the whole globe, our cities etc., and it is extremely clear that landscape architecture has something to contribute. Landscape architecture therefore cannot take only an aesthetic approach; it also has ethical implications.

Landscape architecture relates to sustainability, climate adaptation and social sustainability, and the possible impacts of landscape architecture are obvious and direct. Hence this is a big opportunity for landscape architecture, maybe not to put its foot down, but at least to gain a little bit more of a voice. The situation has actually moved substantially forwards from when I went to school in the 1990s, when landscape architecture was stuck in an aesthetic approach. Even though neoliberalism probably makes things more difficult, I think the whole sustainability agenda has put the ethical dimension very much into landscape architecture, in a way that makes it possible to have an impact on a much larger scale. It goes back to connecting landscapes, as I mentioned before, and this is exactly why I think it is important to work on that question.

Inge Bobbink (Technische Universiteit Delft, the Netherlands)

At TU Delft, our landscape architecture track (which started in 2010) is one of five within a master's programme on architecture and the built environment, leading to a Master of Science degree in Architecture, Urbanism and Building Sciences. As much as landscape architecture relies on concepts from architecture and urbanism, it is an independent discipline. It is specific to landscape architecture that the landscape is the object of design as well as the starting point. Landscape architectonic design and research methods differ from those of other spatial design disciplines in the extent to which natural (e.g. geomorphological or hydrological) patterns and processes determine the form and operation of the spatial system. Students working through scales from 'hard to horizon' need to understand the transformation of the site through time, and to incorporate future consequences into their work.

We teach our students in the first place to be designers of spaces and processes, and to become researchers by designing. For us it is in the nature of things that each project is a unique reflection of geomorphology and geometry, nature and artefact, form and function. General principles are and can be distilled from the research and used to build up our body of knowledge. As long as the specificity of the site is taken as the starting point of the design, and is understood and analysed as part of a system, we are confident that our students (depending on their talent) can deliver good work all over the world.

Today's challenges need professionals who are able to think integrally, in order to design a more sustainable environment. The landscape architect can become the 'spider at the centre of the web': a person who is able to process and bring together knowledge from civil engineering, hydrology, ecology, architecture, urbanism etc. What troubles me is that the aesthetic dimension of our discipline, as well as of architecture and urbanism, seems to be losing ground. Problem-solving is currently the most used term in our faculty, based on economic value and prompt output. I feel that this way of thinking is the result of too great a focus on evidence-based research, where facts and figures seem to speak for themselves, are 'easy' to communicate and are expected to solve problems. Being part of a technical university does not make it easier to emphasize the value of achievements in terms of aesthetics. That is why in our programme, we explicitly stress the importance of form and composition as an expression of culture and identity.

Last year, the executive board obliged all faculties to address ethical issues more explicitly—after all, we do transform people's lives—and to formulate them into learning goals. This is a very good initiative which I strongly support, and hopefully it will provoke a discussion of our responsibility for a more beautiful, meaningful and sustainable environment, since we educate students from all over the world.

David Grahame Shane (Columbia University, USA)

I was always most interested in the 'urban turn' in landscape architecture in the late 1980s and early 1990s, when the theories and practices of 'landscape urbanism' first emerged in the East Coast schools in the USA. As an urban designer, I appreciated how landscape could accommodate multiple scales of intervention and the temporal dimension, including long-term planning for ecological and climate change issues ignored by many urbanists and planners. This active accommodation opened up many questions that remain in play and will undoubtedly affect our future as a species on the planet. Within this framework, landscape urbanists as a profession have tended to work with other professionals in industrialized contexts, while much of the present and future urban growth will be informal and self-built without an industrial base. Perhaps in future our educational system should turn to address the scenario of this massive informal urbanization of the landscape.

Traditional landscape architecture as the space between buildings or within walled enclosures, even encompassing large picturesque landscaped estates, always operated as a closed system. Landscape urbanism as a hybrid opened up the practice to engage with wastelands, brownfields, abandoned industrial sites etc., tracing human occupation across vast areas and creating a new form of landscaped city. One of the most valuable lessons I learnt from my interactions with landscape architects was to re-evaluate the role of the countryside, the peri-urban landscape and my definition of the city. In many cultures, cities and landscape coexist as a complex, interwoven structure, based on hydrology and farming practices at first, but then reinforced by transport and modern communication systems. This rur-urban hybrid has been one of the most interesting fields of landscape research in recent years in Europe and Asia.

We are living in the midst of one of the most rapid global urbanization processes in human and planetary history, on a scale never seen before. While the UN highlights megacities, these cities will host only eight per cent of the urban population; ninety-two per cent will live in networks of towns of between one and four million, spread across a rur-urban territory. Huge global corporations are engaged in finding the resources to feed and house the megacities, while some developing countries leave their poor to live on the street. Unless designers come from one of the developing nations, it is extremely unclear what the ethical urban or landscape designer's role will be in this situation, especially in non-industrialized nations. Finding a proper ethical balance in a complex, non-state based, non-industrial system can be very difficult. What design skills are appropriate? The issue of design aesthetics perhaps seems simpler, as professional designers and landscape architects can package and deliver much-needed infrastructural services and reforms, even without a local industrial base. But even here, the success of these efforts long term, without proper training, support and maintenance, is an open question. Should we train designers for collapse and failure?

Catharina Dyrssen (Chalmers Sweden)

Theory today relies less on philosophy and is instead more anchored in practice. Abilities to deal with complexity and heterogeneity to generate knowledge are increasingly supported by non-hierarchical modes of discourse and theory, lateral strategies, multiple perspectives and relational conceptualizations, such as developed within feminist approaches, environmental systems, sociocultural narratives or art. These more recent knowledge traditions also contribute to broader understandings of 'urbanism', 'landscape' or 'landscape architect'. Fast information flows need alternatives: deep insights into key issues, paired with long-term thinking and listening, to open up profound complementary perspectives on situated knowledge and environmental conditions, underpinned by explorative practice, combinatory modellings of understanding, and multi-modal forms of dialogue and communication. In this situation, we see trends towards both individual expert careers and increasing team-building, urging academic studies towards specialization, increased autonomy, and deep thinking as well as collective work and flexibility. Here, I believe, artistic approaches are needed to expand understandings of knowledge making and pedagogy, including new aspects of aesthetics and ethics.

In education, the training of aesthetic capacities is a key issue in the recognition of situated insights, coherence, and a diversity of solutions and processes. It is important to combine artistic and analytic as well as discursive-modelling-communicative abilities to rethink complexities and situations and to develop teamwork competence, for instance in interprofessional and participative transformation processes—which are all regarded as dynamic, interrelated approaches, but where emphases can shift. In turn, this requires relevant links to related fields. It demands the provision of a generous, explorative environment, with high work expectations and a culture of pleasure, joy, and respect for individuality, diversity, equality and collaboration. Landscape architecture can strengthen its educational processes, recognizing that a dynamic strategic polarity between prominent specialization and thematic diversity is useful in professional cross- and transdisciplinary collaborations. Developments in education and research are linked here, although not necessarily congruent.

In research, landscape architecture should assert its professional competences as unique but not isolated, with compositional-transformational approaches to key issues of great social value. Landscape architecture issues involve both basic research, applied and transdisciplinary knowledge. In light of public-private funding systems, academia needs to assert its authority and reasonable autonomy as a long-term research environment that can go deep, look ahead and move beyond existing conditions.

Challenges here, as I see it, follow two, partly matching lines:

1 Designers as 'professional stars' are involved as key agents in global place- or business-branding, the competitive service economy, and environmental-entrepreneurial collaborations. Global communication among professional networks often requires specialized, authoritative 'exclusive expert' statements, whereas local levels call for broader, more inclusive dialogue formats, with team leaders as 'levers' in transformation processes.
2 Environmental and democratic demands depend on the societal, political evolution of trust. Glocal agency builds on deep ecological thinking, multiple relational perspectives and complementary capabilities. Today, big 'knowledge firms' often build their professional knowledge collectives on diversity in team competence.

The aesthetic and the ethical are closely linked, not least through the issues of ecology and democracy—nature-culture, resource use, equality etc.—and the constant interplay between (re)thinking-(re)making-(re)experiencing-(re)focusing-reflecting and more diversified theory as indicated above. Deep artistic-ethical training of competence helps us to recognize possible coherences, sometimes understood as 'beauty', but may also reveal unexpected logics, alternative perspectives and insights, including links and rhythms in changes, and knowledge movements that can bypass quick statistics and fast information media flows as well as overly complicated solutions, enabling us to rethink complex situations through inventive, well-supported, environmentally well-grounded responsive and shared artistic and deductive logics of judgement.

Maggie Roe (Newcastle University, UK)

In 2012 I was asked to provide the summary chapter for a book edited by three established landscape architecture academics which aimed to examine the boundaries between the field of landscape architecture and cognate disciplinary areas. In that analysis,[9] it became clear to me that while there was considerable regard for landscape architecture and an understanding of potential areas of overlap with other disciplines—from both an academic and a practice point of view—there was also a lot of misunderstanding about the scope of the work. What became clear was that landscape architecture is—as I suggested then—'more than a job; it is a philosophy, a way of life, or a way of living'. People often fall into landscape architecture as a result of meeting someone working in the field. This is precisely what happened to me. I was a classic case of someone interested in both arts and sciences, but I did not want to study one or the other in isolation. This is where the landscape architect comes into her own: in the ability to cross boundaries and work between cultures, whether between arts and natural sciences, or between social sciences and design, or between all of these areas. This is why it is important for those studying not only to become adept in one area, but also to understand how to shift between disciplinary areas and work in the interstices of other fields.

Both practitioners and academics seem to feel hampered by the strictures of working between disciplines—but students in particular should revel in this opportunity. Working between worlds is difficult and challenging, but it is also very rewarding. There is never a dull moment working with landscape studies, and you can define the area that most interests you. Remember that landscape is not just a job. When you talk to anyone—ordinary people, policymakers, professionals, children—they are *always* interested in landscape. You just have to ask someone about their identity; more often than not they will talk to you about their surroundings or their neighbourhood, in short, their landscape.

The tensions of professionalization often lie between regulation, financing and creativity. My own experience of working in practice—and leaving it—was of much frustration. The way schemes were reduced through so-called cost-cutting, which resulted in cheaper solutions and lack of maintenance, left me disillusioned as a young practitioner. What I now realize is that these things are of course part of the creative challenge, and this is one important aspect that we do not actually teach at present. How can you make a meal when the cupboard is bare?

The other key issue is how to deal with a world where the environmental pressures from humans are at a critical point. I have always felt that ecological health is central—or should be a central consideration—to what landscape architects do. Unfortunately this is not always the case, and I am still astonished by students who have spent three years doing landscape architecture and have only a vague inkling of ecological principles and concepts, let alone the detail of how to work with natural processes. In ethical terms—this is where I would put my emphasis. Get a thorough understanding of natural processes before you even try to work with people and design. Perhaps the best thing anyone ever told me in the first year of my own studies was to read Aldo Leopold's *A Sand County Almanac* (1949) and Robert Pirsig's *Zen and the Art of Motorcycle Maintenance* (1974)—I would still support this approach, even if they mark me out as a bit of an old hippie.

Notes

1 Francesco Petrarch and Rodney John Lokaj, *Petrarch's Ascent of Mount Ventoux: The Familiaris* IV, I (Roma: Edizioni dell'Ateneo, 2006).
2 D.W. Meinig, 'The Beholding Eye: Ten Versions of the Same Scene', in *The Interpretation of Ordinary Landscapes: Geographical Essays*, eds. D.W. Meinig and John Brinckerhoff Jackson (New York: Oxford University Press, 1979): 46.
3 James Corner, 'The Agency of Mapping: Speculation, Critique and Invention', in *Mappings*, ed. D. Cosgrove (London: Reaktion Books, 1999), 213–252.
4 J.B. Harley, 'The New Nature of Maps', in *Essays in the History of Cartography*, ed. Paul Laxton (Baltimore: The John Hopkins University Press, 2001). See also D. Cosgrove, 'Mapping Meaning', in *Mappings*, ed. D. Cosgrove (London: Reaktion Books, 1999), 1–23.
5 Jane Hutton, 'Material as Method', in *Landscript 5, Material Culture*, ed. Jane Hutton (Berlin: Jovis Verlag GmpH, 2017), 13–22.
6 Hansen et al., 'The Campfire Design Studio: Design Conversations in Landscape Architecture Education', *Edinburgh Architectural Research Journal* 34 (2016): 66.
7 Sakuteiki, *Visions of the Japanese Garden* (North Clarendon: Tuttle Publishing, 2008), 159.
8 United Nations, 'Sustainable Development Goals: 17 Goals to Transform Our World', http://www.un.org/sustainabledevelopment/sustainable-development-goals/.
9 Maggie H. Roe, 'Crossing the Boundaries?', in *Exploring the Frontiers of Landscape Architecture*, eds. S. Bell, R. Stiles and I. Sarlov-Herlin (Oxford: Routledge, 2012), 299–315.

Bibliography

Hansen et al. 'The Campfire Design Studio: Design Conversations in Landscape Architecture Education'. *Edinburgh Architectural Research Journal* 34 (2016): 63–80.

Leopold, Aldo. *A Sand County Almanac*. Oxford: Oxford University Press, 1968 *c*. 1949.

Meinig, D.W. 'The Beholding Eye: Ten Versions of the Same Scene'. In *The Interpretation of Ordinary Landscapes: Geographical Essays*, edited by D. W. Meinig and John Brinckerhoff Jackson. New York: Oxford University Press, 1979: 46.

Petrarca, Francesco, and Rodney John Lokaj. *Petrarch's Ascent of Mount Ventoux: The Familiaris IV*, I, Roma: Edizioni dell'Ateneo, 2006.

Pirsig, Robert. *Zen and the Art of Motorcycle Maintenance*. NY: Harper Collins, 1974.

Roe, Maggie H. 'Crossing the Boundaries?' In *Exploring the Frontiers of Landscape Architecture*, edited by S. Bell, R. Stiles and I. Sarlov-Herlin, 299–315. Oxford: Routledge, 2012.

Sakuteiki. *Visions of the Japanese Garden*. North Clarendon: Tuttle Publishing, 2008.

INDEX